Molecular Virology

Molecular Virology

Edited by **Drew Farmer**

SYRAWOOD
PUBLISHING HOUSE

New York

Published by Syrawood Publishing House,
750 Third Avenue, 9th Floor,
New York, NY 10017, USA
www.syrawoodpublishinghouse.com

Molecular Virology
Edited by Drew Farmer

International Standard Book Number: 978-1-68286-181-3 (Hardback)

Printed in the United States of America.

Contents

Preface

This book has been a concerted effort by a group of academicians, researchers and scientists, who have contributed their research works for the realization of the book. This book has materialized in the wake of emerging advancements and innovations in this field. Therefore, the need of the hour was to compile all the required researches and disseminate the knowledge to a broad spectrum of people comprising of students, researchers and specialists of the field.

Molecular virology refers to the study of molecular structures and characteristics of viruses. It is aimed at analysing in detail, the specific viral genes and their products. This book provides significant information on molecular virology and provides a comprehensive overview on concepts such as life cycle of viruses, molecular analysis of viruses infecting agricultural crops, use of virus strains for developing vaccines and pharmaceuticals, etc. The book is appropriate for students seeking detailed information in this area as well as for experts.

At the end of the preface, I would like to thank the authors for their brilliant chapters and the publisher for guiding us all-through the making of the book till its final stage. Also, I would like to thank my family for providing the support and encouragement throughout my academic career and research projects.

Editor

Molecular approaches towards analyzing the viruses infecting maize (*Zea mays* L.)

Kamal Sharma[1,2,3] and Raj Shekhar Misra[1]*

[1]International Institute of Tropical Agriculture, PMB 5320, Ibadan, Nigeria.
[2]Central Tuber Crops Research Institute, Thiruvananthapuram, Kerala -695017, India.
[3]International Institute of Tropical Agriculture, PMB 5320, Ibadan, Nigeria.

Information on virus diseases of maize still remains scanty in several maize growing countries. Therefore it is hoped that this description will stimulate more research, which will lead to better understanding of viruses infecting maize in Africa. Plant viruses are a major yield-reducing factor for field and horticultural crops. The losses caused by plant viruses are greater in the tropics and subtropics, which provide ideal conditions for the perpetuation of both the viruses and their insect vectors. Management of viral diseases is more difficult than that of diseases caused by other pathogens as viral diseases have a complex disease cycle, efficient vector transmission and no effective viricide is available. Traditionally, integration of various approaches like the avoidance of sources of infection, control of vectors, cultural practices and use of resistant host plants have been employed for the management of viral diseases of plants. All these approaches are important, but most practical approach is the understanding of seed transmission, symptom development, cell-to-cell movement and virus multiplication and accurate diagnosis of viruses. This update aims to continue on this course while simultaneously introducing additional levels of complexity in the form of microbes that infect plants. Rather than serving as a standard literature review, the objective is to provide a broad conceptual introduction to the field of molecular plant-microbe interactions, virus multiplication, transmission and virus diagnosis and various immunodiagnostic and molecular diagnostic methods such as enzymes linked immunosorbent assay (ELISA), immunosorbent electron microscopy (ISEM), polymerase chain reaction (PCR), nucleic acid hybridization, dot immunoblotting assay (DTBIA) found suitable for diagnosis of viruses infecting maize. These techniques do not only provide information for epidemiological purposes, but also help to develop disease free stock of maize. Therefore, these various techniques with symptoms and history are of immense value to diagnose maize viruses and are the cornerstone of the management of maize viruses. This information will be useful to researchers in understanding of maize viruses. Information on symptomatology, transmission, geographical distribution and properties of viruses is summarized here based on literature review.

Key words: Maize, diagnosis, ELISA, transmission, virus.

INTRODUCTION

Maize (*Zea mays* L.) is one of the major cereal crops; it ranks third in production following wheat and rice with an average of 784,786,580 tons produced annually by 10 countries (FAO, 2007) (Table 1). It is the world's most widely grown crop in almost all tropical areas of the world including tropical highlands over 3000 m in altitude, to temperate areas as far north as the 65th latitude. Maize is one of a few important grasses that humanity has cultivated for centuries to provide food and a considerable number of industrial products (Galinat, 1977). Maize has often been described as "the grain that civilized the New World." Maize, or corn as it is called, has a multitude of uses and ranks third among the world's cereal crops in terms of total production. Also, because of its worldwide distribution and lower prices relative to other cereals, maize has a wider range of uses than any other cereal. Within the developing world, there are a number of

*Corresponding author. E-mail: rajshekharmisra@gmail.com.

Table 1. Top ten maize producers in 2007.

Country	Production (Tonnes)	Area harvested	Yield Hg/ Ha
United States	332,092,180	35022300	94823
China	151,970,000	28074000	54131
Brazil	51,589,721	13827500	37309
Mexico	22,500,000	7800000	28846
Argentina	21,755,364	2838072	76655
India	16,780,000	7770000	21595
France	13,107,000	1481000	88501
Indonesia	12,381,561	3450650	35881
Canada	10,554,500	1361100	77543
Italy	9,891,362	1081680	91444

Source: Food and Agricultural Organization of United Nations: Economic And Social Department: The Statistical Division. http://faostat.fao.org/site/567/ 17.2.09

countries where maize is a major staple food and the per capita human consumption reaches high levels. Maize is a major staple food for most of the indigenous, rural populations in Africa (CIMMYT, 1990). Africans consume nearly one-fourth of the total feed (livestock) maize produced in Africa, and for many countries, the per capita consumption of maize may be as high as 100 kg per year (CIMMYT, 1990) Major maize producing countries in Africa are presented in Table 3. A number of diseases affect the productivity of maize (Bacterial, fungal, viral, virus like, nematode and parasitic).

Because of existence of different ecological conditions between the temperate areas and the tropics, the insect vectors and their disease agents also are different under these different conditions. To date, at least 21 viral plant pathogens affecting maize in the hot and humid tropics have been reported (Table 2). Most of the viruses and molecules referred to as etiological agents of diseases on maize have been found in several parts of Africa. Viruses found infecting maize on the African continent include: Maize streak virus (*geminvirus*), maize mottle/chlorotic stunt virus, maize eyespot virus, guinea grass mosaic virus (*potyvirus*), cynodon chlorotic streak virus (*rhabdoviridae*), maize yellow stripe virus (*tenuivirus*), brome mosaic virus (*bromoviridae* and *bromovirus*), barley stripe mosaic virus (*hordeivirus*), and barley yellow dwarf virus (*luteovirus)* (Thottappilly et al., 1993).

Plant viruses are intracellular pathogens that perform genome replication and encapsidation within the infected cells. In order to use the resources of the host cells efficiently for their genome replication, plant viruses have to interact with the host cells, manipulate host cell pathways, and, ultimately, transform the host cells into "viral factories." The magnitude of physiological and phenotypic changes in the host during viral infection suggests the involvement of a large number of host genes (Golem and Culver, 2003; Whitham et al., 2006). Thus, the intimate interaction between a plant virus and its host is complicated by the systemic nature of infection

and global alterations in host gene expression (Maule et al., 2002; Whitham and Wang, 2004).

All these pathogens represent potential threat to maize productivity. In the event of epidemiological surges, the ability to rapidly and precisely diagnose and identify the causal agents will be necessary to design control measurements.

Genome structure of viruses

The majority of plant viruses have a plus stranded (+)RNA genome compatible with the protein translation apparatus of the host. The infection cycle of these viruses includes entry into the cell, disassembly of the virus capsids, translation of the viral RNA, genome replication and transcription, encapsidation, and cell-to-cell movement. The central event is the genome replication of (+)RNA viruses, which consists of a two-step process: First, the minus strand replication intermediates are produced, which are then used to direct synthesis of excess amounts of (+)RNA progeny by the unique viral replicases (such as RNA dependent RNA polymerase (RdRp), the key enzymes in viral replication. Replication is an asymmetric process leading to a 20- to 100-fold excess of the new (+)RNA progeny over minus-strand RNA. All known plant (+)RNA viruses assemble their own replicase complexes (RC), likely containing both viral- and host-coded proteins (Buck, 1996; Ahlquist, 2002; Ahlquist et al., 2003; Nagy and Pogany, 2006). The assembled viral RCs are associated with cellular membranes, such as the endoplasmatic reticulum and the membranes associated with cell organelles like mitochondria, vacuole, golgi complex, chloroplast, and peroxisome, which serve as sites of viral replication (Laliberte and Sanfacon, 2010).

One intriguing aspect of (+)RNA viruses is that their RNAs must participate in several competing processes, all of which are required for successful viral infections.

Table 2. Maize viruses and virus like diseases.

Diseases	Virus	Virus genus/group	Vectors	Seed transmission	Geographical distribution	Reference
Maize bushy stunt	Mycoplasma like organism (MLO)		Corn leafhopper, *D. maidis*	Unknown		Legrnad and Power,
Maize chlorotic dwarf	Maize chlorotic dwarf virus (MCDV)	IV: (+)sense RNA Viruses (Waikavirus)	Arthropods (*G. nigrifrons, Graminella sonora* and *Exitianus exitiosus*)	No	United States of America	ICTVdB, 2006
Maize chlorotic mottle	Maize chlorotic mottle virus (MCMV)	IV: (+)sense RNA Viruses (Machlomovirus)	Arthropods (*Cicadulina mbila, C. zeae, C. storeyi* and *C. triangula*)	No	Nigeria, Rwanda, Sao Tome and Principe, Tanzania, Togo, Zambia, and Zimbabwe	Rossel and Thottappilly
Maize dwarf mosaic	Maize dwarf mosaic virus (MDMV) strains A, D, E and F	IV: (+)sense RNA (Potyvirus)	Arthropods, insects	Yes	China, South Africa, and the United States of America	ICTVdB, 2006
Maize eyespot virus			Virus transmitted by mechanical inoculation	No	Cote d'Ivoire	Brunt et al., 1996
Maize line	Maize line virus (MLV)	-	Transmitted by an insect; *P. maidis*; Delphacidae	No	Kenya and Tanzania	Kulkarni, 1973
Maize mosaic (corn leaf stripe, enanismo rayado)	Maize mosaic virus (MMV)	V: (-) sense RNA Viruses (Nucleorhabdovirus)	Arthropods, by insects *P. maidis*	No	Australia, Colombia, Costa Rica, Fiji, India, Mauritius, Mexico, Peru, Spain, Tanzania, and the United States of America (and the Caribbean Islands).	ICTVdB, 2006
Maize Iranian mosaic			Transmitted by a vector; an insect; *Unkanodes tanasijevici, Laodelphax striatellus, P. maidis*; Delphacidae	No	Iran	Brunt et al., 1996
Maize rayado fino (fine striping disease)	Maize rayado fino virus (MRFV)	-	Transmitted by a vector; an insect; *D. maidis*; Cicadellidae	No	Argentina, Brazil, Colombia, Costa Rica, Mexico, Peru, Venezuela, and the USA (in the south).	Brunt et al., 1996
Maize mottle/chlorotic stunt virus			Transmitted by a vector; an insect; *C. mbila, C. zeae, C. storeyi* and *C. triangula*; No Cicadellidae	No	Nigeria, Rwanda, Sao Tome and Principe, Tanzania, Togo, Zambia, and Zimbabwe	Brunt et al., 1996

Table 2. Contd.

Maize streak dwarf		Transmitted by a vector; an insect; *Laodelphax striatellus* (both adults and nymphs); Delphacidae	No	China	Brunt et al., 1996
Maize streak monogeminivirus		Transmitted by a vector; an insect; *C. mbila, C. arachidis, C. bipunctella, C. triangula, C. bimaculata, C. similis, C. latens, C. ghaurii, C. parazeae;* Cicadellidae	No	African region; India, Madagascar, Reunion, and Yemen	Brunt et al., 1996
Maize rough dwarf virus (MRDV) (nanismo ruvido)	III: dsRNA Viruses	Virus is transmitted by a vector (*Delphacodes propinqua, Dicranotropis hamata, L. striatellus, Javasella pellucida, Sogatella vibix*). Virus is transmitted by mechanical inoculation	No	Argentina, Czechoslovakia (former), France, Israel, Italy, Norway, Spain, Sweden, and Yugoslavia	ICTVdB Management, 2006
Maize yellow stripe		Transmitted by a vector; an insect; *C. chinai;* Cicadellidae	No	Egypt	Brunt et al., 1996
Maize streak (MSV)	Group II (ssDNA) (*Mastrevirus*)	African leathopper, *C. mbila* Naudé	Unknown	Sub-Saharan Africa	ICTVdB Management, 2006
Maize stripe (maize chlorotic stripe, maize hoja blanca)	V: (-) sense RNA Viruses (Tenuivirus)	Virus is transmitted by arthropods, by insects *P. maidis.*	No	Australia, Botswana, Guadeloupe, India, Kenya, Mauritius, Nigeria, Peru, the Philippines, Reunion, Sao Tome and Principe, the United States of America, and Venezuela	ICTVdB Management, 2006
Maize white line mosaic virus (MWLMV)	Virus unclassified	Insect	No	France, Italy, and the United States of America	ICTVdB Management, 2006

These highly regulated, coordinated, and compartmentalized processes include translation of viral RNA, replication, transcription to produce subgenomic RNA for some viruses, encapsidation, and cell-to-cell movement. Each process can be further divided into distinct steps based on recent detailed analyses of a single replication cycle of (+)RNA viruses. During genome replication, the steps include the following: (1) The recruitment/selection of the viral (+)RNA template for replication, including a requirement for switching of the genomic RNA from translation to replication; (2) Targeting of viral replication proteins to the site of replication; (3) Preassembly of the viral replicase components;

Table 3. Top twenty maize producers country in Africa'.

Country	Production (Tonnes)	Area harvested	Yield Hg/Ha
Nigeria	7800000	4700000	16595
South Africa	7338738	2551800	28759
Egypt	7045000	868000	81163
Ethiopia	4000000	1468000	27248
Malawi	3444700	1688500	20400
Tanzania	3400000	3000000	11333
Kenya	3240000	1600000	20250
Mozambique	1579400	1505400	10491
Zambia	1366158	872800	15652
Uganda	1262000	844000	14952
DR Congo	1155000	1480000	7804
Ghana	1100000	750000	14666
Zimbabwe	952600	1445800	6588
Cameroon	923000	480000	19229
Benin	900000	700000	12857
Angola	570000	1115000	5112
Togo	500000	380000	13157
Chad	200000	200000	10000
Rwanda	90000	110000	8181
Sudan	60000	80000	7500

Source: Food and Agricultural Organization of United Nations: Economic And Social Department: The Statistical Division. http://faostat.fao.org/site/567/ 17.2.09.

(4) Activation/final assembly of the viral RC containing the (+)RNA template on intracellular membranous surfaces; (5) Synthesis of the viral RNA progeny by the RC, including minus- and plus-strand synthesis; (6) Release of the viral (+)RNA progeny from the RC to the cytosole, and (7) Disassembly of the viral RC (Nagy and Pogany, 2006).

Geminiviruses are DNA viruses which are known to infect plants and have a small genome which encode only a few proteins. The genome encodes for both structural and non-structural proteins. Geminiviruses have circular single-stranded DNA. Therefore, their DNA replication cycle relies largely on the use of cellular DNA replication proteins. The genome is either in two segments or not segmented at all. The non-segmented genome is 2500 to 3000 nucleotides long, and the segmented genome is 4800 to 5600 nucleotides long. The strategy used by geminiviruses to replicate their single-stranded DNA (ssDNA) genome consists of a first stage of conversion of ssDNA into double-stranded DNA (dsDNA) intermediates and, then, the use of dsDNA as a template to amplify viral dsDNA and to produce mature ssDNA genomes by a rolling-circle replication mechanism (Gutierrez, 1999).

The characteristic twinned or "geminate" particles, which consist of two joined, incomplete $T = 1$ icosahedra, are unique among viruses. Structure of a geminivirus particle, the Nigerian strain of Maize streak virus (MSV-N), was determined. The particle, of dimensions 220 ×

380 Å, has an overall 52-point-group symmetry, in which each half particle "head" consists of the coat protein (CP) arranged with quasi-icosahedral symmetry (Zhang et al., 2001). Whereas the structure of the maize streak virus genome (Kenyan isolate, MSV-K), as determined from the sequence of clones obtained from DNA isolated from virus particles, is composed of one major DNA component of about 2.6 kb. MSV virion DNA was partially double-stranded, composed of a full-length virion (V) strand and a short (70 to 80 b) primer (P) strand. The primer strand has a fixed 5'-end capped with alkaline labile material, presumably 1 to 2 ribonucleotides. The MSV genome has two major coding regions oriented on opposite strands and flanked by two small intergenic regions. The coding region on the P strand is composed of two major open reading frames (ORFs), arranged in tandem and in the same reading frames (Howell, 1984). The genomes of plant (+)RNA viruses code for 4 to 10 proteins, including 1 to 4 proteins involved in viral RNA replication. The replication proteins include one RNA-dependent RNA polymerase (RdRp) protein and auxiliary proteins with helicase, RNA capping or other functions. Host factors contribute to these functions, provide additional functions, and likely participate in each step of the (+)RNA virus replication. Based on recent genome-wide studies (Kushner et al,, 2003, Jiang et al,, 2006), the emerging picture is that the identified host factors are mostly highly conserved genes, suggesting that (+)RNA viruses might selectively target conserved host functions

as opposed to species factors. This strategy can help viruses to have a broader host range and to expand infections to new host species.

Host factors play crucial roles in plant (+)RNA virus replication and infection. Host factors affect host-virus interactions, virus pathology, virus evolution, and they are also key determinants of host range of a given virus (Nagy and Pogany, 2006).

Maize stripe virus (MStV) is a member of newly recognized *tenuivirus* group (Family: Tospoviridae). This group exhibits several unique properties different from other characterized RNA plant viruses (Gingery, 1988). Thin, filamentous, sometime circular infectious nucleo-protein particles have been associated with *tenuivirus* infected plants (Chen et al., 1993; Tsai and Falk, 1993). The nucleoprotein particles are composed of a ca. 35,000 Mr nucleocapsid protein and 4 to 5 species of RNA (Toriyama, 1982; Chen et al., 1993; Tsai and Falk, 1993). When analyzed by denaturing gel electrophoresis, the 5 RNAs have molecular weights of 0.52, 0.78, 0.81, 1.18 Kd and Mr 3.01×10^6. The 16,000 Mr (16 K) protein which has been referred to as the nuclear coat protein (NCP) is found abundantly in MStV-infected plants (Falk et al., 1987) and it can readily be found in sap from infected plants as crystals by phase-contrast light microscopy and the crystals react with antiserum to MStV NCP in immunofluorescence microscopy (Bradfute and Tsai, 1990). Recent studies showed that fibrous intracellar inclusions can be readily found in paradermal sections of the leaf sheath of MStV infected maize (Overman et al., 1992). The nucleotide sequence of MStV NCP gene has also been determined. The purified Maize mosaic virus (MMV, Nucleorhabdovirus) virions contained a single-stranded RNA of Mr 4.2×10^6 (Falk and Tsai, 1983). MMV virions contain three major structural proteins of Mr 75,000, 54,000, and 30,000 as analyzed by SDS-PAGE. Various sizes of MMV virions have been reported. Dimensions of 255×90 nm for negatively-stained partially-purified preparations, and 242×48 nm for particles in thin sections of MMV-infected tissues have been reported for the Venezuelan isolate (Lastra, 1977); 224×68 nm and 234 to 325×63 nm have been reported for purified virions and those in MMV-infected cells, respectively, for the Florida isolate (Bradfute and Tsai, 1983; Falk and Tsai, 1983); and dimensions of 204×67 nm for bullet-shaped particles, 245×80 nm for bacilliform particles have been reported for the Hawaiian isolate of MMV (McDaniel et al., 1985). Both the perinuclear accumulation of virus particles in the infected cells (Bradfute and Tsai, 1983) and the presence of particle in the cytoplasm of epidermal, mesophyll, and vascular parenchyma cells and phloem and xylem elements of infected plants have been reported (Lastra, 1977; McDaniel et al., 1985). The granular masses were found to surround the nuclei of the epidermal strips of MMV infected leaves and roots using light microscopy (Nault et al., 1984). The viral genome of maize dwarf

mosaic virus (MDMV) was 9515 nt and contained an open reading frame encoding 3042 amino acids, flanked by 3'- and 5'-UTRs of 139 and 250 nucleotides, respectively. The RNA of maize dwarf mosaic virus (Potyvirus) strain A (MDMV-A) was characterized and compared with that of strain B (MDMV-B). Glyoxal-treated MDMV-A RNA has a Mr of 3.32×10^6 measured in agarose gels, compared with that of MDMV-B which is 3.41×10^6 under the same conditions. MDMV-A RNA has a Tm of 50.7°C and a hyperchromicity (increase in absorbance during unfolding of a higher structure) of 23.3%, which are higher than those reported for MDMV-B RNA (Berger et al., 1989). MDMV-Bg was more conserved in the coding region (52.9%) than in the UTRs (45.8%) when compared to the 15 other potyviruses (Kong and Steinbiss, 1998). In Maize chlorotic mottle virus (MCMV), sub-genomic RNA is present in infected cells; encoding the coat protein. The genome expression is based on RNA production which can be analyzed by the dsRNA patterns found in the infected tissues. Usually, there are 2 virus specified dsRNA species found in infected cells. Size of largest virus specified dsRNA 4.4 kb and 2nd largest 1.1 kb. The genome has four ORFs encoding proteins of 32 and 50 kDa (possibly the polymerase), 9 and 25.1 kDa (coat protein) (Lommel et al., 1991). The complete nucleotide sequence of the single-stranded RNA genome of maize rayado fino virus (MRFV), the type member of the genus *Marafivirus* (Family: Tymoviridae), is 6305 nts in length and contains two putative open reading frames (ORFs). The largest ORF (nt 97 to 6180) encodes a polyprotein of 224 kDa with sequence similarities at its N-terminus to the replication-associated proteins of other viruses with positive-strand RNA genomes and to the papain-like protease domain found in tymoviruses (Hammond and Ramirez, 2001). The C-terminus of the 224-kDa ORF also encodes the MRFV capsid protein. A smaller, overlapping ORF (nt 302 to 1561) encodes a putative protein of 43 kDa with unknown function but with limited sequence similarities to putative movement proteins of tymoviruses (Hammond and Ramirez, 2001). Morphologically, Maize chlorotic dwarf virus (MCDV) virions are 30 nm diameter icosahedrons with a buoyant density of 1.507 g/ml (Gingery, 1976). The genome consists of a single-stranded RNA molecule with an estimated molecular mass of 9.4 kDa (Gingery et al., 1981). These properties of MCDV resemble those of rice tungro spherical virus (RTSV) within the family of Sequiviridae (Shen et al., 1993). These two viruses and anthriscus yellow virus have been placed in the genus *Waikavirus* (Mayo et al., 1995). Maize necrotic streak virus (MNeSV) has 32 nm isometric particles that encapsidate a single stranded RNA genome *ca.* 4.3 kb in size, and is a tentative member of the genus *Tombusvirus*. Maize necrotic streak virus has isometric particles 32 nm in diameter, with a poorly resolved surface structure, that encapsidate a single-stranded positive-sense RNA genome of *ca.* 4.3 kb (Louie et al., 2000). Viral RNA has a genome structure and organization that are similar to those of tombusviruses. However, MNeSV is classified as a tentative species in the genus *Tombusvirus* because it is not transmissible by leaf-rub inoculation, has a low rate of vectorless transmission through the soil, and has a small coat protein (CP) that is more similar in size (29 kDa), structure, and sequence to those of necroviruses and sobemoviruses than to those of tombusviruses (de Stradis et al., 2005). The molecular masses of maize fine streak virus (MFSV) the proteins by SDS-PAGE of purified virions were reported 82±2, 50±3 and 32± 2 KDa with mean of ± SD, n = 9. A ribonuclease A-sensitive nucleic acid of more than 10 kb was isolated from purified MFSV. No other RNA species observed in nucleic acid preparations. The open reading frame (ORF) encoded by the MFSV G3A cDNA homologous to the N terminus of *rhabdoviral* L proteins. An ORF encoded by the 5' 1.4 kb of the MFSVG6A cDNA had similarity with the nucleocapsid (N) protein of rhabdoviruses (Redinbaugh et al., 2002). MSV has twin quasi-icosahedral particles (18 × 30 nm) with a coat protein of Mr 26000. The genome is composed of a single 2.7 kb single-stranded circular DNA (Mullineaux et al., 1984).

MOLECULAR ASPECTS OF MULTIPLICATION OF MAIZE VIRUSES

All viruses are intracellular molecular parasites on eukaryotic cells since they possess minimum of essential genetic information. Consequently, they have developed the capacity to utilize metabolic machinery of the host cell for production of their DNA and RNA during viral nucleic acid multiplication and viral proteins during translation process. The invading virus genome thus takes control of and subverts the normal cellular processes and usurps the natural cellular machinery as well as cellular factors for replication and transcription of their DNA and RNA genomes and synthesis of viral proteins.

Propagation is the fundamental aspect of biology of any organism. So it is in plant viruses also. Apart from its basic role in increasing population and perpetuating a virus species, it has great importance in genome recombination, in generating hybrid genomic molecules, in producing defective-interfering RNAs/DNAs, in pathogenesis, and in several other viral functions (Zaitlin and Palukaitis, 2000). It involves four chronologically overlapping fundamental steps: (1) Decapsidation resulting in unmasking of genomic RNA and making it available for various viral functions; (2) Translation during which viral genomic RNA serves as mRNA and produces structural and non-structural proteins coded by viral RNA; (3) Replication of viral RNA genome to yield progeny RNA molecules, and (4) Encapsidation leading to assembly of progeny RNA molecules with the cognate

capsid protein molecules to produce the progeny virus particles. Each of these stages is a complex phenomenon requiring specific conditions and certain essential proteins that perform vital functions during these stages (Salonen et al., 2005). For viral RNA replication, these essential proteins include polymerases, helicases, DNA-binding proteins, capping enzymes, elements needed for binding of host factors, and others (Ahlquist et al., 2003). Additionally, the cis- and transacting nucleotide sequence motifs and RNA secondary structures within 5'-termini of viral genomic RNAs are central to virus RNA replication. Such events, enzymes and factors are also functional in each of the other three stages. Phylogenetically, replication-associated genes constitute the core elements of RNA virus genomes while other gene modules are considered as accessory elements. Very little was known about molecular aspects of plant viral RNA replication (Teycheney et al., 2000) and its initiation enzymatic studies on virus multiplication constitute the frontier area.

Most of the research focused on RNA viruses evolution, which are generally subject to relatively high rates of mutation due to their dependence on error-prone DNA dependent RNA polymerases. RNA viruses have been shown to evolve at rates between 10^{-3} to 10^{-5} substitutions per site per year (subs/site/year) (Malpica et al., 2002). However recently, DNA viruses evolution study has also gained equally importance, in contrast with the hypothesis that polymerase fidelity influences evolution rates double stranded DNA (dsDNA) papillomaviruses and polyomaviruses evolve at rates in the region of 10^{-9} subs/site/year (Drake, 1991; van der Walt et al., 2008).

The genome replication of most DNA viruses takes place in the cell's nucleus. If the cell has the appropriate receptor on its surface, these viruses enter the cell sometimes by direct fusion with the cell membrane or more usually by receptor-mediated endocytosis. Most DNA viruses are entirely dependent on the host cell's DNA and RNA synthesizing machinery, however, viruses with larger genomes may encode much of this machinery themselves.

Geminiviruses utilize three replication modes: Complementary-strand replication (CSR), rolling-circle replication (RCR) and recombination-dependent replication (RDR). Here RCR mechanism is briefly explained

Rolling-circle replication

A characteristic feature of RCR is the involvement of a replication initiator protein (Rep) with a nicking-closing activity similar to that found in topoisomerases. RCR occurs in three stages. In the first stage (SS→RF synthesis), viral ssDNA (+ strand) enters the cell and is converted into a covalently closed dsDNA replicative form (RF) in a process involving host-directed, RNA-primed synthesis of a complementary (-) template for further replication. The purpose of the second stage of RCR (RF→RF synthesis) is to generate additional RF DNA (Saunders et al., 1991). This step is initiated by viral Rep protein, gene A protein (gpA) in the case of φX174, whose function is to nick the plus strand at a specific sequence. Following phosphodiester bond cleavage, Rep protein covalently binds to the 5' terminus via a phosphotyrosine linkage. The 3' -OH terminus is used as a primer for the synthesis of nascent plus strand, which displaces the parental plus strand from the intact minus-strand template. Synthesis again is carried out by host replication proteins (DePamphilis, 1988). Completion of the nascent plus strand regenerates the origin of replication, which again is nicked by Rep, this time acting as a terminase to release the displaced unit-length plus strand, which is simultaneously ligated to circular form by the closing activity. Rep is transferred to the newly created 5' terminus. Early in the replication cycle, the circularized ssDNA is used as template for synthesis of minus-strand DNA, resulting in the amplification of RF. The third stage of RCR (RF→SS synthesis), which occurs late in the replication cycle, is responsible for the accumulation of viral genomes for encapsidation. This stage is similar to RF→RF synthesis, except that priming is prevented and ssDNA is the predominant product (DePamphilis, 1993; Bisaro, 1994).

Multiplication/propagation of maize viruses

Propagation is well studied for several economic important viruses for ex. maize stripe virus (Nault and Gordon, 1988), maize chlorotic mottle virus (Lommel et al., 1991). Here it is briefly explained with another economically important virus maize dwarf mosaic virus (MDMV). The incidence of this virus disease is usually less than 5%, but levels as high as 65% been reported (Seifer and Hackerott, 1987). The virus is spread from plant to plant and field to field by several species of aphids. The most common carriers (vectors) are the greenbug and the corn leaf aphid. Initial infections may occur when over wintering aphids feed on infected weed hosts and then move into a field. Additionally, the virus can spread to great distances when virus-carrying aphids are moved by strong winds associated with weather fronts (Seifer and Hackerott, 1987). MDMV is also readily transmissible by aphids in a non-persistent manner which means that both virus acquisition and inoculation by aphids can occur in a few seconds. At least 25 species of aphids have been reported to be vectors of plant viruses (Knoke et al., 1983a). The transmission efficiency varies greatly depending upon aphid species, environmental conditions, virus strains and host plants. The virus can survive in perennial grasses or in the seed of annual or perennial grasses which represent important sources for both MDMV and the aphids that transmit it. The aphid species known to be efficient vectors of MDMV are:

The green bug, *Schizaphis graminum* (Rondani), the corn root aphid, *Aphis maidiradicis* Forbes, the cowpea aphid, *Aphis craccivora* Koch, the bean aphid, *Aphis fabae* Scopoli, the melon aphid, *Aphis gossypii* Glover, the boat gall aphid, *Hyalopterus atriplicis* (L.), the pea aphid, *Acyrthosiphon pisum* (Harris), the green peach aphid, *Myzus persicae* (Sulzer), the English grain aphid, *Macrosiphum avenae* (F.), the blue grass aphid, the corn leaf aphid, *Rhopalomyzus poae* (Gillette) and *Rhopalosiphum padi* (L.) (Knoke et al., 1983a). Maize mosaic virus (MMV) is solely transmitted by *Peregrinus maidis* in a persistent-propagative manner. The rate of MMV transmission by *P. maidis* by means of plant acquisition ranged from 5 to 42% (Lastra, 1977; Falk and Tsai, 1985). *P. maidis* was able to acquire MMV in less than 15 min and the patterns of transmission were often erratic (Falk and Tsai, 1985). The efficiency of MMV transmission by *P. maidis* could be increased from 20 to 43% by injection with either purified MMV or with sap from MMV-infected corn plants (Falk and Tsai, 1985).

INCIDENCE AND ECONOMIC IMPORTANCE OF VIRUSES INFECTING MAIZE

An economically devasting maize disease throughout the southeastern part of the United States, commonly known as maize chlorotic dwarf virus (MCDV) is caused by a complex of strains of MCDV. It is considered to be the second major corn virus disease in the USA (Knoke and Louie, 1981). Maize dwarf mosaic virus (MDMV) is one of the most important widely distributed virus diseases of corn (*Z.mays* L.) in the temperate regions of the world especially in U.S.A. and Hawaii. This disease has not been reported as a serious disease in the Tropics and Subtropics (Tsai and Brown, 1989). This disease caused severe yield losses in the early 1960's particularly in dent corn (Williams and Alexander, 1965). Yield losses as high as 40% have been attributed to MDMV wherever maize and sugarcane are cultivated, however, occur predominantly in the United States and Australia. Distribution of the diseases caused by other viruses generally reflects the geographic distribution of their host. However, the crop losses varied greatly depending on the susceptibility of the corn genotype, virus strains, plant age and environmental factors. MDMV was named by Williams and Alexander (1965) and is closely related to sugarcane mosaic virus (SCMV) which has at least 13 strains. The leafhopper-borne maize yellow stripe virus (MYSV) is a tentative member of the *tenuivirus* group. MYSV symptom-types include fine stripe, coarse stripe, and chlorotic stunt; these symptom-types usually appear on different leaves of the same plant (Ammar et al., 1990). Maize streak virus (MSV) remains an economically important disease of maize in much of Africa. Yield loss as high as 100% have been reported under favorable conditions for disease development (susceptible varieties and favorable climatic conditions for leafhopper development). MSV is an economically significant pathogen in maize, cereals and sugar-cane throughout Africa (Damsteegt, 1983; Rose, 1978) and has also been isolated from grasses such as *Coix* spp., *Panicum* spp., *Paspalum* spp. and *Setaria* spp. in Africa (Storey and McClean, 1930; Rose, 1978). Maize stripe virus (MStV) was first described in 1936 in East Africa by Storey who recognized two types of symptoms, one with narrow yellow stripes on the leaves, the other with broad stripes (Storey, 1936). Kulkarni (1973) demonstrated that two symptoms of maize stripe were associated with two distinct pathogens and were transmitted persistently by the corn delphacid, *P. maidis* (Ashmead). Later Bock et al. (1974) proved that the narrow yellow stripe was caused by a *rhabdovirus*. To date, maize stripe has been reported from Venezuela, Florida the Philippines, Mauritius, Australia, Peru and Taiwan (Tsai and Falk, 1993). Maize mosaic virus (MMV) was first reported in 1914 in Hawaii (Kunkel, 1921). It is considered a serious disease in the tropics and subtropics, and has been speculated as a possible cause of the collapse of Mayan civilization (Brewbaker, 1980). MMV is also transmitted by *P. maidis* in a persistent manner. Maize mosaic has often been confused with maize stripe in the literature because of their similarity. MMV has been reported in Central and South America, Mexico, India, Mauritius, Reunion, Madagascar, and Tanzania (Tsai and Falk, 1993). Maize rayado fino virus (MRFV) was first reported in El Salvador in the 1960's (Ancalmo and Davis, 1961). Later, Gamez (1983) demonstrated a Costa Rican isolate of MRFV transmission by the corn leafhopper, *Dalbulus maidis*. This disease has also been found in Uruguay, Brazil, Colombia, Panama, Guatemala, Honduras, Nicaragua, Mexico, Peru, Venezuela, Ecuador and the U.S. (Nault et al., 1980; Toler et al., 1985). Yield losses in Central America may be up to 40 to 50% of early infected plants. Losses and incidences may reach 100% for newly introduced cultivars (Gamez, 1983). The isolation, culture and characterization of Maize fine streak virus (MFSV; *rhabdovirus*) obtained from leaf samples of Decatur County in Southwestern Georgia that exhibited chlorotic vein streaking. Based on symptoms incited by the virus, it has been previously named maize fine streak virus. The initial symptoms included chlorotic spots and short streaks on small veins that were unevenly distributed. Symptoms were fully developed with continuous streaks on intermediate and small veins, on leaves four to seven, between 3 and 4 weeks post-vascular puncture inoculation (VPI). The streaks enlarged as the plants matured, such that after approximately 6 weeks post-VPI the leaves appeared white with isolated green spots. Alternatively, some plants showed a partial recovery with only scattered chlorotic spots and streaks on later leaves. The virus has been found only in limited areas of Southwesteren Georgia in two fall-growing seasons, suggesting the virus is currently of limited agronomic

importance (Redinbaugh et al., 2002).

MAIZE VIRUS TRANSMISSION AND VIRUS-VECTOR INTERACTIONS

Viruses are transmitted predominantly by several genera of aphids, leafhoppers for ex. *Cicadulina arachidis* but can also be transmitted mechanically and through seed. Yield losses due to viral disease may be extensive. In economic terms, viruses are only of importance if it is likely that they will spread to crops during their commercial lifetime, which of course varies greatly between very short extremes to long extremes from crop to crop. To date, no plant virus is known to use a specific cellular receptor of the type that animal and bacterial viruses use to attach to cells. Rather, plant viruses rely on a mechanical breach of the integrity of a cell wall to directly introduce a virus particle into a cell. This is achieved either by the vector associated with transmission of the virus or simply by mechanical damage to cells. After replication in an initial cell, the lack of receptors poses special problems for plant viruses in recruiting new cells to the infection (Hull, 1989).

Most plant viruses are absolutely dependent on a vector for plant-to-plant spread. Although a number of different types of organisms are vectors for different plant viruses, phloem-feeding hemipterans are the most common and transmit the great majority of plant viruses (Ng and Falk, 2006). The complex and specific interactions between hemipteran vectors and the viruses they transmit have been studied intensely, and two general strategies, the capsid and helper strategies, are recognized. Both strategies are found for plant viruses that are transmitted by aphids in a non persistent manner. Evidence suggests that these strategies are also found for viruses transmitted in a semipersistent manner. Recent applications of molecular and cell biology techniques have helped to elucidate the mechanisms underlying the vector transmission of several plant viruses (Harris et al., 1981). The apparent absence of sites for virus retention and accumulation in the non-vector, *D. maidis,* provides a plausible explanation for Maize chlorotic dwarf virus (MCDV) leafhopper transmission specificity. MCDV is restricted to the phloem of infected plants (Harris and Childress, 1983), and virions retained in the foregut would be bathed by fluids ingested from the phloem. Overall, the characteristics of semi-persistent, non-circulative virus transmission seem compatible with an internal vector retention site. The vector's loss of the ability to inoculate through moulting (non-transstadial) is a characteristic of non-circulative transmission; shedding of the intima during moulting would result in loss of MCDV and, hence, its transmission (Childress and Harris, 1989).

MStV is transmitted by *P. maidis* in a persistent-propagative manner. Nymphs of *P. maidis* transmitted

MStV with ca. twice the efficiency after a 24, 48, 68, 96 and 192 h acquisition access period (AAP) as did adults. Macropterous adults were slightly more efficient transmitters than brachypterous adults (Tsai and Falk, 1993). MRFV is transmitted by *D. maidis* in a persistent manner. A protracted incubation period in the vector is required. The rate of MRFV transmission by *D. maidis* was usually low ranging from 10 to 34% (Nault et al., 1980; Gamez, 1983). Nymphs were more efficient transmitters than the adults (Gamez, 1983). The average incubation period (IP) in *D. maidis* varied from 12.5 to 16 days. The average retention period in *D. maidis* ranged from 16.5 to 20.2. The infectivity of partially purified MRFV was demonstrated by vector injection and membrane feeding (Gamez, 1983). This author demonstrated that the transmission rate for *D. maidis* injected with partially purified MRFV was dosage dependent. Using ELISA tests, MRFV was shown to multiply in *D. maidis* in a time course study (Gingery et al., 1982; Rivera and Gamez, 1986). The Texas isolate of MRFV has also been experimentally transmitted by *D. elimatus, Stirellus bicolor*, and *Graminella nigrifrons* (Gamez, 1983). Symptoms caused by the severe strain of MCDV include severe stunting, leaf discoloration (reddening and yellowing), and leaf-tearing of maize (Bradfute et al., 1972; Gordon and Nault, 1977). A consistent, diagnostic symptom of MCDV-S infection is chlorosis of the tertiary leaf veins (vein banding) (Gordan and Nault, 1977; Pratt et al., 1994). Transmission is by the detocephaline leafhopper, *G. nigrifrons* (Forbes), in a semi-persistent manner (Gingery et al., 1981). Maize dwarf mosaic virus, a subgroup of the sugarcane mosaic virus (SCMV) complex of potyviruses (Gingery, 1981), has a wide host range in the Gramineae and is nonpersistently transmitted by more than 20 aphid species (Knoke and Louie, 1981). MDMV consists of several strains initially characterized by host range and later by serological differences (Hill et al., 1973; Hill and Benner, 1976). The maize stripe virus (MStV) is a member of the newly described *tenuivirus* group (Gingery, 1988). Maize streak virus (MSV) is transmitted by leafhoppers of the genus *Cicadulina* (Rose, 1978; Van Rensburg, 1981).).Maize fine streak virus (MFSV) is not transmitted by any of the insects tested under nonpersistent or semipersistent conditions. Because electron microscopy indicated the pathogen was a *rhabdovirus*, vector transmission under persistent conditions was tested with species of the Aphididae, Delphacidae, and Cicadellidae. Transmission by the known maize rhabdovirus vectors *P. maidis* (MMV) and *Endria inimical* (WAMSV) was unsuccessful. Only the leafhopper, *G. nigrifrons*, transmitted MFSV. G. nigrifrons also transmitted MFSV to barley (*Hordeum vulgare*), wheat (*Triticum aestivium*), oat (*Avena sativa*), gaint foxtail (*Seteria faberi*), and rye (*Secale cereale*), but did not transmit the virus to sorghum (Sorghum bicolor 'Atlas') or Johnsongrass (*Sorghum halpense*)

(Redinbaugh et al., 2002).

COMMONLY USED DIAGNOSTIC TOOLS IN MAIZE VIRUS DETECTION

PCR based tools play a vital role in diagnosis, detection and identification of viruses in plants. Traditional diagnosis of plant viruses requires bioassay, an indicator plant, determination of host range, symptomatology, virus particle morphology (size and shape), and vector relations. A single diagnostic test or assay may provide adequate information on the identity of a virus but a combination of methods is generally needed which are specific, sensitive and inexpensive (Naidu and Hughes, 2003). However, progress in molecular biology, biochemistry and immunology has led to the development of many new, accurate, rapid and less labour-intensive methods of virus detection. Technologies for the molecular detection of plant pathogens have already undergone two major breakthroughs well over the past three decades. The first was the advent of antibody based detection, in particular monoclonal antibodies and enzyme-linked immunosorbent assay (Kohler and Milstein, 1975; Clark and Adams, 1977). There are various immuno-diagnostic and molecular-diagnostic techniques presently available in the field of virology and these are divided into two: Protein based techniques which include precipitation/agglutination tests, enzymes linked immunosorbent assay (ELISA), Immunosorbent electron microscopy (ISEM), fluorescent antibody test, dot immunoblotting assay (DTBIA). Viral nucleic acid based techniques are dot blot hybridization/slot blot hybridization, polymerase chain reaction (PCR), nucleic acid hybridization with radio labelled and nonradio-labelled probes, DNA/RNA probes. Appropriate screening procedures have been conducted in order to certify any plant free of certain pathogen using ELISA, PCR, DNA probes. Occurrence of virus in maize in several African, Asian and American countries has been reported (Thottappilly et al., 1993). ELISA and other modified forms e.g. direct antigen coating enzymes linked immunosorbent assay (DAC-ELISA), double antibody sandwich ELISA (DAS-ELISA), antigen-coated plate (ACP-ELISA), plate trapped antigen (PTA-ELISA), triple antibody sandwich (TAS-ELISA).

Consequently, several laboratories have developed methods either based on DNA detection using the polymerase chain reaction (PCR) technique, or based on protein detection using ELISA (Anklam et al., 2002). However, the methods vary in their reliability, robustness and reproducibility; in combination with different levels of cost, complexity, and speed etc. Moreover, there is no one method that is applicable in all circumstances. A further consideration is the claim of very high sensitivity reached in the analysis even in absence of clearly proven detailed performance studies.

Serology was the first method adopted in the evolution of rapid plant pathogen detection and identification (Clark, 1981; Miller and Martin, 1988). This technique is based on the recognition of antigens with antibodies produced to them. In its initial application by plant virologists, serology had been used routinely to identify virus species and strains but was not amenable to high throughput assays. ELISA is based on a nearly decade earlier demonstration by Avrameas (1969) that glutaraldehyde cross-linked enzyme-antibody conjugates retained both the specificity of the IgG molecule and the catalytic properties of the enzyme. ELISA allows qualitative and quantitative analysis, high throughput, and high sensitivity and was adopted rapidly and widely (Miller and Martin, 1988).

Virions of viruses that are sap-transmissible to herbaceous hosts usually can be purified in milligram amounts and high purity for serial injection into rabbits or goats and recovery of polyclonal antibody from serum, or into chickens, for recovery from yolk. Examples are members of the genera *Nepovirus, Ilarvirus, Trichovirus* and *Vitivirus*. These preparations have been used for the production of polyclonal antibodies and also to inject into mice to produce monoclonal antibodies (Mabs), the production of which requires special facilities and protocols (Halk and DeBoer, 1985; Hu et al., 1990; Torrance, 1995; Schieber et al., 1997; Boscia et al., 2001). Mabs often have less avidity than polyclonal antibodies, but the high specificity of Mabs allows strain differentiation and eliminates the problem of cross-reaction with host material (Permar et al., 1990; Nemchinov et al., 1996). If strain specificity is not desired, a broad-spectrum reagent can be produced by combining Mabs generated from several cell lines. Where sequence information is available but purified virions have not been obtained, antibodies with reactivity to the virion may be raised against synthetic peptides (Robinson et al., 1997; Ling et al., 2000).

Enzyme-linked immunosorbent assay (ELISA)

Two classes of ELISA protocols are used for surveillance (Koenig and Paul, 1982). Direct methods such as double antibody sandwich (DAS)–ELISA involves enzyme attachment to the antibody probe (Bar-Joseph and Salomon, 1980; Koenig, 1978; Rochow and Carmichael, 1979; Uyemoto, 1980). In the indirect method [(DASI)-ELISA], the antibody probe remains unlabeled. Instead, the enzyme is attached to a second antibody or protein, a reactive specifically to the probe antibody (Koenig and Paul, 1982; Rowhani et al., 1985). DASI-ELISA is favored over DAS-ELISA for its greater sensitivity, broader reactivity and convenience. Only a single enzyme conjugate is needed for assays of different viruses, and usually a suitable conjugate is available commercially. However, many factors may influence the sensitivity and

reliability of ELISA assay, among these are quality of antibodies, preparation and storage of reagents, incubation time and temperature, selection of appropriate parts of sample and the use of suitable extraction buffer (McLaughlim et al., 1981). ELISA is an excellent technique for detection of seed borne viruses (Bashir and Hampton, 1996). Generally a sample is regarded as positive if the absorbance value exceeds the mean value of a negative control by 2 to 3 standard deviations (Naidu and Hughes, 2003). An optimized DTBIA is as sensitive as ELISA, simple, relatively inexpensive and the DTBIA result can be scored visually, but differs from ELISA as the plant extracts are spotted on to a membrane rather than using a microlitre plate as the solid support matrix. Abdullahi et al. (2001) evaluated the detection capacity of ELISA to prove its reliability using a reverse transcriptase PCR assay, thus, PCR confirmed ELISA.

Two virus isolates associated respectively with the maize stripe (MStV) and the maize chlorotic stripe (MCStV) syndromes in Mauritius, have been purified and characterized by using antisera. The most sensitive diagnosis was achieved using $F(ab')_2$ ELISA (Roca de Doyle et al., 2007). Antisera to the 32,000 Mr (32 K) capsid and 16,500 Mr (16 K) In immunological assay the antiserum against the noncapsid protein was found very useful for detecting MStV infections in plants caused by *P. maydis* by indirect ELISA (Falk et al., 1987; Tsai and Falk, 1993). In the DAS-ELISA with purified virus of Taiwan isolate as antigen, specific reactions to two maize streak virus (MStV-FL) antisera were used. The 19.8 kDa antiserum of MStV-FL gave very strong reactions with crude sap and the noncapsid protein of MStV-T by indirect ELISA (Chen et al., 1993). MYSV was detected, through symptomatology and insect transmission, through ELISA and dot-blot methods, for detecting MYSV in several host plants and weeds as well as in the vector leafhopper *C. chinai* (Mahmoud et al., 1996). Using DAC-ELISA, MYSV antiserum and dot-blot methods the virus was detected in clarified extracts of *C. chinai* leafhoppers at a dilution of 10^3. However, DAC-ELISA is more economic and less complicated than other ELISA methods and is thus more useful in field surveys. However, the dot-blot method may be more sensitive for detecting MYSV in single leafhoppers (Ammar et al., 1990). The ELISA was very useful in demonstrating the virus titer and effects of factors for MDMV (Jenson et al., 1985). The serological and molecular diagnostic tools (ELISA and Western blots) were very useful tools for determining the virus distribution in maize and other hosts. The serological relationships among maize-infecting rhabdovirus have been reported using western blot analysis (Redinbaugh et al., 2002). Assays of corn for MDMV-A and MDMV-B and of Johnsongrass for MDMV-A detection was performed by enzyme linked immunosorbent assay (EIA) and distribution of MDMV and MCDV was studied (Knoke et al., 1983b). The information on the serological relationships between coat proteins of MSV isolates has been obtained using polyclonal antisera (Dekker et al., 1988). Comparison of different ELISA methods was performed for MSV to determine which method is most suitable for serotyping of MSV group, that is, DAS-ELISA, Direct ACP ELISA, Indirect ACP ELISA, Indirect DAS F(ab')2 ELISA and PAS-ELISA (Pinner and Markham, 1990). Among these indirect ELISA procedures proved to be the most useful methods for serotyping the MSV isolates (Pinner and Markham, 1990). There is also report of MStV detection by ELISA in *P. maidis* vector and virus was successfully detected from the midgut of vector (Nault and Gordon, 1988). The ELISA method also proved useful for early detection of maize rayado fino virus and epidemiological studies (Gordon et al., 1985). Maize viruses, that is, MSV, MStV, MDMV-A were detected by using serological tests in Zimbabwe (Bonga and Cole, 1997). ELISA technique proved to be very useful for maize virus detection, distribution and epidemiology study.

Polymerase chain reaction (PCR)

PCR was developed in the mid-1980s (Saiki et al., 1988) and was rapidly adopted to identify pathogens through their DNA genetic materials. PCR assays are extremely sensitive, reliable, fast, and highly versatile. An alternative approach to virion purification is immuno-capture (IC), referred to as IC-RT-PCR (Wetzel et al., 1991; Nolasco et al., 1993; Minafra and Hadidi, 1994; Nemechinov et al., 1995; Rowhani et al., 1995).

The PCR DNA amplification technique was used to detect and typing of maize streak virus isolates with degenerate oligonucleotide primers based on short sequence of genomic DNA. The amplification was specific and extremely sensitive. In addition, the techniques was compared with ELISA and reported that endpoint dilution used in PCR was 10^4-fold lower than that routinely obtained in ELISA with purified virus (Rybicki and Hughes, 1990). MSV strain A has been associated with maize streak virus in Africa and only known strain reported in Africa. In Cameroon, maize streak virus is most important disease. PCR based method (rolling circle amplification) was used to identify the new strain MSV-A and first time reported in Cameroon (Leke et al., 2009). PCR approach are not only used for the diagnostic in maize but also used in genetic mapping of lines against maize streak virus resistance (Pernet et al., 1999), to study the transient and transgenic expression of MSV replication-associated protein mutants and demonstrated that rep[1-219Rb-] transgene is effective against a range of MSV strains (Shepherd et al., 2007). The presence of transgenes in landraces in local maize varieties was determined and transgene elements were detected with highly sensitive PCR-based markers (Ortiz-García et al., 2005). PCR-RFLP applied for genetic analysis of MSV isolates in

Uganda. Sixty-two full-genome sequences were determined, 52 of which were detectably recombinant in which two recombinants contained predominantly MSV-A(1)-like sequences. Interestingly, it was demonstrated that its characteristics in MSV are quite different from those observed in related African cassava-infecting geminivirus species (Owor et al., 2007).

Multiplex-PCR has a distinct advantage in that it allows the concurrent identification of viruses in plants with mixed infections, all in a single PCR experiment (Routh et al., 1988; Saldarelli et al., 1998; Saade et al., 2000; Wetzel et al., 2002; Dovas and Katis, 2003). Multiplex PCR procedures were applied for simultaneously detecting multiple target sequences in genetically modified (GM) soybean, maize, and canola. Simultaneous amplification profiling (SAP), rather than target specific detection was used for the identification of four GM maize lines. For maize nonspecific amplification was utilized as a tool for specific and reliable identification of one line of GM maize. SAP proved simple and has the potential to identify both approved and non approved GM lines in maize (James et al., 1999).

In real time-PCR, a fluorescent-labeled oligonucleotide (e.g., TaqMan fluorescent probe) in the reaction mixture and a laser-excited fluorescence detection monitor are utilized to assess the quantity of PCR product at the end of each PCR cycle. The TaqMan probe set consists of a pair of oligonucleotide primers and a TaqMan probe designed to hybridize to a site between the two primer binding sites. This method eliminates the need for product detection by gel electrophoresis. It is quantitative and highly sensitive (Korimbocus et al., 2002; Mackay et al., 2002; Marbot et al., 2003). In a recent study, application of real-time TaqMan RT-PCR was demonstrated. Maize chlorotic mottle virus is transmitted through infected maize seeds. It becomes difficult to detect this virus in the seeds to prevent its introduction and infection. A real-time TaqMan RT-PCR procedure for efficient detection of MCMV was developed and its sensitivity was tested. The sensitivity of the method was 4 fg of total RNA or 25 copies of RNA transcripts, which was approximately ten-fold higher than conventional RT-PCR gel electrophoresis method (Zhang et al., 2010).

Recently, a novel real-time quantitative PCR assay was developed for the detection and quantification of plant viruses (Heid et al., 1996). Polymerase chain reaction (PCR) is a molecular biology method for enzymatically copying target nucleic acid sequence without using a living organism, in which repeated replication of a given sequence forms millions of copies within a few hours. PCR technique is a DNA based technology that permits a small sample of target nucleic acid to be copied multiple times for analysis (Mullis and Faloona, 1987).

Flinders technology associates (FTA)

FTA is a paper-based technology designed for the

collection and archiving of nucleic acids, either in their purified form or within pressed samples of fresh tissue. Proprietary chemicals impregnated into the paper act to lyze cellular material and fix and preserve DNA and RNA within the fibre matrix (Whatman, 2004). After a short drying period, pressed samples can be stored at room temperature for extended periods and processed when required. Nucleic acids are recovered by removing small punches from the pressed area and washing with simple reagents. RNA and smaller DNA molecules, such as plasmids and viral genomic components, are eluted by a simple extraction buffer and used as template for amplification by PCR. Genomic DNA remains attached to the paper matrix but are available for amplification by PCR when the paper punch is included in the PCR reaction mix (Ndunguru et al., 2005). Predicted PCR products were obtained in 100 and 80% of the cassava leaf samples collected from the greenhouse and field respectively; with the entire MSV-infected field grown maize plants sampled yielding viral sequences. The studies described here demonstrate that FTA offers a simple, sensitive and specific tool appropriate for the diagnosis and molecular characterization of plant viral pathogens isolated from plant tissues and transgene sequences integrated into the plant genome (Ndunguru et al., 2005).

Conclusions

The ultimate goal in microbial testing is the ability to accurately and sensitively detect pathogens in real-time or as quickly as possible. Nucleic acid diagnostics (NAD) offer many advantages over traditional microbiological and immunological methods for the detection of infections microorganisms. These include faster processing time as well as greater potential for intra-species identification and identification of antibiotic susceptibility and strain typing based upon unique sequences. The original techniques of PCR and gel electrophoresis are being superseded by real-time PCR while the development of integrated sample preparation and amplification devices with a simplified user interface will allow for true point-of-care disease detection and suitably tailored treatments. This chapter describes the principles of maize virus incidence, host-pathogen interaction, ELISA and nucleic acid diagnostics including an overview of the technology's history as well as the general properties of diagnostics target. Special emphasis is placed upon the detection of pathogens relevant to maize. Ongoing developments in molecular detection platforms including nucleic acid based provide potential for new test methods that will enable multi-parameter testing and at-line monitoring for viral contaminants.

Mostly disease problems are first noticed in the field and depending on the scope and urgency, this will immediately be followed by applied efforts to contain, control, or eradicate the pathogen. If the problem is sufficiently relevant from an economic perspective and interesting from a fundamental viewpoint, research on molecular diagnostic most likely will be performed to find the actual causal agent. This can be the natural host or an alternative suitable plant species. Aided by genomics and functional genomics data, genes will be identified and/or manipulated to eliminate the pathogen, or protect the plant from either the infection or the symptoms.

Accurate identification and early detection of the viral diseases is the cornerstones of the management of maize cultivar. Maize viruses are difficult to identify using morphological criteria, which can be time consuming and challenging and requires extensive knowledge in taxonomy. Molecular and immunological detection such as ELISA and other modified forms, precipitation/agglutination, fluorescent antibody, PCR, nucleic acid hybridization are best suitable techniques to detect the various viruses, which include, MMV, MCMV, MSV, MStV, MMV and MDMV infecting maize. Until now, ELISA and other modified forms have been extensively used, because these are quick. However PCR has been widely used with the varying degree of modification for detection of viral genomes in infected plant in the last two decades. The disadvantage of PCR is that it requires sophisticated equipment like thermocycler which is expensive, where as ELISA/DAC-ELISA can be used for diagnoses even in field conditions and are very cost effective. Viruses and virus strains cannot be distinguished on the basic of common sources of resistance. These various immunological and molecular diagnostic tests with symptoms and history are of immense value to diagnose maize viruses, thus, these diagnostic techniques can become a routine in plant pathology research. This review will substantially accelerate to understand host-pathogen interaction, multiplication and their diagnostic assay. This should, in turn, lead to development of very effective and durable control measures against harmful pathogens.

Abbreviations: MStV, maize stripe virus; **MMV,** Maize mosaic virus; **MDMV,** maize dwarf mosaic virus; **MCMV,** maize chlorotic mottle virus; **MRFV,** maize rayado fino virus; **MCDV,** Maize chlorotic dwarf virus; **MNeSV,** maize necrotic streak virus; **MFSV,** maize fine streak virus; **SCMVN,** sugarcane mosaic virus; **MBSM,** maize bushy stunt mycoplasma; **MYSV,** maize yellow stripe virus; **MSV,** maize streak virus; **CSS,** corn stunt spiroplasma; **RSV,** rice stripe virus; **MCStV,** maize chlorotic stripe; **WASMV,** wheat american striate mosaic virus; **SSMV,** Sorghum stunt mosaic virus; **RBSDV,** rice black-streaked dwarf virus; **RTSV,** rice tungro spherical virus; **ISEM,** Immunosorbent electron microscopy; **DIBA,** dot immunoblotting assay; **DAC-ELISA,** direct antigen coating enzymes linked immunosorbent assay; **DAS-ELISA,** double antibody sandwich ELISA; **ACP-ELISA,** antigen-coated plate; **PTA-ELISA,** plate trapped antigen; **TAS-ELISA,** triple antibody sandwich; **DTBIA,** direct tissue blot immunoassay; **NAD,** nucleic acid diagnostics; **IC-RT-PCR,** immuno-capture RT-PCR; **NCP,** nuclear coat protein; **UTR,** untranslated region; **ORFs,** open reading frames; **CP,** coat protein; nts, nucleotides; **IAP,** inoculation access period; **AAP,** acquisition access period; **Mabs,** monoclonal antibodies; **VPI,** vascular puncture inoculation ; **IP,** incubation period.

REFERENCES

Abdullahi I, Ikotun T, Winter S, Thottappilly G, Atiri A (2001). Investigation on seed transmission of cucumber mosaic virus in cowpea. Afr. Crop Sci. J., 9(4): 677-684.

Ahlquist P (2002). RNA-dependent RNA polymerases, viruses, and RNA silencing. Science, 296: 1270-1273.

Ahlquist P, Noueiry AO, Lee WM, Kushner DB, Dye BT (2003). Host factors in positivestrand RNA virus genome replication. J. Virol., 77: 8181-8186.

Ammar ED, Gingery RE, Gordon DT, Aboul-Ata AE (1990). Tubular helical structures and fine filaments associated with the leafhopper-borne maize yellow stripe virus. Phytopathology, 80: 303-309.

Ancalmo O, Davis WC (1961). Achaparramiento (corn stunt). Plant Dis., Rep., 45: 281.

Anklam E, Gadani F, Heinze P, Pijnenburg H, Eede GVD (2002). Analytical methods for_detection and determination of genetically modified organisms in agricultural crops and plant-derived food products. Eur. Food Res. Technol., 214: 3-26.

Avrameas S (1969). Coupling of enzymes to proteins with glutaraldehyde. Immunochemistry, 6: 43-52.

Bar-Joseph M, Salomon R (1980). Heterologous reactivity of tobacco mosaic virus strains in enzyme-linked immunosorbent assays. J. Gen. Virol., 47: 509-512.

Bashir M, Hampton RO (1996). Detection and identification of seedborne viruses from cowpea (*Vigna unquiculata* (L.) walp) germplasm. Plant Pathol., 45: 54-58.

Berger PH, Luciano CS, Thornbury DW, Benner HI, Hill JH, Zeyen RJ (1989). Properties and *in vitro* translation of maize dwarf mosaic virus RNA. J. Gen. Virol., 70: 1845-1851.

Bisaro DM (1994). Recombination in geminiviruses: Mechanisms for maintaining genome size and generating genomic diversity. In homologous recombination and gene silencing in plants (ed. J. Paszkowski), Kluwer, Dordrecht, pp. 39-60.

Bock KR, Guthrie EJ, Woods RD (1974). Purification of maize streak virus and its relation-ship to viruses associated with streak diseases of sugarcane and *Panicum maximum*. Ann. Appl. Biol., 77: 289-296.

Bonga J, Cole DL (1997). Identification of viruses infecting maize in Zimbabwe. Afr. Plant Protect., 3(1): 1-9.

Boscia D, Digiaro M, Safi M, Garau R, Zhou Z (2001). Production of monoclonal antibodies to grapevine virus D and contribution to the study of its aetiological role in grapevine diseases. Vitis, 40: 69-74.

Bradfute OE, Tsai JH (1983). Identification of maize mosaic virus in Florida. Plant Dis., 67: 1339-1342.

Bradfute OE, Tsai JH (1990). Rapid identification of maize stripe virus. Phytopathology, 80: 715-719.

Bradfute OE, Gingery RE, Gordon DT, Nault LR (1972). Tissue ultrastructure, sedimentation and leafhopper transmission of a virus associated with a maize dwarfing disease (Abstract). J. Cell Biol., 55: 25a.

Brewbaker JL (1980). Diseases of maize in the wet lowland tropics and the collapse of the Maya civilization. Econ. Bot., 33: 101-118.

Buck KW (1996). Comparison of the replication of positive-stranded RNA viruses of plants and animals. Adv. Virus Res., 47: 159-251.

Chen CC, Tsai JH, Chiu RJ, Chen MJ (1993). Purification, characterizaiton, and serological analysis of maize stripe virus in Taiwan. Plant Dis., 77: 367-372.

Childress SA, Harris KF (1989). Localization of virus-like particles in the foreguts of viruliferous *Gramineila nigrifrons* leafhoppers carrying the semi-persistent maize chlorotic dwarf virus. J. Gen. Virol., 70: 247-251.

CIMMYT (1990). 1989/90 CIMMYT World Maize: Facts and trends: Realizing the potential of maize in sub-Sahara Africa. Mexico, DF, Mexico.

Clark MF, Adams AN (1977). Characteristics of the microplate method of enzyme-linked immunosorbent assay for the detection of plant viruses. J. Gen. Virol., 34: 475-483.

Clark MF (1981). Immunosorbent assays in plant pathology. Ann. Rev. Phytopathol., 19: 83-106.

Damsteegt VO (1983). Maize streak virus: I. Host range and vulnerability of maize germplasm. Plant Dis., 67: 734-737.

de Stradis A, Redinbaugh MG, Abt JJ, Martelli GP (2005). Ultra structure of maize necrotic streak virus infections. J. Plant Pathol., 87(3): 213-221.

Dekker EL, Pinner MS, Markham PG, Van-Regenmortel MHV (1988). Characterization of maize streak virus isolates from different plant species by polyclonal and monoclonal antibodies. J. Gen. Virol., 69: 983-990.

DePamphilis ML (1988). Transcriptional elements as components of eukaryotic origins of replication. Cell, 52: 635-638.

DePamphilis ML (1993). Eukaryotic DNA replication: Anatomy of an origin. Annu. Rev. Biochem., 62: 29-63.

Dovas CI, Katis NI (2003). A spot-nested RT-PCR method for the simultaneous detection of members of the *Vitivirus* and *Foveavirus* genera in grapevine. J. Virol. Meth., 170: 99-106.

Drake JW (1991). A constant rate of spontaneous mutation in DNA-based microbes. 1991. Proc. Natl. Acad. Sci. USA, 88: 7160-7164.

Falk BW, Tsai JH (1983). Physicochemical characterization of maize mosaic virus. Phytopathology, 73: 1536-1539.

Falk BW, Tsai JH (1985). Serological detection and evidence for multiplication of maize mosaic virus in the planthopper, *Peregrinus maidis*. Phytopathology, 75: 852-855.

Falk BW, Tsai JH, Lommell SA (1987). Differences in levels of detection for maize stripe virus capsid and major noncapsid proteins in plant and insect hosts. J. Gen. Virol., 68: 1801-1811.

FAO (2007). http://faostat.fao.org/site/567/default.aspx.

Galinat WC (1977). The origin of corn. In: Corn and Corn Improvement. (eds: G.F. Sprague). American Society of Agronomy. Madison, Wisconsin, USA, pp. 1-47.

Gamez R (1983). Maize rayado fino disease: The virus-host-vector interaction in neotropical environments. In: Proc. Int'l. Maize Virus Dis. Colloq. and Workshop, 2-6 August 1982. (eds: D.T. pp. 62-68.

Gordon JK, Knoke LR, Nault RMR (YYYY). The Ohio State University, Ohio Agricultural Research and Development Center, Wooster, pp. 266.

Gingery R (1988). The rice stripe virus group. In: The Plant Viruses 4: The Filamentous Plant Viruses. (eds: R.G. Milne). Plenum, New York, pp. 297-329.

Gingery RE (1976). Properties of maize chlorotic dwarf virus and its ribonucleic acid. Virology, 73: 311-318.

Gingery RE (1981). Chemical and physical properties of maize viruses pp. 38-39. In: viruses and virus-like diseases of maize in the United States. (Eds: D.T. Gordon, J.K. Knoke and G.E. Scott). Wooster: Ohio Agricultural. Research Development Center.

Gingery RE, Gordon DT, Nault LR (1982). Purification and properties of an isolate of maize rayado fino virus from the United States. Phytopathology, 72: 1313-1318.

Gingery RE, Gordon DT, Nault LR, Bradfute OE (1981). Maize chlorotic dwarf virus pp.19-32. In: (eds: E.Kurstak). Plant Virus Infection and Comparative Diagnosis Elsevier/North Holland Biochemical Press, Amsterdam.

Golem S, Culver JN (2003). Tobacco mosaic virus induced alterations in the gene expression profile of *Arabidopsis thaliana*. Mol. Plant-Microbe Interact., 16: 681-688.

Gordon DT, Nault LR (1977). Involvement of maize chlorotic dwarf virus and other agents in stunting diseases of *Zea mays* in the United States. Phytopatholgy, 67: 27-36.

Gordon DT, Nault LR, Gordon NH, Heady SH (1985). Serological detection of corn stunt spiroplasma and maize rayado fino virus in field-collected *Dalbulus* spp. from Mexico. Plant Dis., 69: 108-111.

Gutierrez C (1999). Geminivirus DNA replication. Cell. Mol. Life Sci., 56(3-4): 313-329.

Halk EL, DeBoer SH (1985). Monoclonal antibodies in plant-disease research. Ann. Rev. Phytopathol., 23: 321-330.

Hammond RW, Ramirez P (2001). Molecular characterization of the genome of *maize rayado fino virus*, the type member of the genus *Marafivirus*. Virology, 282(2): 338-347.

Harris KF, Childress SA (1983). Cytology of maize chlorotic dwarf virus infection in corn. Int. J. Trop. Plant Dis., 1: 135-140.

Harris KF, Treur B, Tsai J, Toler R (1981). Observations of leaf hopper ingestion-egestion behavior: its likely role in the transmission of noncirculative viruses and other plant pathogens. J. Econ. Entomol., 74: 446-453.

Heid CA, Stevens J, Livak KJ, Williams PM (1996). Real time quantitative PCR. Genome Res., 10: 986-994.

Hill JH, Benner HI (1976). Properties of potyvirus RNAs: turnip mosaic, tobacco etch, and maize dwarf mosaic viruses. Virology, 75: 419-432.

Hill JH, Ford RE, Benner HI (1973). Purification and partial characterization of maize dwarf mosaic virus strain B (sugarcane mosaic virus). J. Gen. Virol., 20: 327-339.

Howell SH (1984). Physical structure and genetic organisation of the genome of maize streak virus (Kenyan isolate). Nucleic Acids Res., 12(19): 7359-7375.

Hu JS, Gonsalves D, Boscia D, Namba S (1990). Use of monoclonal antibodies to characterize grapevine leafroll associated closteroviruses. Phytopathology, 80: 920-925.

Hull R (1989). The Movement of Viruses in Plants. Ann. Rev. Phytopathol., 27: 213-240.

ICTVdB Management (2006). 00.000.4.00.008. Maize white line mosaic virus. In: ICTVdB - The Universal Virus Database, version 4. Büchen-Osmond, C. (Ed), Columbia University, New York, USA.

ICTVdB Management (2006). 00.057.0.01.039. Maize dwarf mosaic virus. In: ICTVdB - The Universal Virus Database, version 4. Büchen-Osmond, C. (Ed), Columbia University, New York, USA.

ICTVdB Management (2006). 00.060.0.07.003. Maize rough dwarf virus. In: ICTVdB - The Universal Virus Database, version 4. Büchen-Osmond, C. (Ed), Columbia University, New York, USA.

ICTVdB Management (2006). 00.065.0.02.003. Maize chlorotic dwarf virus. In: ICTVdB - The Universal Virus Database, version 4. Büchen-Osmond, C. (Ed), Columbia University, New York, USA.

ICTVdB Management (2006). 00.069.0.01.002. Maize stripe virus. In: ICTVdB - The Universal Virus Database, version 4. Büchen-Osmond, C. (Ed), Columbia University, New York, USA.

ICTVdB Management (2006). 01.062.0.05.004. Maize mosaic virus. In: ICTVdB - The Universal Virus Database, version 4. Büchen-Osmond, C. (Ed), Columbia University, New York, USA.

James D, Jelkmann W, Upton C (1999). Specific detection of cherry mottle leaf virus using digoxygenin-labeled cDNA probes and RT-PCR. Plant Dis., 83: 235-239.

Jenson SG, Palomar MK, Ball EM, Samson R (1985). Factors influencing virus titer in maize dwarf mosaic virus-infected sorghum. Phytopathology, 75: 1132-1136.

Jiang Y, Serviene E, Gal J, Panavas T, Nagy PD (2006). Identification of essential host factors affecting tombusvirusRNAreplication based on the yeastTet promoters Hughes Collection. J. Virol., 80: 7394-404.

Knoke JK, Louie R (1981). Epiphytology of maize virus diseases. In: Gordon DT, Knoke JK, Scott GE (eds): Virus and virus-like diseases of maize in the United States. Southern Coop. Ser. Bull., 247: 92-102.

Knoke JK, Anderson RJ, Louie R, Modelen LV, Findley WR (1983a). Insect vectors of maize dwarf mosaic virus and maize chlorotic dwarf virus. pp. 130-138. In: (eds: Gordon DT, Knoke JK, Nault LR, Ritter RM). Proc. Int'l. Maize Virus Disease Calloq. and Workshop. 1982. Ohio Agriculture Research and Development Center, Wooster.

Knoke JK, Louie R, Madden LV, Gordon DT (1983b). Spread of maize dwarf mosaic virus from Johnsongrass to corn. Plant Dis., 67: 367-370.

Koenig R (1978). ELISA in the study of homologous and heterologous reactions of plant viruses. J. Gen. Virol., 40: 309-318.

Koenig R, Paul HI (1982). Variants of ELISA in plant virus diagnosis. J. Virol. Methods, 5: 113-125.

Kohler G, Milsten C (1975). Continuous culture of fused cells secreting antibody of predefined specificity. Nature, 256: 495-497.

Kong P, Steinbiss HH (1998). Complete nucleotide sequence and analysis of the putative polyprotein of maize dwarf mosaic virus

genomic RNA (Bulgarian isolate). Arch. Virol., 143(9): 1791-1799.

Korimbocus J, Coates D, Barker I, Boonham N (2002). Improved detection of Sugarcane yellowleaf virus using a real-time fluorescent (TaqMan) RT-PCR assay. J. Virol. Methods, 103: 109-120.

Kulkarni HY (1973). Comparison and characterization of maize stripe and maize line viruses. Ann. Appl. Biol., 75: 205-216.

Kunkel LO (1921). A possible causative agent for the mosaic disease of corn. Hawaii. Sugar Plantation Association Experimental Station. Bull. Bot. Ser., 3: 44-58.

Kushner DB, Lindenbach BD, Grdzelishvili VZ, Noueiry AO, Paul SM, Ahlquist P (2003). Systematic, genome-wide identification of host genes affecting replication of a positivestrand RNA virus. Proc. Natl. Acad. Sci. USA, 100: 15764-15769.

Laliberte JF, Sanfacon H (2010). Cellular Remodeling During Plant Virus Infection. Ann. Rev. Phytopathol., 48: 69-91.

Lastra RJ (1977). Maize mosaic and other maize virus and virus-like diseases in Venezuela. In: Proc. Int'l. Maize Virus Dis. Colloq. and Workshop, 16-19 Aug., 1976. (eds: Williams LE, Gordon DT and Nault LR). Ohio Agriculture Research Developmental Center, Wooster, pp. 30-39.

Leke WN, Njualem DK, Nchinda VP, Ngoko Z, Zok S, Ngeve JM, Brown JK, Kvarnheden A (2009). Molecular identification of Maize streak virus reveals the first evidence for a subtype A1 isolate infecting maize in Cameroon. Dis. Rep., 19: 36.

Ling K, Zhu HY, Jiang Z, Gonsalves D (2000). Effective application of DASELISA for detection of grapevine leafroll associated closterovirus-3 using a polyclonal antiserum developed from recombinant coat protein. Eur. J. Plant Pathol., 106: 301-309.

Lommel SA, Kendall TL, Siu NF, Nutter RC (1991). Characterization of maize chlorotic mottle virus. Phytopathology, 81: 819-823.

Louie R, Redinbaugh MG, Gordon DT, Abt JJ, Anderson RJ (2000). Maize necrotic streak virus, a new maize virus with similarity to species of the family Tombusviridae. Plant Dis., 84: 1133-1139.

Mackay IM, Arden KE, Nitsche A (2002). Survey and summary, real-time PCR in virology. Nucl. Acid Res., 30: 1292-1305.

Mahmoud A, Hotjvenel JC, Abol-Ela SE, Sewify GH, Ammar ED (1996). Detection of maize yellow stripe tenui-like virus by ELISA and dot-blot tests in host plants and leafhopper vector in Egypt. Phytopathol. Mediterranea, 35: 19-23.

Malpica JM, Fraile A, Moreno I, Obies CI, Drake JW, Garcia-Arenal F (2002). The rate and character of spontaneous mutation in an RNA virus. Genetics, 162: 1505-1511.

Marbot S, Salmon M, Vendrame M, Huwaert A, Kummert J (2003). Development of real-time RT-PCR assay for detection of *Prunus necrotic* ringspot virus in fruit trees. Plant Dis., 87: 1344-1348.

Maule A, Leh V, Lederer C (2002). The dialogue between viruses and hosts in compatible interactions. Curr. Opinion Plant Biol., 5: 279-284.

Mayo MA, Murant AF, Turnbull-Ross AD, Reavy B, Hamilton RI, Gingery RE (1995). Family-Sequiviridae In: (eds: Murphy F.A., Fauquet C.M., Bishop D.H.L., Ghabrial S.A., Jarvis A.W., Martelli G.P., Mayo M.A., Summers M.D): Virus Taxonomy. 6th Rep. Int. Committee on Taxonomy of Viruses. Springer-Verlag, Vienna and New York, pp. 337-340.

McDaniel LL, Ammar ED, Gordon DT (1985). Assembly, morphology, and accumulation of a Hawaiian isolate of maize mosaic virus in maize. Phytopathology, 75: 1167-1172.

McLaughlin MR, Barnett OW, Burrows PM, Bavm RH (1981). Improved ELISA conditions for detection of plant viruses. J. Virol. Methods. 3: 13-25.

Miller SA, Martin RR (1988). Molecular diagnosis of plant diseases. Ann. Rev. Phytopathol., 26: 409-432.

Minafra A, Hadidi A (1994). Sensitive detection of grapevine virus A, B, or leafroll associated III from viruliferous mealybugs and infected tissue by cDNA amplification. J. Virol. Methods, 47: 175-188.

Mullineaux PM, Donson J, Morris-Krsinich BAM, Boulton MI, Davies JW (1984). The nucleotide sequence of maize streak virus DNA. EMBO J., 3: 3063-3068.

Mullis KB, Faloona FA (1987). Specific synthesis of DNA in vitro via a polymerase catalyzed chain reaction. Methods Enzymol., 155: 335-350.

Nagy PD, Pogany J (2006). Yeast as a model host to dissect functions of viral and host factors in tombusvirus replication. Virology, 344: 211-20.

Naidu RA, Hughes JDA (2003). Methods for the detection of plant viral diseases in plant virology in sub-Saharan Africa, Proceedings of plant virology, IITA, Ibadan, Nigeria. Eds. Hughes JDA, Odu B, pp. 233-260.

Nault LR, Gordon DT (1988). Multiplication of maize stripe virus in *Peregrinus maidis*. Phytopathology, 78: 991-995.

Nault LR, Gingery RE, Gordon DT (1980). Leafhopper transmission and host range of maize rayado fino virus. Phytopathology, 70: 709-712.

Nault LR, Madden LV, Styer WE, Triplehorn BW, Shambaugh GF, Heady SE (1984). Pathogenicity of corn stunt spiroplasma and maize bushy stunt mycoplasma to its vector *Dalbulus longulus*. Phytopathology, 74: 977-979.

Ndunguru J, Taylor NJ, Yadav J, Aly H, Legg JP, Aveling T, Thompson G, Fauquet CM (2005). Application of FTA technology for sampling, recovery and molecular characterization of viral pathogens and virus-derived transgenes from plant tissues. Virol. J., 2: 45 doi:10.1186/1743-422X-2-45.

Nemchinov L, Hadidi A, Maiss E, Cambra M, Candresse T (1996). Sour cherry strain of plum pox potyvirus (PPV): molecular and serological evidence for a newsubgroup of PPV strains. Phytopathology, 86: 1215-1221.

Nemechinov L, Hadidi A, Candresse T, Foster JA, Verdervskaya TD (1995). Sensitive detection of Apple chlorotic leafspot virus from infected apple or peach tissue using RT-PCR, IC-RT-PCR, or multiplex IC-RT-PCR. Acta Hortic., 386: 51-62.

Ng JCK, Falk BW (2006). Virus-vector interactions mediating nonpersistent and semipersistent transmission of plant viruses. Ann. Rev. Phytopathol., 44: 183-212.

Nolasco G, de Blas C, Torres V, Ponz F (1993). A method combining immunocapture and PCR amplification in a microtiter plate for the detection of plant viruses and subviral pathogens. J. Virol. Methods, 45: 201-218.

Ortiz-García S, Ezcurra E, Schoel B, Acevedo F, Soberón J, Snow AA, Schaal BA (2005). Absence of Detectable Transgenes in Local Landraces of Maize in Oaxaca, Mexico (2003-2004). Proc. Natl. Acad. Sci. USA, 102(35): 12338-12343.

Overman MA, Ko NJ, Tsai JH (1992). Identification of viruses and mycoplasmas in maize by light microscopy. Plant Dis., 76: 318-322.

Owor BE, Martin DP, Shepherd DN, Edema R, Monjane AL, Rybicki EP, Thomson JA, Varsani A (2007). Genetic analysis of maize streak virus isolates from Uganda reveals widespread distribution of a recombinant variant. J. Gen. Virol., 88(3): 154-165.

Permar TA, Garnsey SM, Gumpf DJ, Lee RF (1990). A monoclonal antibody that discriminates strains of *Citrus tristeza* virus. Phytopathology, 80: 224-228.

Pernet A, Hoisington D, Franco J, Isnard M, Jewell D, Jiang C, Marchand JL, Reynaud B, Glaszmann JC, de Leon DG (1999). Genetic mapping of the maize streak virus resistance from the Mascarene source.I. Resistnace in line D211 and stability against different virus clones. Theor. Appl. Genet., 99: 524-539.

Pinner MS, Markham PG (1990). Serotyping and strain identification of maize streak virus isolates. J. Gen. Virol., 71: 1635-1640.

Pratt RC, Anderson RJ, Louie R, McMullen MD, Knoke JK (1994). Maize responses to a severe isolate of maize chlorotic dwarf virus. Crop Sci., 34: 635-641.

Redinbaugh MG, Seifers DL, Meulia T, Abt JJ, Anderson RJ, Styer WE, Ackerman J, Salomon R, Houghton W, Creamer R, Gordon DT, Hogenhout SA (2002). Maize fine streak virus: a new leafhopper-transmitted rhabdovirus. Phytopathology, 92: 1167-1174.

Rivera C, Gamez R (1986). Mustiplication of maize rayado fino virus in the leafhopper vector *Dalbulus maidis*. Intervirology, 25: 76-83.

Robinson E, Galiakparov N, Radian S, Sela I, Tanne E (1997). Serological detection of grapevine virusA using antiserum to a non structural protein, the putative movement protein. Phytopathology, 87: 1041-1045.

Roca de Doyle MM, Autrey LJC, Jones P (2007). Purification, characterization and serological properties of two virus isolates associated with the maize stripe disease in Mauritius. Plant Pathol., 41(3): 325-334.

Rochow WF, Carmichael LE (1979). Specificity among barley yellow

dwarf viruses in enzyme-linked immunosorbent assays. Virology, 95: 415-420.

Rose OJW (1978). Epidemiology of maize streak virus. Ann. Rev. Entomol., 23: 259-282.

Routh G, Zhang YP, Saldarelli P, Rowhani A (1988). Use of degenerate primers for partial sequencing and RTPCR-based assays of grapevine leafroll associated viruses 4 and 5. Phytopathology, 88: 1238-1243.

Rowhani A, Maningas MA, Lile LS, Daubert SD, Golino DA (1995). Development of a detection system for viruses of woody plants based on PCR analysis of immobilized virions. Phytopathology, 85: 347-352.

Rowhani A, Mircetich SM, Shepherd RJ, Cucuzza JD (1985). Serological detection of cherry leafroll virus in English walnut trees. Phytopathology, 75: 48-52.

Rybicki EP, Hughes FL (1990). Detection and typing of maize streak virus and other distantly related geminiviruses of grasses by polymerase chain reaction amplification of a conserved viral sequence. J. Gen. Virol., 71: 2519-2526.

Saade M, Aparicio F, Sanchez-Navarro JA, Herranz MC, Myrta A (2000). Simultaneous detection of the three ilarviruses affecting stone fruit trees by nonisotopic molecular hybridization and multiplex reverse-transcription polymerase chain reaction. Phytopathology, 90: 1330-1336.

Saiki RK, Gelfank GH, Staffel S, Scharf SJ, Higuchi R (1988). Primer-directed enzymatic amplification of DNA with a thermostable DNA polymerase. Science, 239: 487-491.

Saldarelli P, Rowhani A, Routh G, Minafra A, Digiaro M (1998). Use of degenerate primers in a RT-PCR assay for the identification and analysis of some filamentous viruses, with special reference to clostero- and vitiviruses of the grapevine. Eur. J. Plant Pathol., 104: 945-950.

Salonen A, Ahola T, Kaariainen L (2005). Viral RNA replication in association with cellular membranes. Curr. Topics Microbiol. Immunol., 285: 139-173.

Saunders K, Lucy A, Stanley J (1991). DNA forms of the geminivirus African cassava mosaic virus consistent with a rolling circle mechanism of replication. Nucl. Acids Res., 19: 2325-2330.

Schieber O, Seddas A, Belin C, Walter B (1997). Monoclonal antibodies for detection, serological characterization and immunopurification of grapevine fleck virus. Eur. J. Plant Pathol., 103: 767-774.

Seifer DL, Hackerott HL (1987). Estimates of yield loss and virus titre in sorghum hybrids infected with maize dwarf mosaic virus strain B. Agric. Ecosyst. Environ., 19: 81-86.

Shen P, Kaniewska M, Smith C, Beachy RN (1993). Nucleotide sequence and genomic organization of rice tungro spherical virus. Virology, 193: 621-630.

Shepherd DN, Mangwende T, Martin DP, Bezuidenhout M, Thomson JA, Rybicki P (2007). Inhibition of maize streak virus (MSV) replication by transient and transgenic expression of MSV replication-associated protein mutants. J. Gen. Virol., 88: 325-336.

Storey HH (1936). Virus diseases of East African plants. IV. A Survey of the viruses attacking the Gramineae. E. Afr. Agric. J., 1: 333-337.

Storey HH, McClean APD (1930). The transmission of streak disease between maize, sugarcane and wild grasses. Ann. Appl. Biol., 17: 691-719.

Teycheney PY, Aaziz R, Dinant S, Salanki K, Tourneur C, Balazs E, Jacquemond M, Tepfer M (2000). Synthesis of (-)-strand RNA from the 3' untranslated region of plant virus genomes expressed in transgenic plants upon infection with related viruses. J. Gen. Virol. 81: 1121-1126.

Thottappilly G, Bosque-Perez NA, Rossel HW (1993). Viruses and virus diseases of maize in tropical Africa. Plant Pathol., 42: 494-509.

Toler RW, Skinner G, Bockholt AJ, Harris KF (1985). Reactions of maize (*Zea mays*) accessions to maize rayado fino virus. Plant Dis., 68: 56-57.

Torrance L (1995). Use of monoclonal antibodies in plant pathology. Eur. J. Plant Pathol., 101: 351-63.

Tsai JH, Brown LG (1989). Maize dwarf mosaic virus. Division of Plant Industry. Fla. Dept. Agric. & Consumer Serv. Plant Path. Circular, No. 320.

Tsai JH, Falk BW (1993). Viruses and mycoplasmal agents affecting maize in the Tropics. In: (eds: R.J. Chiu and Y.Yeh) Proc. Symp. Plant Virus Virus-like Dis., pp. 43-48.

Uyemoto JK (1980). Detection of maize chlorotic mottle virus serotypes by enzyme-linked immunosorbent assay. Phytopathology, 70: 290-292.

van der Walt E, Martin DP, Varsani A, Polston JE, Rybicki EP (2008). Experimental observations of rapid Maize streak virus evolution reveal a strand-specific nucleotide substitution bias. Virol. J., 5: 104.

Van-Rensburg GDJ (1981). Effect of plant age at the time of infection with maize streak virus on yield of maize. Phytophylactica, 13: 197-198.

Wetzel T, Candresse T, Ravelonandro M, Dunez J (1991). A polymerase chain reaction assay adopted to plum pox virus detection. J. Virol. Methods, 33: 355-365.

Wetzel T, Jardak R, Meunier L, Ghorbel A, Reustle GM (2002). Simultaneous RT/PCR detection and differentiation of arabis mosaic and grapevine fanleaf nepoviruses in grapevine with a single pair of primers. J. Virol. Methods, 101: 63-69.

Whatman (2004). Application of FTA-based technology for sample collection, transport, purification and storage of PCR-ready plant DNA. [http://www.whatman.co.uk/repository/docu ments/s3/usFtaPlantDna.pdf].

Whitham SA, Wang Y (2004). Roles for host factors in plant viral pathogenicity. Curr. Opinion Plant Biol., 7: 365-371.

Whitham SA, Yang C, Goodin MM (2006). Global impact: Elucidating plant responses to viral infection. Mol. Plant-Microb. Interact., 19: 1207-1215.

Williams LE, Alexander LJ (1965). Maize dwarf mosaic: A new corn disease. Phytopathology, 55: 802-804.

Zaitlin M, Palukaitis P (2000). Advances in understanding plant viruses and virus diseases. Ann. Rev. Phytopathol., 38: 117-143.

Zhang W, Olson NH, Baker TS, Faulkner L, Agbandje-McKenna M, Boulton MI, Davies JW, McKenna R (2001). Structure of the Maize Streak Virus Geminate Particle. Virology, 279(2): 471-477.

Zhang Y, Zhao W, Li M, Chen H, Zhu S, Fan Z (2010). Real-time TaqMan RT-PCR for detection of maize chlorotic mottle virus in maize seeds. J. Virol. Methods, doi:10.1016/j.jviromet.2010.11.002.

2

Serological and RT-PCR detection of *cowpea mild mottle* carlavirus infecting soybean

M. Tavasoli[1], N. Shahraeen[2] and SH. Ghorbani[1]

[1]Division of Microbiology, College of Science, Alzahra University, Tehran, Iran.
[2]Plant Virus Research Department, Iranian Research Institute for Plant Protection (IRIPP), Tehran, P.O. Box- 19395-1454, Iran

During 2006-2007 growing seasons, survey was carried to identify a disease of possible viral etiology causing mosaic of soybean in the soybean field. Leaf samples showing symptoms of mild mosaic were collected from soybean fields in Dezful region (Khozestan province, Southern Iran). Electron microscopy, DAS-ELISA serological tests and RT-PCR assays with specific pairs of primers were applied to identify and determine the viral etiology of the agent. Flexuous particles of ca. 15 x 650 nm were present in the leaf dip preparations examined by electron microscopy. On the basis of serology, RT-PCR assays and reaction of indicator host plants, the causal virus of these mild mosaic symptoms on soybean in the Southern Iran was identified as *Cowpea mild mottle virus*- CPMMV, a whitefly-transmitted carlavirus.

Key words: Cowpea mild mottle virus, serology, RT-PCR, detection, Southern Iran

INTRODUCTION

As soybean (*Glycine max* (L.) Merril) cultivation increases in Iran (106000 ha in 2005) (Anonymous 2005) occurrence of virus and virus-like diseases can limit production (Golnaraghi et al., 2002a,b, 2004; Tavasoli et al., 2007). Soybean (*Glycine max*) is susceptible to infection by several viruses, which substantially reduce yield and quality and at the moment it is known that it is a natural host for 35 potentially important viruses (Edwardson et al., 1991).

Cowpea mild mottle virus (CPMMV) was first reported on cowpea (*Vigna unguiculata*) in Ghana (Brunt and Kenten, 1973). Subsequently it was reported from several tropical regions of Africa (Brunt and Philips, 1981; Thouvenel et al.,1982; Anno-Nyako ,1984; 1986; Mink and Keswani, 1987) , Asia (Antignus and Cohen ,1987; Nolt and Rajeshwari, 1987; Shahraeen , 1989; Reddy, 1991;), Brazil and Argentine (Laguna et al., 2006; Almeida et al., 2005) and from Ivory Coast in a diverse range of plant species that include leguminous and Solanaceous food crops (Hartman et al., 1999). CPMMV is re-

ported to be transmitted by the whitefly, *Bemisia tabaci* (Homoptera: *Aleyrodidae*), in a non-persistent manner (Jeyanandarajah and Brunt, 1993; Memelink et al., 1990). CPMMV has filamentous particles of approximately 650 ×15 nm in size with a coat protein of 32-36 KDa (Demski et al., 1989). CPMMV is a member of genus *Carlavirus* which has recently been classified under the plant virus family Flexiviridae (Memelink et al.1990; Giovanni et al., 2007). CPMMV causes mosaic, chlorosis, necrosis and distortion in a range of indicator host plants (Iwaki et al., 1986; Demski and Kuhn, 1989). Soybean, (*Glycine max* L Merill, groundnut (*Arachis hypogaea* L.), cowpea (*Vigna unguiculata* (L.) Walp.), tomato (*Lycopersicon esculen0tum* Mill. , broad bean (*Vicia faba* L.) and *Nicotiana clevelandii* Gray. have been reported as diagnostic hosts of CPMMV (Demski and Kuhn., 1989; Reddy,1991). The CPMMV genome consists of a single-stranded RNA of Mr 2.5 × 10^6 with six open reading frames (ORF) encoding for the following putative proteins: methyltransferase (Mt), papain-like protease (P-Pro), helicase (Hel), RNA-dependent RNA polymerase (Pol), coat protein (CP), nucleic acid binding protein (NB), plus three triple gene block (TGB) (Nolt et al., 1997). CPMMV-S and CPMMV-M, are

---

*Corresponding author. E-mail: shahraeen@yahoo.com.

Table 1. Primers used in RT-PCR and predicted amplicon size for detection of CPMMV

Tm(°C)	Primer position	primer sequence
56.8	CPMMV Upstream1778-1799 bp	5' - CAC TTG GAA TTT TAT GTT GAC - 3'
58.8	CPMMV Downstream1982-2002bp	5' - TCA TTT CGA TTG GAC CTA TC - 3'

two reported strain of the virus. Mild to severe systemic symptom appearance is reported by CPMMV in different hosts (Laguna et al., 2006; Tavassoli et al., 2007). There have bean a few previous studies on soybean viral diseases and their distribution in Iran, where a few important viruses from nepo-, poty-, bromo-, cucumo- and tospoviruses group have been reported (Shahraeen et al., 2005; Golnaraghi et al., 2000, 2002a,b, 2004, 2007; Ghorbani et al., 2007).

Field symptoms associated with virus infection includeing mosaic, mottling, vein clearing and leaf crinkle were observed in soybean fields in Dezful region. CPMMV has been reported to be transmitted by the whitefly, *Bemisia tabaci*(Laguna et al.,2006). Recently a disease causing stem necrosis of soybean in Brazil and Argentina has also been attributed to CPMMV or a related carlavirus (Almeida et al., 2005; Laguna et al., 2006). In Iran CPMMV has been reported to infect cowpea in Guilan (Northern) province of Iran (Ghorbani et al., 2007) where mix cropping pattern (pulses and solanaceous food crop is a common practice and a large population of the whitefly vector *Bemisia tabacci* Genn. may be responseble for the transmission and spread of pathogen to other crop .

Investigation of incidence and distribution of soybean viruses are very important in developing appropriate control measures. During this study limited number of infected sample were collected and tested using DAS-ELISA serology, RT-PCR and electron microscopic studies for the presence of CPMMV in the sample.

Material and methods

During 2006-2007 growing seasons, infected leaf samples of soybean (*G. max* cv. Clark) plants with mild mosaic and leaf defoliation symptoms were collected from fields in Dezful territory (Khozestan province). For virus detection and host range studies, the sap extracted from infected soybean leaf samples using 0.1 M phosphate buffer (pH 7.2) containing 2% PVP and 1% Na-DIECA (Golnaraghi et al., 2004) were mechanically inoculated to several host plants: soybean, cowpea, groundnut, bean, (*Phaseolus vulgaris* L.) , broad bean tomato, *Chenopodium amaranticolor* Coste & Reyn, *C. quinoa* Willd., *Datura stramonium* L., *Gomphrena globosa* L., *Vigna radiata* (L) R. Wilcz , *Vigna aconitifolia* (Jacq.) Marechal and *Nicotiana rústica* L., tobacco (*N. tabacum* L.) and *Cassia tora* L.. Seeds from infected filed plants were sown and grown under greenhouse conditions.

DAS-ELISA (Clark and Adams, 1977; Lister, 1987) serological test was performed in order to identify CPMMV in the samples. CPMMV antibody kit (plus positive control) was gifted by Dr. Winter (DSMZ, Braunschweig, Germany) and used as prescribed. RT-PCR test was performed (Tavasoli et al., 2007), where total RNA was extracted by TRI-reagent solution as described (Sigma Company).

30-50 mg. of infected tissue was grounded to a fine powder using a wooden applicator stick, in a 1.5 ml Eppendorf tube cooled in liquid nitrogen. Grounded leaf tissue was extracted with 500 µl of TRI-reagent buffer, after 10 min, 250 µl of phenol-chloroform was added, and mixed, followed by centrifuged at 30000 rpm for 15 min. The supernatant was transferred to a new micro tube and 300 µl of isopropanol was added. Samples were kept at -20 °C for at least 20 min, then centrifuged as above; supernatant removed and the pellet was washed by ethanol 75%, and finally dissolved in 20 µl distilled water. For RT-PCR reaction the specific pair of CPMMV primers were designed (Table 1) applying Gene Runner and Blast programs in 1778-2002bp sequences region of two CPMMV strains in GenBank (Accession number AF024628 and AF024629, pair of primers were synthesized (Isogene company , Netherland, Figure 1). Reverse transcription reaction was followed using RevertAid TM first strand cDNA synthesis Kit (BioNeer Co. Korea, Kit-1621), following the specific protocol by applying reverse primer. PCR amplification was performed using lyophilized PCR micro tubes (Accupower PCR PreMix (BioNeer Co., Korea) under the following conditions: (denaturation: 94°C , 30 s ; annealing: 56.5°C , 30 s ; extension: 72°C , 30 s) for 35 cycles, finally 5.5 min of extension at 72°C. Amplified DNA fragments were separated by eletrophoresis in 1% (w/v) agarose gels in 0.5 x TBE and visualized at 302nm after staining in ethidium bromide (0.5 ug/ml).

For virus particle morphology observation (Edwardson et al., 1991), applying leaf dip preparation method (Milne ,1984) young soybean symptomatic trifoliate leaf sample infected by CPMMV were selected and prepared using 400 mesh carbon coated grids, stained with 2% uranyl acetate and examined under a transmission electron microscope (TEM-Philips-301G Model).

RESULTS

The virus isolated from soybean in the Southern part of Iran was identified as CPMMV based on serological reaction, particle morphology, experimental host range (Edwardson et al., 1991) and RT-PCR test with specific pair of primers. Soybean CPMMV infected few indicator plants and produced necrotic local lesions followed by systemic mosaic, mild mottling and veinal necrosis on soybean cv. Clark, chlorotic local and systemic vein clearing and mosaic on cowpea (local cultivar), bean cv. Tender Green reacted with mosaic and leaf distortion while cv. Top Crop with mild mottling, Yellow vein clearing downward leaf rolling in groundnut *(*cv.NC-2*)*, chlorotic local lesion on *Chenopodium amaranticolor* and *C. quinoa*. The virus isolate did not produce any visible symptoms in *Vigna radiata, V. aconitifolia,* and tomato but they were shown to be infected and gave positive results in the in ELISA recovery test. On the other hand, CPMMV could not infect broad bean, *Datura stramonium* and tobacco (Table 2). The symptoms induced following mechanical inoculation of herbaceous host plants resembled those described for CPMMV-S (Laguna et al., 2006),

Figure 1. Upstream and downstream specific primers by matching CPMMV-S and CPMMV-M sequences from GenBank using GeneRunner software. (*) regions are indicating of similar sections.

Table 2. Reaction of selected host plants to inoculation with CPMMV the soybean isolate from Southern Iran.

| Host plants | Symptoms | | ELISA Test results |
	Local reaction	Systemic reaction	
C. amaranticolor	CLL	-	+
C. quinoa	CLL	-	+
A. hypogaea (cv. Local)	Ylvc,Cr,Ld		+
Glycine max	NLL	M, Mot,Vn	+
P. vulgaris (cv. Top Crop)	-	MMot	-
P. vulgaris (cv. Tender Green)	CLL	M,Ld	+
V. unguiculata	CLL	M,Ld	+
V. radiatae	(-)	(-)	+
V. aconitifolia	(-)	(-)	+
L. esculentum	-	MMot	+
V. faba	-	-	—
D. stramonium	-	-	—
N. tabacum (cv. Rustica)	-	-	—
G. globosa	-	-	-

CLL= Chlorotic local lesion, NLL=Necrotic local lesion, YlVc=Yellow vein clearing, Cr=Crinkle, M=Mosaic, Mot=Mottling, Vn=Veinal necrosis, Ld=Leaf distortion, MMot=Mild mottle, (-) =Negative result, + = Positive reaction, (-)=doubtful
V. unguiculata and *Vigna radiata* were without any conspicuous symptoms, but ELISA result was positive.

no obvious symptoms were seen upon inoculation of the virus isolates on *Gomphrena globosa*, *Cassia tora* .*Nicotiana rustica*, bean (local cultivar Chitee), *D. stramonium*, and broad bean. None of 800 planted seeds from infec-

ted field soybean plants showed symptoms upon emergence. Inoculation test results were further rechecked by DAS-ELISA. The expected band of 206 bp (Figure 2) was obtained in gel electrophoresis using a 1 kb ready to use

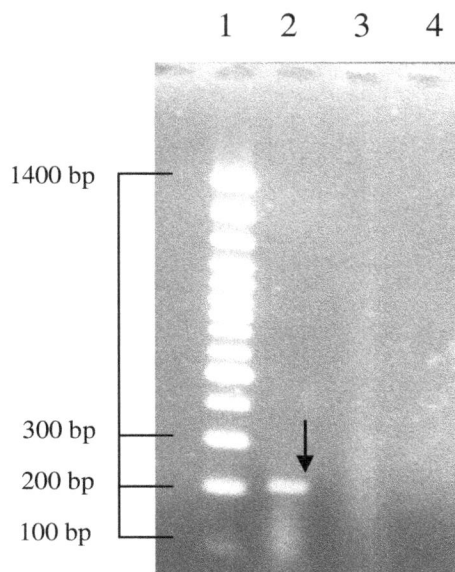

Figure 2. 1% agarose gel. Electrophoresis of CPMMV.
1- 100 bp DNA ladder (Fermentas).
2- CPMMV band in 206 bp region.
3- Negative control consisting of distilled water and loading buffer
4 - Negative control using healthy soybean leaves.

Kenten, 1973; Thouvenel et al., 1982 and Iwaki et al., 1986) reported seed transmission of CPMMV is soybean, while others (Lizuka et al. 1984; Anon. 1987; Horn et al., 1991 and Rossel and Thottappilly, 1993) failed to produce such transmission. There are reports in other parts of the world that *Solanum incanum* L. may act as a perennial reservoir of CPMMV during the summer when region is virtually free from the soybean (Iwaki et al., 1986; El-Hammady et al.,2004). A survey on the plants growing near soybean field must be made to identify possible natural reservoirs of the CPMMV. Symptoms caused by this Iranian isolate of CPMMV on soybean is quite mild and similar to previous reports of infection of this crop by CPMMV and seem to be less severe that those described in Brazil which involve stem necrosis (Almeida et al., 2005). There is not yet an assessment of the yield losses on soybean caused by CPMMV associated to the plant stage when infection occurs nor the demonstration that it is being transmitted by the white flies present in the fields, which still have to be identified. The presence of CPMMV in the soybean in Southern Iran must be considered as an additional threat to this culture as well as to other legume crops since white flies populations are present everywhere. A thorough screening for resistance to CPMMV must be made on the available soybean germplasm collection of Iran and a cooperation with researchers form other parts of the world dealing with this virus on soybean must be considered to face this menace.

On the basis of serological studies the isolate designnated as similar to CPMMV-S it is not yet clear if the cowpea isolate (Northern region, Ghorbani et al., 2007) and CPMMV isolate described in this paper are strain of one virus or distinct but serologically related viruses. In order to discriminate and determine sequence similarities between the CPMMV isolates from Iran with other reported isolates and their tentative relation to carlavirus or as a distinct virus require further molecular studies.

ACKNOWLEDGMENT

The authors thank Dr. Winter S (DEMZ, Braunschweig Germany) for the generous gift of antibodies used for serological detection of CPMMV in this study. This study is a part of M.Sc. thesis of the first author submitted to Alzahra University Tehran.

marker (100 bp.) DNA ladder (Fermentas Co, Germany) and the primers designed by us. In this study for PCR reaction, the lyophilized microtube of AccuPower PCR premix (Bioneer Corporation Korea) was found more convenient replacing the needs for loading buffer compared to a common protocol. Particles of approximately 650 nm in length were observed (micrograph not shown). In leaf dip preparations of young symptomatic leaves examined in the transmission electron microscope Particle length measurement for CPMMV and of other related carlaviruses has been described elsewhere (Demski and Kuhn, 1989; Edwardson et al., 1991).

DISCUSSION

Our experimental data clearly point out that the soybean plants with mild mosaic symptoms found in the Dezful territory was infected by an isolate of CPMMV. Possibly it was being transmitted by the whiteflies present in the culture. It is not known yet how the virus came into the region. Since no seed transmission could be demonstrated under our conditions, it is likely that the virus was present in the region in some natural host or it is was introduced from the Northern region where CPMV was already identified infecting cowpea fields (Ghorbani et al., 2007; Tavasoli et al., 2007).. However, seed transmission of CPMMV is still a controversial subject. As discussed by El-Hammady et al., (2004), some authors (Brunt and

REFERENCES

Almeida AMR, Piuga FF, Marin SR, Kitajima EW, Gaspar JO, Moraes TG (2005). Detection and partial characterization of a carlavirus causing stem necrosis of soybeans in Brazil. Fitopatologia Brasileira 30: 191-194.

Anno-Nyako FO (1984). Identification and partial characterization of a mild mottle disease in soybean (*Glycine max* (L.) Merril) in Nigeria. University of Science and Technology, Kumasi, Ghana, Ph.D. Thesis.

Anno-Nyako FO (1986). Semi-persistent transmission of an extra mild isolate of *cowpea mild mottle virus* on soybean by whitefly *Bemisia tabaci* in Nigeria. Trop. Agric. 63:193-194.

Antignus Y, Cohen S (1987). Purification and some properties of a new strain of *cowpea mild mottle virus* in Israel. Annals Applied Biol. 110: 563-569.

Anonymous (2005). Montly report .Iranian oilseed industry magazine. Status of soybean cultivations in January and February 2005. N0 (35-36): 14-18.

Anonymous (1987). *Cowpea mild mottle virus* (CPMMV) is not seed-borne in soybean. IITA , Annual Report and Research Highlights 1986-1987, Ibandania, Niger, p.127.

Brunt AA, Kenten RH (1973). *Cowpea mild mottle*, a newly recognized virus infecting cowpeas (*Vigna unguiculata*) in Ghana. Annals of Applied Biol. 74: 67-74.

Brunt AA, Philips S (1981). "Fuzzy vein", a disease of tomato (*Lycopersicon esculentum*) in Western Nigeria induced by *Cowpea mild mottle virus*. Trop. Agric. 58: 177-180.

Clark MF, Adams AN (1977). Characteristic of the microplate method of enzyme-linked immunosorbent assay for the detection of plant viruses. J. General Virol. 34: 475-483.

Demski JW, Kuhn CW (1989). *Cowpea mild mottle virus*. In: Compendium of soybean diseases (3rd edition), American Phytopathological Society , St. Paul, USA, pp. 60-61.

Edwardson JR, Christie GR, Raton B (1991). Handbook of viruses infecting legumes. Vol.II. CRC Press, Boca Raton, Florida USA, P. 354

El-Hammady M, Albrechtsen SE, Abdelmonem AM, Abo El-Abbas FM, Gazalla W (2004). Seed-borne *cowpea mild mottle virus* on soybean in Egypt. Amin University, Cairo. Arab J. Agric. Sci. 12 (2): 838-850.

Ghorbani SGM, Shahraeen N, Elahinia A, Bananej K (2007). Identification of mixed viral infection on cowpea (*Vigna unguiculata*) in Guilan province. Proceedings of the second Iranian Microbiology Congress, Alzahra University, Tehran, Iran, p. 110.

Giovanni P, Martelli P, Michael JA, Kreuze JF, Valerian VD (2007). Family flexiviridae: a case study in virion and genome plasticity. Annual Review of Phytopathology 45: 73-100.

Golnaraghi AR, Pourrahim R, Farzadfar Sh, Ohshima K , Shahraeen N, Ahoonmanesh A (2007). Incidence and distribution of *tomato fruit ring virus* on soybean in Iran. Plant Pathology J. 6(1): 14-21.

Golnaraghi AR, Shahraeen N, Pourrahim R, Farzadfar Sh, Ghasemi A (2002c). First report of natural occurrence of eight viruses affecting soybeans in Iran. Plant Pathology 51(6): 794.

Golnaraghi AR, Shahraeen N, Pourrahim R, Farzadfar Sh, Ghasemi A (2004). Occurrence and relative incidence of viruses infecting soybeans in Iran. Plant Disease 88: 1069-1074.

Golnaraghi AR, Shahraeen N, Pourrahim R (2000). Detection of *tomato* spotted wilt virus (TSWV), Bean yellow mosaic virus (BYMV) and peanut mottle virus (PeMoV) from soybean fields in Iran. Proceedings of the 14th Iran and Plant Protection Congress, University of Isfahan, Iran, p. 243.

Golnaraghi AR, Shahraeen N, Pourrahim R, Farzadfar Sh (2002a). Comparative host reactions of three Tomato spotted *wilt tospovirus* isolates from soybean in Iran. Proceedings of the 1st Iranian Plant Virology Congress, Tehran Medical University, Iran, p. 120.

Golnaraghi AR, Shahraeen N, Pourrahim R, Farzadfar Sh (2002b). Soybean seed-borne viruses in Mazandaran and Golestan provinces of Iran. Proceedings of the 15th Iranian Plant Protection Congress, University of Razi, Iran, p. 342.

Hartman GL, Sinclair JB, Rupe JC (1999) Compendium of soybean diseases. Fourth edition. The American Phytopathological Society, St. Paul, Minnesota, USA, p. 182.

Horne NM, Saleh N, Balidi Y (1991). Cowpea mild mottle virus could not be detected by ELISA in soybean and groundnut seeds in Indonesia. Netherlands J. Plant Pathol. 97(2): 125-127.

Iwaki M, Thongmeearkom T, Honda Y, Prommin M, Deema N, Tibi T, Lzuka N, Ong CA, Salaeh N (1986). Cowpea mild mottle virus occurring on soybean and peanut in South East of Asian countries. Tropical Agricultural Research Centre, Tech. Bull. 21:106-120.

Jeyanandarajah P, Brunt AA (1993). The natural occurrence, transmission, properties and possible affinities of Cowpea mild mottle virus. J. Phytopathol. 137: 148-156.

Laguna IG, Arneodo JD, Rodriguez-Pardina P, Fiorona M (2006). Cowpea mild mottle virus infecting soybean crops in North-Western Argentina. Fitopatologia Brasilia 31(3): 317.

Lister RM (1978). Application of the Enzyme-Linked Immunosorbent Assay for detecting viruses in soybean seed and plants. Phytopathology 68: 1393-1400.

Lizuka N, Rajeshwari R, Reddy DVR, Goto T, Muniyappa V , Bharathan N, Ghanekar AM (1984). Natural occurrence of a strain of cowpea mild mottle virus on groundnut (*Arachis hypogaea*) in India. Phytopathologishe Zeitschrift 109: 245-253.

Memelink J, Van der Vlug CIM, Linthorst HJM, Derks AFLM, Asjes CJ, Bol JF (1990). Homologies between the genomes of a carlavirus (Lily symptomless virus) and a potexvirus (*Lily virus X*) from lily plants. J. General Virol. 71: 917-924.

Milne RG (1984). Electron microscopy for the identification of the plant viruses in *in vitro* preparation. In: Maramorosch K, Koprowski ,H (eds) Methods of Virology, (7). Academic Press, Orlando, USA, pp. 87-120.

Mink GI, Keswani CL (1987). First report of Cowpea mild mottle virus on bean and mung bean in Tanzania. Plant Dis. 71: 557.

Nolt BL, Rajeshwari R (1987). Properties of a Cowpea mild mottle virus isolate from groundnut. Indian Phytopathol. 40(1): 22-26 .

Reddy DVR (1991). Crop profile. Groundnut viruses and virus diseases: distribution, identification and control. Rev. Plant Pathol. 70: 665-678.

Rossel HW, Thottappilly G (1993). Seed transmission of viruses on soybean (*Glycine max*) in relation to sanitation and international transfer of improved germplasm. Seed Sci. Technol. 21: 25-30.

Shahraeen N (1989). Studies on a virus causing mosaic diseases of soybean (cv. MACS-13). Indian Phytopathol. 42: 338.

Shahraeen N, Bananaj K (1996). Occurrence of Peanut mottle virus in Gorgan Province. Iran J. Plant Pathol. 32(1-2): 21.

Shahraeen N, Ghotbi T, Salati M, Sahandi A (2005). First report of Bean pod mottle virus in soybean in Iran. Plant Dis. 88: 1109.

Tavasoli M, Shahraeen N, Ghorbani SH, Hashemi SH (2007). Identification of *cowpea mild mottle virus*-CPMMV in soybean from Iran. Proceedings of the 4th Iranian Virology Congress, Medical University of Tehran, Iran, p. 342.

Thouvenel JC, Monsarrat A, Fauquet C (1982). Isolation of cowpea mild mottle virus from diseased soybean in Ivory Cost. Plant Dis. 66: 336-337.

3

Epidemiology and molecular characterization of chikungunya virus involved in the 2008 to 2009 outbreak in Malaysia

Mohd Apandi Yusof*, Lau Sau Kuen, Norfaezah Adnan, Nur Izmawati Abd Razak, Liyana Ahmad Zamri, Khairul Izwan Hulaimi and Zainah Saat

Virology Unit, Infectious Disease Research Centre, Institute for Medical Research, Kuala Lumpur, Malaysia.

The 2008 to 2009 outbreak of chikungunya was considered as the huge and worst outbreak of CHIKV (Chikungunya virus) infections in history of the country affecting all states in both Peninsular and East Malaysia. This was unlike the first outbreak in late 1998, which was only restricted to Klang district in Selangor and six years later the second outbreak which only involved the state of Perak. The objective of the study was to detect the presence of chikungunya antibody and antigen by immunofluorescence technique and RT-PCR from the sera samples. A total of 2,692 sera samples were received in 2008, in which 19.2% were positive by antibody detection and 42.6% were positive by RT-PCR. The following year in 2009, the samples size increased to 3,592, only 16.3% sample were positive by antibody detection and 31.7% were positive by RT-PCR. Majority of the hospitalized cases were adults between 30 to 60 years of age and the highest incidence rate was amongst patients' age between 40 to 49 year old. In 2008, most of the confirmed CHIKV infection cases were female but the opposite was seen in 2009, where more male cases were reported. In this outbreak, the prominent ethnic group affected was the Malays. CHIKV involved in the 2008 to 2009 outbreaks was the new Central/East African genotype which was found to be similar with strains causing the outbreaks in the India Ocean and main CHIKV genotype circulating in the European countries from 2006 to 2009.

Key words: Chikungunya virus, genotype, epidemiology, outbreaks in Malaysia.

INTRODUCTION

Chikungunya virus (CHIKV) is responsible for an acute infection of an abrupt onset of high fever, arthralgia, myalgia, headache and rash (Johnson and Peters, 1996). This small envelope and single stranded positive sense RNA virus is in the genus of "alphavirus" under the Togaviridae family. It was first isolated nearly six decades ago in Tanzania in 1953 (Ross, 1956). Until now, only *Aedes* mosquitoes either *Aedes aegypti* or *Aedes albopictus* were documented as the responsible vectors in transmitting the disease. However, several cases of CHIKV infections as the result of maternal- foetal

transmission have been reported (Geradin et al., 2008). During the past 50 years, emergence and re-emergence of CHIKV infection has occurred globally (Rao, 1971; Thuang et al., 1975; Thaikruea et al., 1997). Numerous CHIKV re-emergences have been reported in both Africa and Asia with the intervals of 2 to 20 years between the outbreaks (Powers and Logue, 2007). In the late 50s to 1970s, many African countries experienced the outbreaks of CHIKV and the new Central/East African strains were isolated in 2000. These viruses were believed to have originated from the outbreak in the Democratic Republic of Congo (Pastorini et al., 2004). Now, CHIKV disease has become extremely important in public health control programme as the disease could spread to several countries around the world in a short period of time. In the last decade,

*Corresponding author. E-mail: apandi@imr.gov.my.

CHIKV infection had caused many outbreaks across the globe. Starting from the emergence of CHIKV in Kenya in 2004 (Chretien et al., 2007; Pialoux et al., 2007) then the outbreaks spread to Comoros and Seychelles in early 2005, followed by Mauritius (Beeson et al., 2008). Later the outbreaks were reported to have spread to other islands in the Indian Ocean, including La Reunion Islands which is part of France in 2005 to 2006 (Schuffenecker et al., 2006; Bonn, 2006) and causing major outbreaks in these regions (Renault et al., 2007).

Based on the viral genetic analysis, there was a strong link between the infection in La Reunion Islands in 2005 to 2006 with the outbreak in Kenya in 2004 (Kariuki et al., 2008) that were caused by Central/East African genotype. Then epidemic of CHIKV rapidly spread to Europe and the United States of America. Their activities were seen in countries such as France (Parola et al., 2006), Germany, Italy, Norway and Switzerland due to imported cases from people returning from endemic areas (WHO, 2006). The best example was the transmission of CHIKV in Italy in 2007 where the index case was a traveller from Kerala, India and the virus was isolated from local *Aed. Albopictus* (Europen Centre for Disease Prevention and Control, 2007; Watson, 2007; Vazeille et al., 2008). In 2008, CHIKV fever was listed by US National Institute of Allergy and Infectious Disease (NIAID) under category C priority pathogen (Powers and Logue, 2007; Staples et al., 2009). In Malaysia, the first ever outbreak of CHIKV was recorded in 1998 to 1999 in Port Klang, Selangor affecting more than 51 people (Lam et al., 2001). Six years later, CHIKV re-emerged in Bagan Panchor, Perak in March 2006 (Kumarasamy et al., 2006; Abubakar et al., 2007) and Kinta district, Perak in December 2006 (Noridah et al., 2007). The third and largest outbreak so far, was first detected in Tangkak, Johor in April 2008 and over short period of time the number of cases drastically increased and spread beyond the states of Johor. By the end of 2008 a total of 2,692 cases were investigated and CHIKV activities were recorded in all states in Malaysia including Sabah and Sarawak. The trend of infections continued till end of 2009 with 3,592 cases being investigated.

Currently, only two laboratories in the Ministry of Health Malaysia are able to perform laboratory investigations for CHIKV, namely: the Institute for Medical Research (IMR) in Kuala Lumpur and the National Public Health Laboratory (NPHL) in Sungai Buloh, Selangor. IMR carries out laboratory tests for hospitalized cases, whereas NPHL carries out tests for outpatients, surveillance and outbreaks cases. Since this is third and the largest outbreak of CHIKV infection in Malaysia, in this study we attempt to identify, analyze and characterise all CHIKV strains circulating in each different states of Malaysia and also study the epidemiology of the CHIKV infection during the outbreak.

MATERIALS AND METHODS

In year 2008 and 2009, a total of 2,692 and 3,592 samples respectively were received and tested in IMR. The RT-PCR, using primers E1-C and E1-S (Hasebe et al., 2002) was used to detect the nucleic acid of the E1 gene and the indirect immunofluorescence technique Nagasaki University (1999) was used to detect the presence of IgM antibody against CHIKV. Date of disease onset (such as fever, arthalgia, arthritis and rash) was the criteria used to determine the types of test to be conducted. Samples from patients with onset of disease of ≤5 days duration were subjected to RT-PCR whereas those with >5 days of disease duration were screened for CHIKV IgM. CHIKV IgM detection was also performed on samples tested negative by RT-PCR test. Table 1 shows the number of samples received and tested at IMR in 2008 and 2009. There are 13 states and 3 federal territories (FT) in Malaysia. Representative samples from each state that were found positive by RT-PCR were inoculated into BHK (baby hamster kidney) cells and observed daily for cytopathic effect (CPE). RT-PCR, using E1-S and E1-C primers which amplified 294 bp glycoprotein E1 gene of CHIKV, was performed on the supernatant from cell culture samples that showed CPE (Hasebe et al., 2002). RNA was first extracted using QIAamp® Viral RNA Mini Kit from Qiagen (Hilden, Germany) prior RT-PCR amplification.

The resulting PCR product was subjected to agarose gel electrophoresis and the QIAquick® Gel Extraction kits from Qiagen (Hilden, Germany) was then used to extract the DNA from the gel. DNA sequencing was performed using both forward (E1-S) and reverse (E1-C) primers.

RESULTS

In 2008, from a total of 2,692 sera samples received at IMR from patients showing signs and symptoms of CHIKV infection, 42.6% (1146/2692) were found positive for CHIKV infection by RT-PCR and 19.2% (517/2692) positive by antibody detection. The following year in 2009, there was a 33% (3592/2692) increased in number of samples received of which, 31.7% (1140/3592) were PCR positive for CHIKV and 16.3% (585/3592) were antibody positive. Overall, positivity detected by RT-PCR was higher compared to IgM detection (Table 1). Based on the number of samples received in IMR, FT of Kuala Lumpur recorded the highest cases of CHIKV infection with 1,134 suspected cases in 2008 followed by Melaka (475 cases), Johor (411 cases) and Selangor (331 cases). The trend of infection was slightly different in 2009 with Sarawak (566 cases) and Kelantan (566 cases) recorded a significant increased in number of cases whereas some states such as Johor and Melaka which recorded high number of cases in 2008, showed a significant decreased in number of suspected cases in 2009. However, FT of Kuala Lumpur still maintained as the state with the highest number of suspected chikungunya cases for both 2008 and 2009. A total of 54 CHIKV isolates, 14 isolates from 9 states received form suspected chikungunya cases in 2008 and 35 isolates from all states in Malaysia in 2009 were sequenced. Phylogenetic tree of these isolates was constructed together with isolates retrieved from GenBank for partial E1 gene.

Molecular analysis showed that all the isolates from

Table 1. Number of samples received and tested at IMR in 2008 and 2009.

State	No. of sample tested in 2008			No. of sample tested in 2009		
	No. of sample	Positive by RT-PCR	Positive by IgM	No. of sample	Positive by RT-PCR	Positive by IgM
Johor	411	263	77	44	13	6
Kedah	58	31	10	308	174	36
Kelantan	28	18	5	566	230	136
FT Kuala Lumpur	1134	308	240	811	172	109
Melaka	475	301	58	43	6	3
Negeri Sembilan	44	24	8	2	0	0
Pahang	82	33	24	85	41	16
Perak	79	45	8	142	58	13
Perlis	4	1	0	23	9	2
Pulau Pinang	24	6	7	166	46	39
FT Putrajaya	0	0	0	21	5	2
Sabah and FT Labuan	8	2	1	47	27	5
Sarawak	3	1	0	566	262	89
Selangor	331	109	78	757	94	128
Terengganu	11	4	1	11	3	1
Total	2692	1146	517	3592	1140	585

each states in Malaysia including Sabah and Sarawak for both years were of the Central/Eastern African genotype and belongs in the same cluster with isolates from Thailand in 2009 and other Malaysian isolates in 2008 (Figure 1). These isolates were totally different from the CHIKVs isolated in the Malaysian 1998 and 2006 outbreaks which were of the Asian genotype. Throughout the outbreak which started in April in 2008, based on the number of samples received, there were 3 peaks of CHIKV infections as shown in Figure 2. The first peak was in July 2008 to August 2008 which recorded more than 500 samples received from suspected chikungunya cases; the second peak which was the highest from December 2008 to January 2009 with nearly 1,000 suspected cases and the third peak with 300 suspected cases from October 2009 to November 2009. Majority of the CHIKV infections were in adults between the ages of 30 to 60 years with highest incident in the 40 to 49 years age group (Figure 3). There were more female cases in 2008 but the reverse was recorded in 2009, where more male was affected compared to female (Figure 4). Among the racial groups in Malaysia, the Malays were the highest racial group affected in this outbreak (Figure 5).

DISCUSSION

The first outbreak of chikungunya in Malaysia was in Klang, Selangor in late 1998 with more than 51 people infected (Lam et al., 2001). Six years later, the second outbreak occurred in Bagan Panchor, Perak in March 2006 (Kumarasamy et al., 2006; Abubakar et al., 2007) and Kinta district, Perak in December 2006 (Noridah et al., 2007). The third outbreak initially started in early April 2008, where increasing number of CHIKV infections were first detected in state of Johor. Following this, neighbouring states Kuala Lumpur and Melaka were also affected with increased in number of chikungunya cases. By July 2008, the numbers of cases drastically increased and later spread to other states in Peninsular Malaysia where more than 2,692 cases were reported (IMR, 2009). Highest activities were recorded in the Federal Territory Kuala Lumpur, Malacca, Johor and Selangor; but lower in Kelantan and East Malaysia states of Sabah and Sarawak. However, the scenario was different the following year, where states like Kelantan, Sarawak and Kedah which had low number of cases in 2008, recorded high number of cases in 2009. Of the 5,430 chikungunya cases reported by Control of Disease Division, Malaysia in 2009 (MOH, 2010), 2,505 cases were from Sarawak. The two earliest outbreaks in 1998-1999 and 2006 were due to CHIKV belonging to Asian genotype strain (Lam et al., 2001; Kumarasamy et al., 2006; Abubakar et al., 2007). All the CHIKV isolates in 1988 and 2006 shared high sequence similarities. However, the recent CHIKV circulating in 2008 and 2009 were totally different from the 1998 to 1999 or 2006 isolates. The recent outbreak in Malaysia recorded the emergence of new Central/East

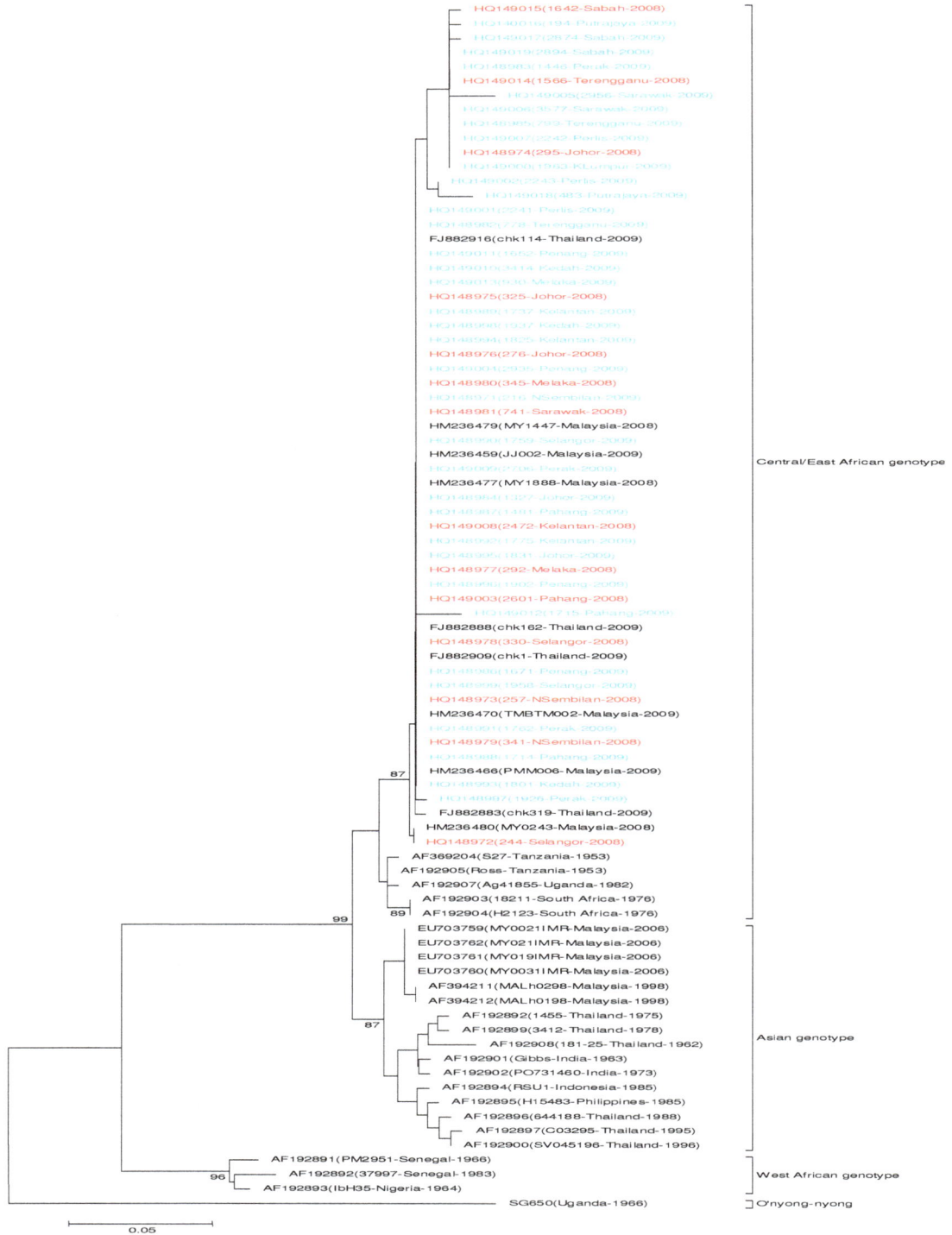

Figure 1. Phylogenetic tree of partial glycoprotein E1 sequences (257 bp) of CHIKV inferred using the Neighbor-Joining method from the software MEGA 4. The evolutionary distances were computed using the "maximum composite likelihood" method. Genotype Asian, Central/East African and West African are indicated by square brackets with O'nyong-nyong virus as anoutgroup. 49 CHIKV isolates from Malaysia in 2008 and 2009 are indicated in red and blue underline words respectively. Representative strains of each genotype obtained from GenBank are labeled using the following format: 'Accession number'- 'isolate'-'Country of origin'-'Year isolation'. Bootstrap values (>75%) for 1,000 pseudoreplicate dataset are indicated at branch nodes.

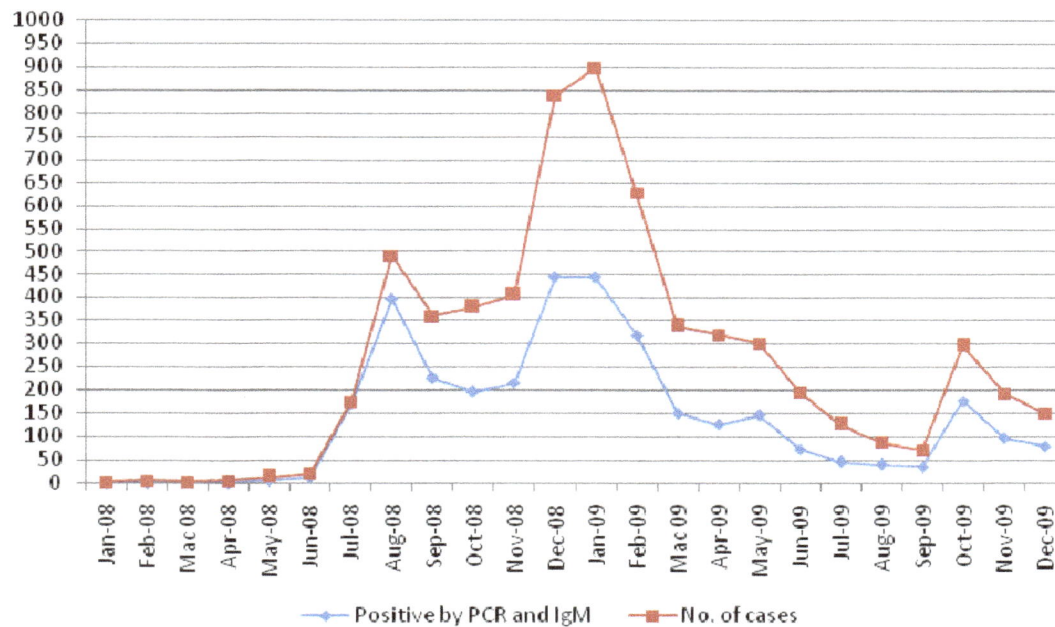

Figure 2. Distribution of CHIKV infection and positive cases by month from 2008 to 2009.

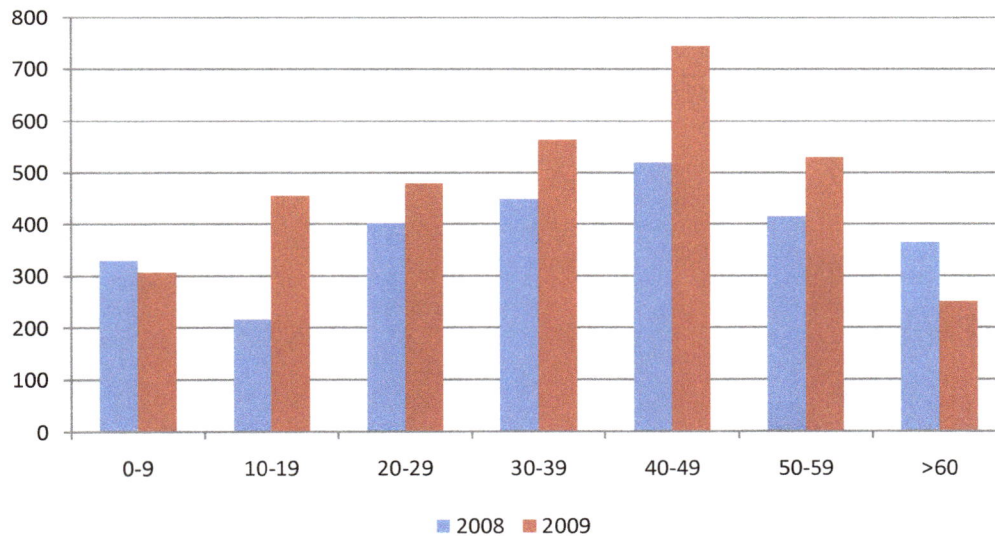

Figure 3. CHIKV infections during 2008 to 2009 outbreaks by age groups.

African genotype (Noridah et al., 2007; Sam et al., 2009; Maizatul., 2009), and was found to be similar with strains causing outbreaks in the Indian Ocean (Vidya et al., 2007) and also the same strains found to be circulating in the state of Kelantan in 2009 (Apandi et al., 2010). The phylogenetic tree based on partial E1 gene of the isolates involved in 2008 to 2009 outbreaks in Figure 2 showed all were from Central/East African genotype.

The different strains of CHIKV involved in the three outbreaks in Malaysia raised concern regarding the epidemiology and the capability of CHIKV to change from one genotype to another. The Asian strains involved in 1998 to 1999 and re-emerged in 2006 was thought to be probably due to CHIKV which was already endemic in Malaysia as suggested by Abubakar et al. (2007). But prior to 1998, CHIKV has never been isolated from humans or animals

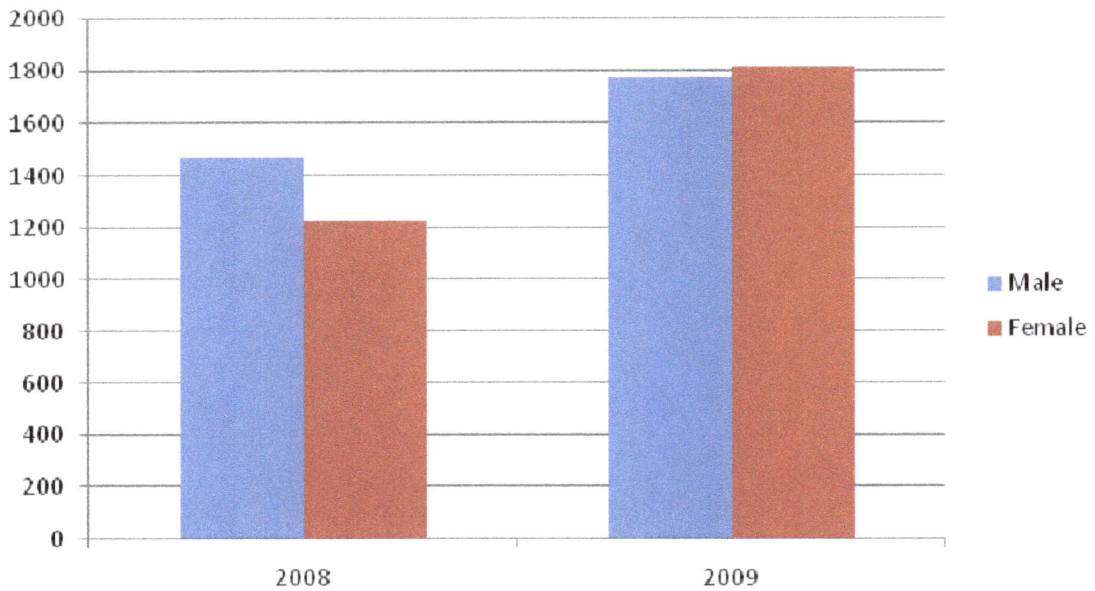

Figure 4. CHIKV infections during 2008 to 2009 outbreaks by gender.

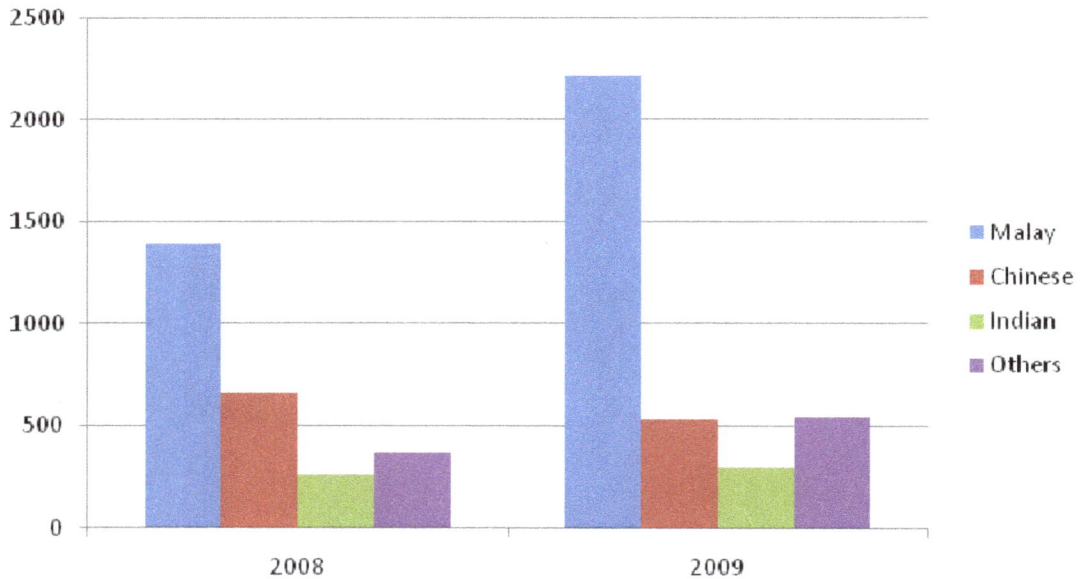

Figure 5. CHIKV infections during 2008 to 2009 outbreaks by ethnic groups.

in Malaysia. Neither was there any clinical disease due to CHIKV infection reported in Malaysia during that time period, even though earlier serological survey of human sera collected from 1965 to 1969 in West Malaysia showed neutralizing antibodies to CHIKV among adults, especially those inhabiting the rural northern and eastern states bordering Thailand (Marchette et al., 1980). The spread of CHIKV appears to be maintained by human-mosquito-human transmission (urban cycle). However, the presence of neutralizing antibodies in wild monkeys, pigs and chickens (Marchette et al., 1980) and recently isolated CHIKV from non-human primates (Apandi et al.,

2009) suggested that a CHIKV sylvatic transmission cycle may exists in Malaysia and could possibly contributes to the outbreaks. The emergence of Central/East African genotype in the third outbreak could probably originated from strain circulating in the Indian Ocean outbreaks in 2007 and responsible for other outbreaks in several regions (Vidya et al., 2007; Xavier et al., 2008; D'Ortenzia et al., 2009; CDC, 2010) including Thailand, (Theamboonlers et al., 2009). It was noted that before 2006, all the three CHIKV genogroups; namely: Asian, Central/East African and West African genotypes (Powers et al., 2000; Schuffenecker et al., 2006) were restricted to the geographical areas denoted by their names.

The recent explosive epidemics of African genotype in Indian Ocean Islands and India; and other parts of Asia, Africa and Europe have indicated that international travellers have disseminated new strain of the virus, some into region from which CHIKV has been absent (Townson and Nathan, 2008). This scenario has changed the geographical distribution of CHIKV worldwide and endemicity of CHIKV as previously stated was not the only factor involved in the epidemics. Other factors include point mutation of the virus with the presence of A226V (Tsetsarkin et al., 2007; Bordi et al., 2008; Rianthavorn et al., 2010) resulting in the virus becoming more susceptible to the new host of *Aedes* spp. especially *A. albopictus* which is found in high density in Malaysia. This mutation enables them to be more adaptive to new vector (Xavier et al., 2008) and could facilitate the spread of CHIKV. Another possibility was probably due to the presence of a large naive population who do not have prior exposure to and immunity against the virus. These could probably be the main factors causing re-emerging of CHKV in Malaysia especially in rural areas. Also, the present of high viral load in patients travelling from epidemic areas could be an added factor (Parola et al., 2006) that been seen in Italy (Bordi et al., 2008).

The disease is almost self limiting and rarely fatal (Fields et al., 1996) and so far in Malaysia, no mortality was ever reported since its first appearance in 1998. However, mortality has been reported in Reunion Island (Michault and Staikowsky, 2009) and suspected in India (Mavalankar et al., 2008).

Conclusion

The CHIKV strains circulating in all states of Malaysia during the outbreak in 2008 to 2009 were from the Central/East African genotype and were different from CHIKV strains previously isolated in 1998 to 1999 and 2006 outbreaks. Majority of the cases were adults between the age of 30 to 60 years with highest incident at amongst the 40 to 49 age group and the Malays were the major race affected in the 2008 to 2009 outbreak.

ACKNOWLEDGEMENTS

The authors would like to thank the Director General of Health Malaysia and Director of the Institute for Medical Research for permission to publish this paper. This research project was funded by the MRG 08-002 fund vote.

REFERENCES

Abubakar S, Sam IC, Wong PF, Mat Rahim NA, Hooi PS, Roslan N (2007). Reemergence of endemic chikungunya, Malaysia. Emerg .Infec. Dis. Vol., 13(1): 147-149.

Apandi Y, Lau SK, Izmawati N, Amal NM, Faudzi Y, Wan M, Hani MH, Zainah S (2010). Identification of chikungunya virus strains circulating in Kelantan, Malaysia in 2009. SEA J. Trop Med. Pub. Healt., 41(6): 1374-1380.

Apandi Y, Nazni WA, Noor Azleen ZA, Vythilingan I, Noorazian MY, Azahari AH, Zainah S, Lee HL (2009). The first isolation of chikungunya virus from non-human primates in Malaysia. J. Gen. Mol. Virol., 1(3): 035-039.

Beeson S, Funkhouser E, kotea N, Spielman A, Robich RM (2008). Chikungunya fever, Mauritius, 2006. Emer. Inf. Dis., 14: 154.

Bonn D (2006). How did chikungunya reach the Indian Ocean? Lancet. Infec. Dis., 6: 543

Bordi L, Carletti F, Castiletti C, Chiappini R, Sambri V, Cavrini F, Ippolito G, Di CA, Capobianchi MR (2008). Presence of the A226V mutation in autochthonous and imported Italian chikungunya virus strains. Corespondence. CID 2008:47 (1 Aaugust).

CDC (2010). Centers for Disease control and prevention. Outbreak notice. Chikungunya fever in Asia and the India Ocean, February 26.

Chretien JP, Anyamba A, Bedno SA, Breiman RF, Sang R, Sergon K, Powers AM, Onyango CO, Small J, Tucker CJ, Linthicum KJ (2007). Drought-associatedchikungunya emergence along coastal East Africa. Amer. J. Trop. Med. Hyg., 76: 405-407.

D'Ortenzio E, Grandadam M, Balleydier E, Dehecq JS, Jaffar-BMC, Michault A, Andriamandimby SF, Reynes JM, Filleul L (2009). Sporadic cases of chikungunya, Réunion Island, August 2009. Eurosurveillance, 14(35), 03 September 2009.

Fields BN, Knipe DM, Howley PM (1996). Alphavuruses. In: Fields of Virology Third Edition; V 1. Lippincott-Raven Publishers. Pg 858-898.

Geradin P, Barau G, Michault A, Bintner M, Randrianaivo H, Choker G, Ienglet Y, Touret Y, Bouveret A, Grivard P, LeRoux K, Blanc S, Schuffenecker I, Couderc T, Seisdos FA, Lecuit M, Robillard PY (2008). Multidisciplinary prospective study of mother to child chikungunya virus infections on the island of La Reunion. Plos. Med. 5: e60.

Hasebe F, Parquet MC, Pandey BD, Mathenge EG, Morita K, Balasubramaniam V, Saat Z, Yusop A, Sinniah M, Natkunam S, Igarashi A (2002). Combined detection and genotyping of Chikungunya virus by a specific reverse transcription polymerase chain reaction. J. Med. Virol., 67: 370-4.

IMR (2008). Institute for Medical Research. Annual Report 2009.

IMR (2009). Institute for Medical Research. Number of Chikungunya case investigated Virology Unit in 2010. Unpiblished data.

Johnson RE, Peters CJ (1996). Alphaviruses associated primarily with fever and polyarthritis. Fields virology Philadelphia: Lippincott-Raven Publishers. Fields BN, Knipe DM, Howley PM. pp 843-898.

Kariuki NM, Nderitu L, Ledermann JP, Ndirangu A, Logue CH, Kelly CH, Sang R, Sergon K, Breiman R, Powers AM (2008). Tracking epidemic chikungunya virus into the Indian Ocean from east Africa. J. Gen. Virol., 89: 2754-2760.

Kumarasamy V, Prathapa S, Zuridah H, Chem YK, Norizah I, Chua KB (2006). Re-Emergence of Chikungunya Virus in Malaysia. Med. J. Malaysia, 61(2): 221-225. ISSN 0300-5283.

Lam SK, Chua KB, Hooi PS, Rahimah MA, Kumari S, Tharmaratnam M, Chua SK, Smith DW, Sampson IA (2001). Chikungunya infection – an emerging disease in Malaysia. Southeast Asian J. Trop. Med. Pub. Healt., 32: 447–51.

Maizatul A (2009). Analisa jujukan gen glikoprotein envelop E1 virus chikungunya Malaysia bagi tahun 2008. Thesis B. Sc (Hons). Universiti Putra Malaysia.

Marchette NJ, Rudnick A, Garcia R (1980). Alphaviruses in Peninsular Malaysia: II. Serological evidence of human infection. Southeast Asian J. Trop. Med. Public Healt., 11: 14–23.

Mavalankar D, Shastri P, Bandyopadhyay T, Parmar J, Ramani KV (2008). Increased mortality rate associated with chikungunya epidemic, Ahmedabad, India Emerg. Infect. Dis.Mar., 14(3): 412-5.

Michault A, Staikowsky F (2009). Chikungunya: First step toward specific treatment and prophylaxis. J. Infec. Dis. Aug; 200:489-491.

MOH (Ministry of Health) (2010). Distribution of CHIKV infections by states in 2008 and 2009. CDC Report.

Nagasaki University (1999). Immuno-fluorescence assay for chikungunya virus.edited by Shingo Inoue (unpublished).

Noridah O, Paranthaman V, Nayar SK, Masliza M, Ranjit K, Norizah I, Chem YK, Mustafa B, Kumarasamy V, Chua KB (2007). Outbreak of chikugunya due to virus of central/East Africn genotype in Malaysia. Med. J. Malaysia, 62(4): 323-8.

Parola P, de LX, Jourdan J, Rovery C, Vaillant V, Minodier P, Brouqui P, Flahault A, Raoult D, Charrel RN (2006). Novel chikungunya virus variant in travellers returning from Indian Ocean islands. Emer. Inf. Dis., 12(10): 1493-1499.

Pastorini B, Muyembe-TJJ, Bessaud M, Tock F, Tolou H, Durand JP, Peyrefitte CN (2004). Epidemic resurgence of Chikungunya virus in Democratic Republic of the Congo: identification of a new central African strain. J. Med. Virol., 74: 277–282.

Pialoux G, Gauzere BB, Jaureguiberry S, Strobel M (2007). Chikungunya, an epidemic arbovirosis. Lancet. Infect. Dis., 7(5): 319-27

Powers AM, Logue CH (2007). Changing patterns of chikungunya virus: re-emergence of a zoonotic arbovirus. J. Gen. Virol., 88: 2363-2377.

Powers AM, Brault AC, Tesh RB, Weaver SC (2000). Re-emergence of chikungunya and o'nyong-nyong viruses: evidence for distinct geographical lineages and distant evolutionary relationships. J. Gen. Virol., 81: 471–479.

Rao TR (1971). Immunological surveys of arbovirus infections in South East Asia, with special reference to dengue, chikungunya and Kyasanur Forest disease. Bull. WHO., 44: 585-591.

Renault P, Solet JL, Sissoko D, Balleydier E, Larrieu S, Filleul L, Lassalle C, Thiria J, Rachou E, de Valk H, Ilef D, Ledrans M, Quatresous I, Quenel P, Pierre V (2007). A major epidemic of chikungunya virus infection on reunion Island, France. Am. J. Trop. Med. Hyg., 77(4): 727-731.

Rianthavorn P, Prianantathavorn K, Wuttirattanakowit N, Theamboonlers A, Poovorawan Y (2010). An outbreak of chikungunya in southern Thailand from 2008 to 2009 caused by African strains with A226V mutation. Int. J. Infect. Dis., Apr 21.

Ross RW (1956). A laboratory technique for studying the insect transmission of animal viruses, employing a bat-wing membrane, demonstrated with two African viruses. J. Hyg. (Lond.), 54: 192–200.

Sam IC, Chan YF, Chan SY, Loong SK, Chin HK, Hooi PS, Ganeswarie R, Abubakar S (2009). Chikungunya virus of asian and central/East African genotypes in Malaysia. J. Clin. Virol., 46(2): 180-3.

Schuffenecker I, Iteman I, Milchault A, Murri S, Frangeul, L, Vaney, MC, Lavenir R, Pardigon N, Reynes JM, Pettinelli F, Biscornet L, Diancourt L, Michel S, Duquerroy S, Guigon G, Frenkiel MP, Brehin AC, Cubito N, Despres P, Kunst F, Rey FA, Zeller H, Brisse S (2006). Genome microevolution of chikungunya viruses causing the Indian Ocean outbreak. Plos Med., 3: e263.

Staples JE, Breiman RF, Powers AM (2009). Chikungunya fever: an epidemiological review of re-emerging infectious disease. Clin. Infect. Dis., 49: 942-948.

Thaikruea L, Cheareansook O, Reanphumkarkit S. Dissomboon P, Phonjan R, Ratchbud S, Kounsang Y, Buranapiyawong D (1997). Chikungunya in Thailand: a re-emerging disease? SEA J. Trop. Med. Pub. Healt., 28: 358-364.

Theamboonlers A, Rianthavorn P, Praianantathavor K, Wuttirattanakowit N, Poovorawan Y (2009). Clinical and molecular characterization of chikungunya virus in south Thailand. Jpn. J. infect. Dis., 62(4): 303-305.

Thuang U, Ming CK, Swe T, Thein S (1975). Epidemiological features of dengue and chikungunya infections in Burma. SEA J. Trop. Med. Pub. Healt., 6: 276-283.

Townson H, Nathan MB (2008). Resurgence of chikungunya. Trans. Royal Society Trop. Med. Hyg., 102(4): 308-309.

Tsetsarkin KA, Dana LV, Charles EM, Stephen H (2007). A single mutation in Chikungunya Virus affects vector specificity and epidemic potential. Plos Pathogen, 3(12): e201

Vazeille M, Jeannin C, Martin E, Schaffner F, Failoux AB (2008). Chikungunya: a risk for Mediterranean countries. Acta. Trop., 105: 200-202.

Vidya AA, Shubham S, Sarah C, Rashmi SG, Atul MW, Santosh MJ, Sudeep AB, Akhilesh CM (2007). Genetic divergence of chikungunya viruses in India (1963-2006) with special reference to the 2005-2006 explosive epidemic. J. Gen. Virol., 88: 1967-1976.

Watson R (2007). Europe witnesses first local transmission of chikungunya fever in Italy. BMJ, 335: 532-533.

Xavier DL, Leroy E, Charrel RN, Ttsetsarkin K, Higgs S, Gould EA (2008). Chikungunya virus adapts to tiger mosquito via evolutionary vonvergence: a sign of things to come? Virol. J., 5: 33

4

Molecular epidemiology of human enterovirus71 (HEV71) strains isolated in Peninsular Malaysia and Sabah from year 2001 to 2009

Mohd Apandi Yusof*, Fazilah Rais, Maizatul Akma Abdullah, Liyana Ahmad Zamri, Hariyati Md Ali, Fauziah Md Kassim and Zainah Saat

Virology Unit, Infectious Disease Research Centre, Institute for Medical Research, Kuala Lumpur, Malaysia.

Human enterovirus71 (HEV71) together with other enteroviruses such as Coxsackie A16 and Coxsackie A10 is known to be responsible for hand, foot and mouth disease. Several of the hand, foot and mouth disease (HFMD) outbreaks caused by HEV71 were associated with neurological manifestations and deaths. In Peninsular Malaysia and Sabah, even though huge outbreaks of HFMD have never been reported; HEV71 strains however, were isolated periodically from HFMD cases throughout the year. From 2001 to 2009, four genetic lineages of HEV71 have been found to be prevalent in Peninsular Malaysia and Sabah. The predominant circulating strain was subgenogroup B4 in 2001 and this was later followed by subgenogroup B5 in 2003. The subgenogroup B5 was dominant between 2005 and 2009. Viruses belonging to subgenogroups C1 and C4 were also detected.

Key words: Human enterovirus, neurological diseases, molecular epidemiology, gene sequences.

INTRODUCTION

Human enterovirus71 (HEV71) is a positive ssRNA virus belonging to the genus enterovirus in the family *Picornaviridae*. It is normally associated with epidemics of hand, foot and mouth disease (HFMD) with typical symptoms including lesions on the palms, soles and oral mucosa (Minor et al., 1995). Since its isolation in 1969 in California, USA from an encephalitis case (Schmidt et al., 1974), this virus has been reported to cause several outbreaks around the world with cases which do not only present with the typical HFMD syndromes but also with neurological diseases such as aseptic meningitis, encephalitis and meningioencephalitis (Blomberg et al., 1974; Shindarov et al., 1979; Lum et al., 1998; McMinn et al., 2001; Kehle et al., 2003). HEV71is also the most common non-polio enterovirus associated with poliomyelitis-like paralysis (Melnick et al., 1984) due to its affinity to anterior horn cell (Chumakov et al., 1979).

In Malaysia, the 1997 cases of HFMD in Sarawak presented with acute myocarditis dysfunction and acute flaccid paralysis, and resulted in 34 paediatric deaths (Cardosa et al., 2003) aside the typical symptoms of HFMD. HEV71was isolated in several of these cases (Chan et al., 2000). This outbreak became a landmark in molecular epidemiology of HEV71 and the study of HEV71 strains in association with disease severity. Numerous reports on the molecular epidemiology of HEV71 strains from the Asia-Pacific region have been published (AbuBakar et al., 1999a; Brown et al., 1999; Shimizu et al., 1999; Wang et al., 1999; Shih et al., 2000; Singh et al., 2000; McMinn et al., 2001; Chu et al., 2001; Wang et al., 2002; Cardosa et al., 2003; Herrero et al., 2003; Shimizu et al., 2004; Li et al., 2005) and all have indicated the existence of three genogroups. Genogroup A represented by prototype BrCr-CA-70 which was isolated in California, USA in 1970 (Schmidt et al., 1974; Brown and Pallanch, 1995) has not been reported elsewhere. Genogroup B has 5 genetic lineages namely subgenogroup B1, B2, B3, B4, B5 and Genogroup C has 5 lineages; C1, C2, C3, C4 and C5.

HEV71strains isolated from the state of Sarawak in East Malaysia have been described by many researchers involving huge number of HEV71 isolates (McMinn et al., 2001; Cardosa et al., 2003; Podin et al., 2006; Ooi et al., 2007). However, only a few data is available on the

*Corresponding author. E-mail: apandi@imr.gov.my.

molecular epidemiology of HEV71 strains from Peninsular Malaysia and Sabah despite the fact that several HEV71 outbreaks have been reported by AbuBakar et al. (1999b), Singh et al. (2000) and Herrero et al. (2003). Work done by AbuBakar et al. (1999b) only involved nine HEV71 isolated in 1997. Singh et al. (2000) reported on four HEV71 isolates in 1997 and 1998; while Herrero et al. (2003) reported on molecular epidemiological data of 43 HEV71 isolates from 1997 to 2000. Even though there was no report on huge HFMD outbreaks in these two regions, HEV71 was constantly isolated from patients with HFMD without CNS infection throughout 2001 to 2009. Therefore, the main objective of this study was to analyze and characterise all HEV71 that have been circulating in both Peninsular Malaysia and Sabah for the past 10 years.

MATERIALS AND METHODS

Samples

Samples such as vesicle swabs, throat swab and rectal swabs were collected from patients showing signs and symptoms of HFMD. All state hospitals in Peninsular Malaysia and Sabah were involved in this study and acted as a specimen collection centre. All samples were cultured in rhabdomyosarcoma cells (RD cells) using Minimum Essential Medium (GIBCO, Invitrogen, USA) supplemented with 10% heat inactivated fetal bovine serum (GIBCO, Invitrogen, USA), 50 U/ml benzyl penicillin and 50 µg/ml streptomycin sulphate (Sigma, St Louis, USA). Cultures were observed daily for cytopathic effect (CPE) and harvested when more than 90% of the cell monolayer showed CPE.

Viruses

A total of 70 HEV71 strains were isolated. There were 4 isolates in 2001, 12 in 2003, 1 in 2004, 7 in 2005, 15 in 2006, 4 in 2007, 23 in 2008 and 4 in 2009. All the HEV71 isolates including information of accession number, type of specimen, age, gender, clinical diagnosis and year of isolation are shown in Tables 1 and 2.

HEV71 screening assays

Cell cultures showing CPE were screened for HEV71 using HEV71-specific RT-PCR assays as described by Perera et al. (2004).

Reverse transcriptase–polymerase chain reaction (RT-PCR)

Viral RNAs were extracted using the QIAamp® Viral RNA Mini Kit from Qiagen (Hilden, Germany). Briefly, samples were added to buffer AVL carrier RNA in a microcentrifuge tubes and incubated at room temperature for 10 min. Later ethanol was added and mixtures were transferred to QI Aamp spin column for centrifugation. Buffer AW1 and AW2 were used to wash viral RNA in the spin column and finally buffer AVE was added to elute viral RNA in the clean microcentrifuge tubes. The VP4 gene was amplified using forward primer EVP2 (5'-CCT CCG GCC CCT GAA TGC GGC TAA-3') (Chua et al., 2001) and reverse primer OL68-1 (5'-GGT AAY TTC CAC CAC CAN CC-3') (Ishiko et al., 2002) in a

one tube reaction (50µl) containing 5µl of RNA, 20 µM of each primer, 2.5U of AMV reverse transcriptase and Promega Access Quick RT-PCR Kit (Cat. No: A1703). Reverse transcription was carried out at 48°C for 45 min followed by 10 min at 70°C to stop the reaction to get the first strand cDNA synthesis. Samples were then subjected to 35PCR cycles, denaturation at 95°C for 45 s, annealing at 55°C for 45 s and extension at 72°C for 60 s.

The VP1 gene was also amplified using the Promega Access Quick RT-PCT Kit with primers VP1F2 (5'-ATA ATA GCA YTR GCG GCA GCC CA-3')-VP1R1 (5'-TGR GCR GTG GTA GAY GAY AC-3') as described by Tu et al. (2007). Polymerase Chain Reaction (PCR) cycling conditions were set up at 51°C for 30 min for reverse transcription followed by 35 cycles of 92°C for 30 s, 51°C for 45 s, and 72°C for 1 min. PCR products (≈1.1 kb) were examined by gel electrophoresis. QIAquick Gel Extraction Kits (QIAGEN Inc, Valencia, CA) was used to extract the DNA from the gel.

Nucleotide sequencing of HEV71 VP4 and VP1 gene

Whereas HEV71 VP4 gene amplicons were sequenced on both strands by using PCR primers, the VP1 gene amplicons were sequenced by PCR primers and in house internal VP1 primers; VP1 Int F (5'-TTC ACY TAY ATG CGY TTT GA-3') and VP1 Int R (5'-ACA AAC ATA TAY TGR AGY AAT TG-3'). Sequencing was performed by using the Big Dye Cycle Sequencing kit version 3.0 and an ABI377 automated DNA sequencer (Applied Biosystems, Foster City, USA). The SeqMan software module in the Lasergene suite of programs (DNASTAR, Madison, USA) was used to format the nucleotide sequences.

HEV71 Sequence data obtained from genbank

The VP4 gene and VP1 gene sequences of HEV71 strains from different genogroups were obtained from genbank for the purpose of generating dendograms.

Phylogenetic analysis

Alignment of the VP1 and VP4 gene sequences was undertaken by using the Megalign software module in Lasergene suite of programs (DNASTAR, Madison, USA). Phylogenetic trees were constructed by using the neighbor-joining method from the Software MEGA4. The CA16 strain G10 was used as an outgroup for phylogenetic analysis of both the VP4 and VP1 sequence data.

RESULTS

From year 2001 to 2007 a total of 43 HEV71 strains were isolated from patients suspected of having HFMD in Malaysia. Phylogenetic tree constructed based on the complete VP4 gene (207) nucleotides, shown in Figure 1, revealed that 4 HEV71 strains isolated in 2001, 2 isolates belonged to subgenogroup C1 and another 2 to subgenogroup B4. Of the 12 HEV71 strains isolated in 2003, 11 isolates belonged to subgenogroup C1 and 1 isolate in subgenogroup B5. All the isolates in 2004 and 2005 were in the subgenogroup B5 whereas in 2006, about 13 isolates were in B5 and 2 were in C1. The scenario was different in 2007 in which for the first time the subgenogroup C4 was isolated in Malaysia and this

Table 1. HEV71 used in the study to sequence VP4 gene.

Isolate	Accession No	Sample type	Sex/Age	Diagnosis	Isolation year
PP046-MAL-04	EU925766	R/S	M/3	HFMD	2004
SAB064-MAL-06	EU925767	V/S	M/1	HFMD	2006
PHG065-MAL-06	EU925768	V/S	M/2	HFMD	2006
SAB070-MAL-06	EU925769	T/S	F/0.5	HFMD	2006
PHG084-MAL-06	EU925770	V/S	M/10	HFMD	2006
J092-MAL-06	EU925771	R/S	M/5	HFMD	2006
PHG1100-MAL-07	EU925772	V/S	M/4	HFMD	2007
J1179-MAL-07	EU925773	T/S	F/4	HFMD	2007
J142-MAL-03	EU925774	T/S	M/1	HFMD	2003
KL161-MAL-06	EU925775	T/S	M/2	HFMD	2006
PP166-MAL-03	EU925776	T/S	M/3	HFMD	2003
PP168-MAL-03	EU925777	T/S	F/3	HFMD	2003
J171-MAL-06	EU925778	R/S	M/6	HFMD	2006
J183-MAL-03	EU925779	V/S, T/S	F/3	HFMD	2003
NS213-MAL-03	EU925780	V/S, R/S	M/3	HFMD	2003
PTJ218-MAL-06	EU925781	T/S	M/3	HFMD	2006
PP28-MAL-01	EU925782	R/S	F/4	HFMD	2001
PP028-MAL-06	EU925783	R/S	F/4	HFMD	2006
PP307-MAL-06	EU925784	T/S	F/2	HFMD	2006
J317-MAL-06	EU925785	T/S	F/1	HFMD	2006
KED379-MAL-05	EU925786	T/S	F/2	HFMD	2005
KED393-MAL-05	EU925787	T/S	M/2	HFMD	2005
KED394-MAL-05	EU925788	T/S	F/4.5	HFMD	2005
J514-MAL-06	EU925789	V/S	F/1	HFMD	2006
PP525-MAL-05	EU925790	R/S	F/4	HFMD	2005
PP533-MAL-05	EU925791	T/S	M/2	HFMD	2005
PP550-MAL-05	EU925792	T/S	F/2	HFMD	2005
PP576-MAL-05	EU925793	R/S	F/1	HFMD	2005
PP648-MAL-06	EU925794	V/S	M/2	HFMD	2006
J78-MAL-03	EU925795	T/S	M/1.5	HFMD	2003
J80-MAL-03	EU925796	R/S	F/2.5	HFMD	2003
TRG828-MAL-07	EU925797	V/S	M/2	HFMD	2007
PP85-MAL-03	EU925798	T/S	M/2	HFMD	2003
PP96-MAL-03	EU925799	T/S	F/2	HFMD	2003
J98-MAL-03	EU925800	V/S	F/3	HFMD	2003
J05-MAL-01	EU925801	R/S	M/2	HFMD	2001
J145-MAL-01	EU925802	T/S	M/3	HFMD	2001
PP76-MAL-03	EU925803	T/S	M/5	HFMD	2003
J085-MAL-06	EU925804	V/S	F/5	HFMD	2006
PP72-MAL-03	EU925805	T/S	M/3	HFMD	2003
J149-MAL-01	EU925806	V/S,T/S,R/S	M/2	HFMD	2001
TRG1381-MAL-07	EU925807	V/S	F/4	HFMD	2007
PHG53-MAL-06	EU925808	R/S	M/3	HFMD	2006

T/S = throat swab, R'S = rectal swab, V/S = vesicle swab, M = Male; F = Female; age in year.

was closely related to the strains from China. Two of the isolates from 2007 were from subgenogroup B5 and one was from C1.

Twenty-seven of HEV71 isolated in 2008 and 2009 were sequenced for the complete VP1 gene which consists of 891 nucleotides. The phylogenetic tree was constructed together with the complete VP1 gene derived from the Genbank (Figure 2). All the HEV71 strains isolated in 2008 and 2009 were from the subgenogroup B5.

Table 2. HEV71 used in the study to sequence VP1 gene.

Isolate	Accession No	Sample type	Sex/Age	Diagnosis	Isolation year
EV1075-Pahang-08	HM358809	V/S	M/7	HFMD	2008
EV0336-Sabah-08	HM358810	V/S	M/3	HFMD	2008
EV0372-Sabah-08	HM358811	T/S	F/2	HFMD	2008
EV0408-Penang-08	HM358812	R/S	F/2	HFMD	2008
EV0466-Johor-08	HM358813	T/S	M/4	HFMD	2008
EV0482-Sabah-08	HM358814	T/S	F/4	HFMD	2008
EV0562-Penang-08	HM358815	T/S, R/S	M/4	HFMD	2008
EV0577-Pahang-08	HM358816	T/S	M/1.5	HFMD	2008
EV0758-Sabah-08	HM358817	V/S	M/6	HFMD	2008
EV0764-Johor-08	HM358818	V/S	M/1	HFMD	2008
EV0811-Penang-08	HM358819	R/S	M/1	HFMD	2008
EV0879-Bintulu-08	HM358820	T/S	M/2	HFMD	2008
EV0884-Johor-08	HM358821	V/S	M/1	HFMD	2008
EV0891-Johor-08	HM358822	V/S	M/6	HFMD	2008
EV0911-Kedah-08	HM358823	V/S	M/7	HFMD	2008
EV0943-Johor-08	HM358824	T/S	F/2	HFMD	2008
EV0972-Johor-08	HM358825	V/S	M/4	HFMD	2008
EV1019-Penang-08	HM358826	T/S	F/5	HFMD	2008
EV1025-Penang-08	HM358827	T/S	M/1	HFMD	2008
EV1035-Pahang-08	HM358828	T/S	M/1	HFMD	2008
EV1078-Johor-08	HM358829	T/S	F/5	HFMD	2008
EV1094-Johor-08	HM358830	T/S	M/1	HFMD	2008
EV0338-Sabah-08	HM358831	V/S	F/3	HFMD	2008
EV0031-Johor-09	HM358832	V/S	M/1	HFMD	2009
EV0076-KLumpur-09	HM358833	R/S	F/1	HFMD	2009
EV1705-Johor-09	HM358834	T/S	M/1	HFMD	2009
EV1945-Kuching-09	HM358835	T/S	F/2	HFMD	2009

T/S = throat swab, R'S = rectal swab, V/S = vesicle swab, M = Male; F = Female; age in year.

DISCUSSION

Before the HFMD outbreak in Sarawak in 1997, HEV71 was considered as an etiological agent of HFMD without any impact together with other enteroviruses especially CA16 and CA10. However, the report of child deaths (Chan et al., 2000; Cardosa et al., 2003) in which HFMD presented with not only the typical symptoms of HFMD but also with acute myocardial dysfunction and acute flaccid paralysis resulted in a major impact in government policy towards the public health including monitoring and control of HFMD cases. The Malaysian government made a ruling that all HFMD cases either with or without neurological manifestations become a notifiable disease in 2007 (MOH, 2007) and since then a surveillance system for monitoring HFMD has been established.

Analyzing the phylogenetic relationship of HEV71 strains in order to determine whether there is some relationship between their genotypes and pathogenic properties has been carried out by many researchers (Shimizu et al., 1999; Brown et al., 1999; Shih et al., 2000; Wang et al., 2002; Munemura et al., 2003). It was

initially thought that the new virulent form of HEV71 was responsible for the outbreaks and deaths among HFMD cases. However, until now, a distinct association of certain genogroups with particular neurovirulence has not been identified from molecular analysis (Shimizu et al., 2004). No significant nucleotide difference has been found between the fatal and non fatal cases by sequence comparison regardless of whether those findings are based on the phylogenetic analysis of VP4 (Shimizu et al., 1999), VP1 (Brown et al., 1999) or 5'UTR (Abubakar et al., 1999b). Thus, it was suggested that virulence might not be determined by a single viral gene (Chua et al., 2001).

In this study, phylogenetic trees were performed using the complete VP4 gene for strains isolated from 2001 to 2007 and VP1 gene for HEV71 in 2008 to 2009. Although, different regions were used, identical clustering was shown regardless of the regions examined (Cardosa et al., 2003 and Shimizu et al., 2004). The VP4 was normally used for molecular typing of human enterovirus (Ishiko et al., 2002) and also for detailed molecular epidemiological analysis (Cardosa et al., 2003). However,

Molecular epidemiology of human enterovirus71 (HEV71) strains isolated in Peninsular Malaysia and Sabah...

35

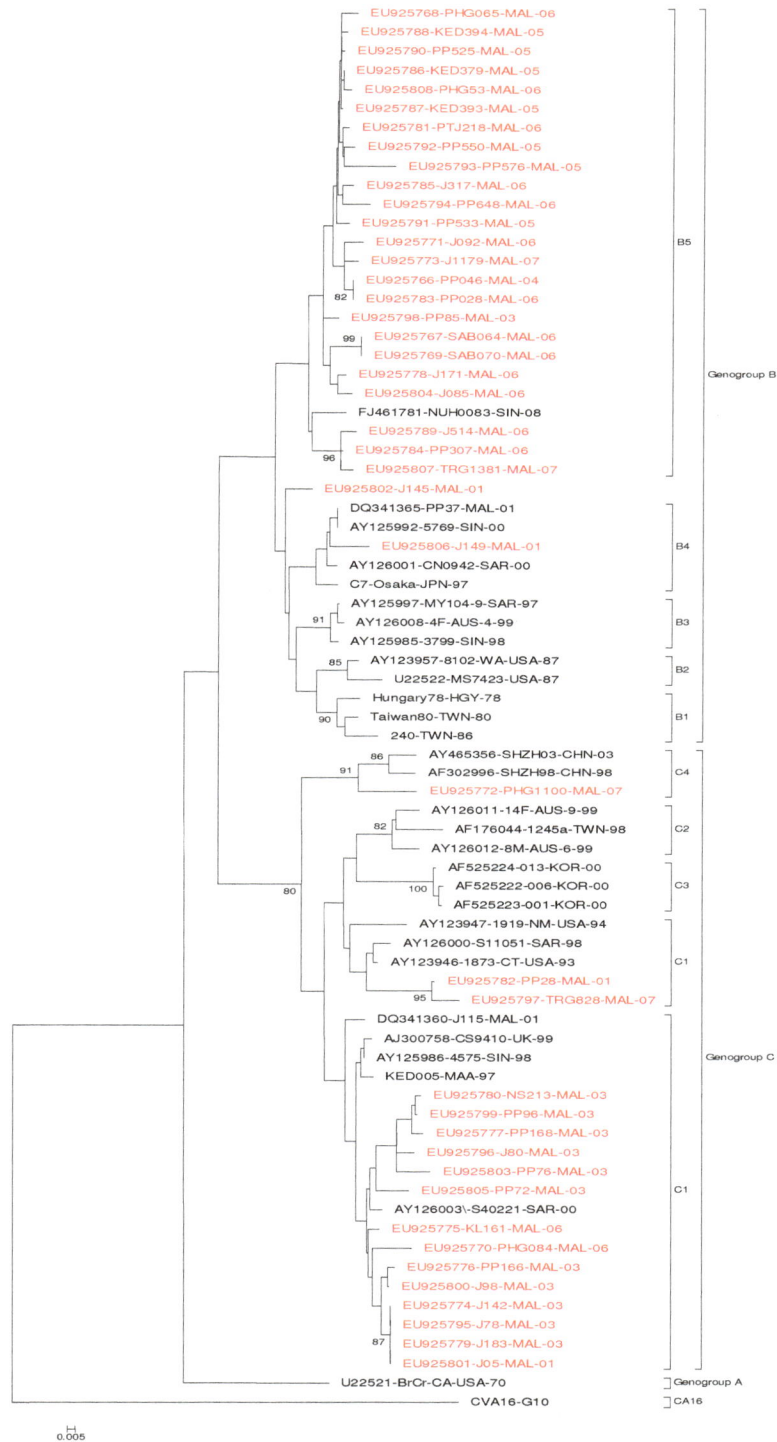

Figure 1. Phylogenetic tree of complete VP4 gene (207) of HEV71 inferred using the Neighbor-Joining method from the software MEGA 4. The evolutionary distances were computed using the Maximum Composite Likelihood method. Genogroups and subgenogroups are indicated by square brackets with CA16-G10 virus as an outgroup. 43 HEV71 isolates in 2001-2007 are indicated in red and underline words. Representative strains of each genogroup obtained from GenBank are labeled using the following format: 'Accession number'- 'isolate'-'Country of origin'- 'Year isolation'. Bootstrap values (>80%) for 1,000 pseudoreplicate dataset are indicated at branch nodes.

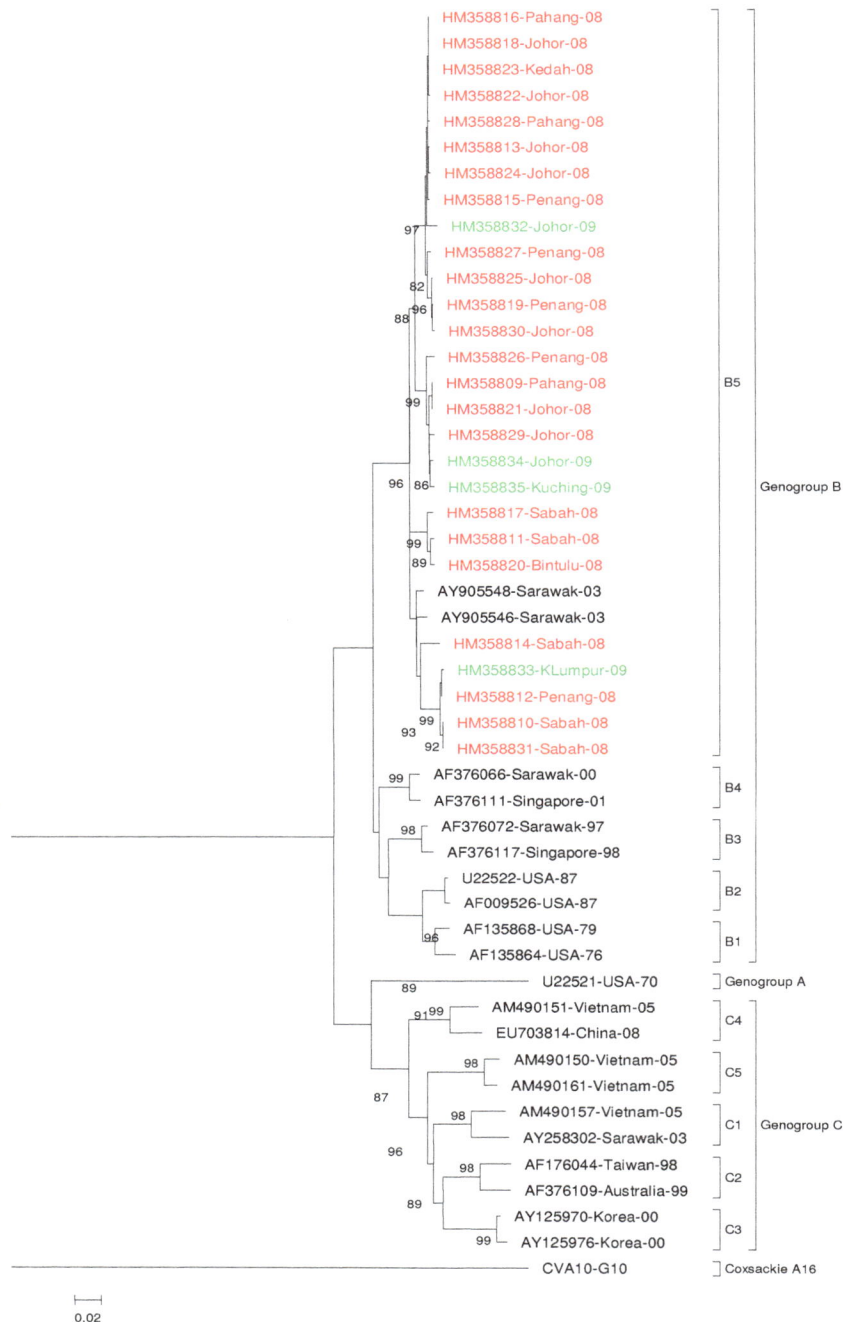

Figure 2. Phylogenetic tree of complete VP1 gene (891) of HEV71 inferred using the Neighbor-Joining method from the software MEGA 4. The evolutionary distances were computed using the Maximum Composite Likelihood method. Genogroups and subgenogroups are indicated by square brackets with CA16-G10 virus as an out group. 27 HEV71 isolates in 2008 and 2009 are indicated in red and green underline words. Representative strains of each genogroup obtained from GenBank are labeled using the following format: 'Accession number'-'Country of origin'- 'Year isolation'. Bootstrap values (>80%) for 1,000 pseudoreplicate dataset are indicated at branch nodes.

VP1 based phylogenetic tree provided more reliable genogrouping of HEV71 because of longer nucleotides and possible correlation of the genogroups with viral antigenicity (McMinn et al., 2001; Cardosa et al., 2003 and

Herrero et al., 2003). Currently there are three genogroups related to HEV71. The first HEV71 isolated in the world was from California, USA. This strain was known as prototype BrCr-CA-70 strain and the only member of the genogroup A (Schmidt et al., 1974; Brown and Pallanch, 1995). It has not been reported since then. The genogroup B has evolved over the years from 2 subgroups described by Brown et al. (1999) into 5 subgenogroups; B1, B2, B3, B4 and B5 while genogroup C has been grouped into C1, C2, C3, C4 and C5.

Molecular epidemiological analysis in this study showed that only 4 subgenogroups namely B4, B5, C1 and C4 were circulating in Peninsular Malaysia and Sabah between 2001 and 2009. The subgenogroup B4 was isolated in 2001 however, analysis 43 HEV71 isolated from 1997 to 2000 in peninsular Malaysia by Herrero et al. (2003) showed that the B4 subgenogroup was the most prevalent subgenogroups in 1997, 1999 and 2000. HEV71 belonging to genogroup B4 were also identified in several HFMD cases in Singapore in 1997 (McMinn et al., 2001), Peninsular Malaysia in 1997 to 1998 (Herrero et al., 2003), Taiwan in 1998 (Shih, et al., 2000) and Sarawak in 2000 (Cardosa et al., 2003). Thus, indicating that this genogroup was widespread, although not predominant, throughout the Asia-Pacific region until it became the focus of large epidemic activities in Peninsular Malaysia and Sarawak, and Singapore in 2000 (Cardosa et al., 2003), apparently replacing subgenogroup B3 viruses. The subgenogroup B3 was reported as the predominant strain in 1997 especially in Sarawak (Cardosa et. al., 2003), but none was found to be circulating in Peninsular Malaysia and Sabah from 2001 to 2009. The subgenogroup B5, which emerged after the subgenogroup B4, was the predominant strain between 2003 and 2009. Thus, indicating that viruses in the genogroup B appeared to be evolving over the years from 2 subgroups described by Brown et al. (1999). To date, 5 subgenogroups have been identified viz. B1, B2, B3, B4 and B5.

The subgenogroup B1 viruses were circulating from 1974 to 1979 in Bulgaria, Hungary, USA and Australia and also in Taiwan in the early eighties. This was followed by subgenogroup B2, which was isolated mostly in the USA from 1987 to 1988. However, B2 has not been isolated in the USA since 1988 (Brown et al., 1999). A decade later in 1998, B2 was isolated from meningitis cases in Germany (Kehle et al., 2003). Between 1997 and 1999, the subgenogroup B3 viruses were the predominant strains in Southeast Asia and were identified as the major cause of epidemics in Sarawak in 1997 (Cardosa et al., 2003). The subgenogroup B3 viruses might be associated with cases of severe encephalitis resulting in fatalities in some cases in Sarawak from 1997 to 1999 (McMinn et al., 2001; Cardosa et al., 2003; Herrero et al., 2003), Singapore in 1998 and Western Australia in 1999 (McMinn et al., 2001). Later, the subgenogroup B4 viruses were isolated in several HFMD

cases in Singapore in 1997 (McMinn et al., 2001), Peninsular Malaysia from 1997 to 1998 (Herrero et al., 2003), Taiwan in 1998 (Shih et al., 2000) and Sarawak in 2000 (Cardosa et al., 2003). In 2003, the subgenogroup B5 was also isolated from Sarawak during a large HEV71 outbreak that started in February 2003 (Cardosa, unpublished data).

Two subgenogroup C were also isolated in this study. The subgenogroup C1 was isolated annually isolated from 2001 to 2007 but none in 2008 and 2009. This finding was parallel to the report by Brown et al. (1999) where the subgenogroup C1, previously prevalent in North America and Eastern Australia, appears to have undergone low-level endemic circulation within Southeast Asia and Western Australia between 1997 and 2003 and were isolated sporadically throughout 1997 to 2003 either from Sarawak, Malaysia, Singapore, Australia, United Kingdom and USA. However, subgenogroup C1 has not been reported as the cause of large outbreaks. Viruses belonging to the subgenogroup C4 were isolated in China in year 2000 (Shimizu et al., 2004; Li et al., 2005) even though no outbreaks were recorded. In Taiwan the subgenogroup C4 emerged and became predominant in 2004 (Lin et al., 2006). However, only one HEV71 in subgenogroup C4 was isolated in the east coast state of Peninsular Malaysia in 2004. No subgenogroups C2, C3 and C5 were isolated in this study.

The subgenogroup C2 was first reported in Malaysia in 1997 (Abubakar et al., 1999) from HFMD cases and this subgenogroup was found mostly during the large Taiwan outbreaks in 1998 (Shih et al., 2000; Wang et al., 2002) which was known to be associated with the emergence of fatal encephalitis. It was also occasionally isolated in Japan in 1997, Australia in 1999 and UK from 1998 to 1999. The subgenogroups C3 and C5, so far have never been reported in Malaysia (AbuBakar et al., 1999b; Cardosa et al., 2003; Herero et al., 2003). In the Asia-pacific region, the subgenogroup C3 was isolated only during the major outbreaks in Korea in 2000. A new subgenogroup C5 has been found circulated widely in Southern Vietnam throughout 2005 and became the predominant virus strain identified during the second half of the year in Vietnam (Tu et al., 2007).

Hand foot and mouth disease and HEV71 infections are still a major problem in Malaysia. However, a good surveillance system for HFMD and monitoring of severe encephalitis due to enteroviruses especially HEV71 has been established in Malaysia resulting in an easier and prompt action in order to stop spread of infection. This system has assisted the government in containing HEV71 outbreaks and has benefited young Malaysians from the disease.

Conclusion

Both genogroups B and C were found to be circulating in

Peninsular Malaysia and Sabah from 2000 to 2009. In the genogroup B, B4 was isolated in 2001 and was replaced by the B5 in 2003 which later became the most predominant HEV71 strain circulating between 2005 and 2009. In the genogroup C, C1 was periodically isolated through 2001 to 2007 but C4 was only isolated in 2007 in Peninsular Malaysia. These four strains were closely related to the strains that caused outbreaks of HFMD in several Asia-Pacific countries; however, no major outbreak has been reported in Peninsular Malaysia and Sabah so far.

ACKNOWLEDGEMENT

The authors would like to thank the Director General of Health and the Director of the Institute for Medical Research for permission to publish this paper. The authors would also like to thank and acknowledge Professor Dr Jane Cardosa and Dr David Perera from UNIMAS for their assistance in sequencing the VP1 gene of the 2008 HEV71 isolates.

REFERENCES

AbuBakar S, Chee HY, Shafee N, Chua KB, Lam SK (1999a). Molecular detection of enterovirus from an outbreak of hand, foot and mouth disease in Malaysia in 1997. Scan. J. Infect. Dis., 31: 331-335.

AbuBakar S, Chee HY, Al-Kobaisi MF, Xiaoshan J, Chua KB, Lam SK (1999b). Identification of enterovirus 71 isolates from an outbreak of hand, foot and mouth disease with fatal cases of encephalomyelitis in Malaysia. Virus Res., 61: 1-9.

Blomberg J, Lycke E, Ahlfors K, Johnson T, Wolontis S, von Ziepel G (1974). New enterovirus type associated with epidemic of aseptic meningitis and /or hand, foot, and mouth disease [letter]. Lancet, 2: 112.

Brown BA, Pallansch MA (1995). Complete nucleotide sequence of enterovirus 71 is distinct from poliovirus. Virus Res., 39: 195-205.

Brown BA, Oberste MS, Alexander JP, Kennett ML, Pallansch MA (1999). Molecular epidemiology and evolution of enterovirus 71 strains isolated from 1970 to 1998. J. Virol., 73: 9969-9975.

Cardosa MJ, Perera D, Brown BA, Cheon D, Chan HM, Chan KP, Cho H, McMinn PC (2003). Molecular Epidemiology of Human Enterovirus 71 Strains and Recent Outbreaks in the Asia-Pacific Region: Comparative Analysis of the VP1 and VP4 Genes. Emerging Infect. Dis., 9(4): 461-468.

Chan LG, Parashar UD, Lye MS, Ong FGL, Zaki SR, Alexander JP, Ho KK, Han LL, Pallansch MA, Suleiman AB, Jegathesan M, Anderson LJ (2000). Deaths of children during an outbreak of hand, foot and mouth disease in Sarawak, Malaysia. Clin. Pathol. Characteristics Dis. Clin. Infect. Dis., 31: 678-683.

Chu PY, Lin KH, Hwang KP, Chou LC, Wang CF, Shih SR, Wang JR, Shimada Y, Ishiko H (2001). Molecular epidemiology of enterovirus 71 in Taiwan. Arch. Virol., 146: 589-600.

Chua BH, McMinn PC, Lam SK, Chua KB (2001). Comparison of complete nucleotide sequences of echovirus 7 strains UMMC and the prototype (Wallace) strain demonstrates significant genetic drift over time. J. Gen. Virol., 82: 2629-2639.

Chumakov M, Voroshilova M, Shindarov L, Lavrova I, Gracheva L, Koroleva G, Vasilenko S, Brodvarova I, Nikolova M, Gyurova S, Gacheva M, Mitov G, Ninov N, Tsylka E, Robinson I, Frolova M, Bashkirtsev V, Martiyanova L, Rodin V (1979). Enterovirus 71 isolated from cases of epidemic poliomyelitis-like disease in Bulgaria. Arch. Virol., 60: 329-340.

Herrero LJ, Lee CS, Hurrelbrink RJ, Chua BH, Chua KB, McMinn PC

(2003). Molecular epidemiology of enterovirus 71 in peninsular Malaysia, 1997-2000. Arch. Virol., 148(7): 1369-1385.

Ishiko H, Shimada Y, Yonaha M, Hashimoto O, Hayashi A, Sakae K, Takeda N (2002). Molecular diagnosis of human enteroviruses by phylogeny-based classification by use of the VP4 sequence. J. Infect. Dis., 185: 744-754.

Kehle J, Roth B, Metzger C, Pfitzner A, Enders G (2003). Molecular characterization of an enterovirus 71 causing neurological disease in Germany. J. Neurovirol., 9: 126-128.

Li L, He Y, Yang H, Zhu J, Xu X, Dong J, Zhu Y, Jin Q (2005). Genetic Characteristics of Human Enterovirus 71 and Coxsackievirus A16 Circulating from 1999 to 2004 in Shenzen, People's Republic of China. J. Clin. Microbiol., 3835-3839.

Lin KH, Hwang KP, Ke GM, Wang CF, Ke LY, Hsu YT, Tung YC, Chu PY, Chen BH, Chen HL, Kao CL, Wang JR, Eng HL, Wang SY, Hsu LC, Chen HY (2006). Evolution of EV71 genogroup in Taiwan from 1998 to 2005: An emerging of subgenogroup C4 of EV71. J. Med. Virol., 78(2): 254-262.

Lum LCS, Wong KT, Lam SK, Chua KB, Goh AYT (1998). Neurogenic pulmonary oedema and enterovirus 71 encephalomyelitis. Lancet, 352: 1391.

McMinn PC, Lindsay K, Perera D, Chan HM, Chan KP, Cardosa MJ (2001). Phylogenetic analysis of enterovirus 71 strains isolated during linked epidemics in Malaysia, Singapore and Western Austr. J. Virol., 75(16): 7732-7738.

Melnick JL (1984). Enterovirus type 71 infections: a varied clinical pattern sometimes mimicking paralytic poliomyelitis. Rev. Infect. Dis., 6(2): S387-S390.

Ministry of Health, Malaysia (MHM) (2007). Prevention and Control of infectious diseases 9amendment of first schedule0 order 2007.

Minor PD, Brown F, Domingo E, Hoey E, King A, Knowles N, Lemon S, Palmenberg A, Rueckert RR, Stanway G, Wimmer E, Yin-Murphy M (1995). The Picornaviridae. In virus taxanomy. Suxth Report of The International Committee on Taxanomy of Viruses, Edited by FA Murphy, C.M. Fauquet, D.H.L. Bishop, S.A. Chabrial, A.W. Jarvis, G.P. Martelli, M.A. Mayo and M.D. Summer. Vienna & New York: Springer-Verlag, pp. 268-274.

Munemura T, Saikusa M, Kawakami C, Shimizu H, Oseto M, Hagiwara A, Kimura H, Miyamura T (2003). Genetic diversity of enterovirus 71 isolated from cases of hand, foot and mouth disease in Yokohama City between 1982 and 2000. Arch. Virol., 148: 253-263.

Ooi MH, Wong SC, Podin Y, Akin W, Syvia del S, Mohan A, Chieng CH, Perera D, Clear D, Wong D, Blake E, Cardosa J, Solomon T (2007). Human Enterovirus 71 Disease in Sarawak, Malaysia: A Prospective Clinical, Virological, and Molecular Epidemiological Study. Clin. Infect. Dis., 44: 646-656.

Perera D, Podin Y, Winnie A, Tan CS, Cardosa MJ (2004). Incorrect identification of recent Asian strains of Coxsackievirus A16 as human enterovirus 71: Improved primers for the specific detection of human enterovirus 71 by RT PCR. BMC Infect. Dis., 4: 1-11.

Podin Y, Edna LMG, Ong F, Leong YW, Yee SF, Apandi Y, Perera D, Teo B, Wee TY, Yao SK, Kiyu A, Arif MT, Cardosa MJ (2006). Sentinel surveillance for human enterovirus 71 in Sarawak, Malaysia: lessons from the first 7 years. BMC Pub. Health, 6: 180.

Schmidt NJ, Lennette EH, Ho HH (1974). An apparently new enterovirus isolated from patients with disease of the central nervous system. J. Infect. Dis., 129: 304-309.

Shih SR, Ho MS, Lin KH, Wu SL, Chen YT, Wu CN (2000). Genetic analysis of enterovirus 71 isolated from fatal and non-fatal cases of hand, foot and mouth disease during an epidemic in Taiwan, 1998. Virus Res., 68: 127-136.

Shimizu H, Utama A, Yoshii K, Yoshida H, Yoneyama T, Sinniah M (1999). Enterovirus 71 from fatal and nonfatal cases of hand, foot and mouth disease epidemics in Malaysia, Japan and Taiwan in 1997–1998. Jpn. J. Infect., 52: 12-15.

Shimizu H, Utama A, Onnimala N, Li C, Li-Bi Z, Yu-Jie M, Pongsuwana Y, Miyamura T (2004). Molecular epidemiology of enterovirus 71 infection in the Western Pacific Region. Paediatr. Int., 46: 1-5.

Shindarov LM, Chumakov MP, Voroshilova MK, Bojinov S, Vasilenko SM, Iordanov I (1979). Epidemiological, clinical and pathomorphological characteristics of epidemic poliomyelitis-like disease caused by enterovirus 71. J. Hyg. Epidemiol. Microbiol.

Immunol., 23: 284-295.

Singh S, Chow VTK, Chan KP, Ling AE, Poh CL (2000). RT-PCR, nucleotide, amino-acid and phylogenetic analyses of enterovirus type 71 from Asia. J. Virol. Meth., 88: 193-204.

Tu PV, Thao NTT, Perera D, Huu TK, Tien NTK, Thuong TC, Ooi MH, Cardosa MJ, McMinn PC (2007). Epidemiologic and virologic investigation of hand, foot, and mouth disease, Southern Vietnam, 2005. Emerg. Infect. Dis., 13(11): 1733-1741.

Wang SM, Liu CC, Tseng HW, Wang JR, Huang CC, Chen YJ, Yang YJ, Lin SJ, Yeh TF (1999). Clinical spectrum of enterovirus 71 infection in chikdren in southern Taiwan, with emphasis on neurological complications. J. Clin. Infect. Dis., 29: 184-190.

Wang JR, Tuan YC, Hsai HP, Yan JJ, Liu CC, Su IJ (2002). Change of major genotype of enterovirus 71 in outbreaks of hand foot and mouth disease in Taiwan between 1998 and 2000. J. Clin. Virol., 40: 10-15.

5

No xenotropic murine leukemia virus-related virus (XMRV) detected in Swedish monozygotic twins discordant for chronic fatigue

Patrick F. Sullivan, Franzcp[1,2], Tobias Allander[3], Cecilia Lindau[3], Kristina Fahlander[3], Andreas Jacks[2], Birgitta Evengård[4], Nancy L. Pedersen[2] and Björn Andersson[5]*

[1]Department of Genetics, University of North Carolina at Chapel Hill, NC, USA.
[2]Department of Medical Epidemiology and Biostatistics, Karolinska Institutet, Stockholm, Sweden.
[3]Laboratory for Clinical Microbiology, Department of Microbiology, Tumor and Cell Biology, Karolinska University Hospital, Karolinska Institutet, Stockholm, Sweden.
[4]Department of Clinical Microbiology, University of Umeå, Umeå, Sweden.
[5]Department of Cell and Molecular Biology, Science for Life Laboratory, Karolinska Institutet, Stockholm, Sweden.

The recent debate regarding the possible involvement of xenotropic murine leukemia virus-related virus (XMRV) in chronic fatigue syndrome (CFS) in humans has been intense due to the conflicting results from different studies and the unclear origin of the XMRV virus. In this study, we have addressed this issue by screening 47 pairs of monozygotic twins discordant for CFS using a sensitive PCR assay. The results are clearly negative for all samples. This is consistent with studies showing no evidence for the involvement of XMRV in CFS and supports the possibility that previous positive studies could be caused by laboratory artifacts.

Key words: Chronic fatigue syndrome (CFS), idiopathic chronic fatigue, infectious disease, xenotropic murine leukemia virus-related virus (XMRV), monozygotic twins.

INTRODUCTION

Chronic fatigue syndrome (CFS) is a severe syndrome characterized by prolonged and impairing fatigue and the absence of other disorders (Fukuda et al., 1994; Reeves et al., 2003). The initial symptoms are often similar to an infection. Immune dysfunction, which could result from a chronic infection or an inappropriate response to an initial infection (Komaroff and Buchwald, 1998; Mihrshahi and Beirman, 2005; Devanur and Kerr, 2006; Hempel et al., 2008; Lorusso et al., 2009), is believed to be a major causative factor. The possible role of a range of specific viruses in CFS has been investigated by searching for case-control differences in past or current viral infection (e.g., cytomegalovirus, Epstein-Barr virus, GBV-C, human herpes virus-6 and parvovirus B19) (Devanur and Kerr, 2006). Several reports of positive associations with a particular virus have been published, but it has not been possible to replicate the findings.

This is exemplified by the recent series of publications and intense debate on the role of the recently discovered xenotropic murine leukemia virus-related virus (XMRV) in CFS. XMRV was claimed to be present in 67% of cases with CFS and 3.7% of controls in a study from 2009 (Lombardi et al., 2009); however, these findings did not replicate in multiple independent samples (McClure and Wessely, 2010) but an NIH study (Lo et al., 2010) also found a significant increase in the frequency of retroviral infection in CSF patients. More recent studies have suggested that the positive XMRV findings were likely to be due to a sample contamination (Hue et al., 2010; Smith, 2010; Sato et al., 2010; Oakes et al., 2010; Robinson et al., 2010; Shin et al., 2011; Knox et al., 2011), which explained the conflicting results, and it is even possible that XMRV in itself was created as a laboratory artifact (Paprotka et al., 2011).

In order to further clarify the issue of XMRV and CFS further, it is useful to search for the virus in well-selected sets of patients and controls, as inconsistent findings

*Corresponding author. E-mail: bjorn.andersson@ki.se.

across case-control studies could be due to bias if controls are inappropriate to cases. For example, in the initial XMRV study, apart from the contamination issue, cases were highly selected (chronically ill patients treated in medical practices specializing in CFS) and controls were described only as "healthy" (Lombardi et al., 2009). The study of discordant monozygotic twins offers substantially improved experimental control (that is, an individual affected with CFS contrasted with their well monozygotic twin) (Byrnes et al., 2009). In this study, 47 pairs of monozygotic twins discordant for chronic fatigue, which have been used for CFS studies, including virus screening in our laboratory (Byrnes et al., 2009; Sullivan et al., 2011), were screened for XMRV using a sensitive PCR assay.

METHODS

The protocol was approved in advance by the ethical review board at UNC-CH and the Karolinska Institutet and all subjects provided written informed consent. The selection and diagnosis of patients was described exhaustively in Byrnes et al. (2009) and Sullivan et al. (2011). Briefly, 61,000 twin pairs from the Swedish twin registry was screened and 47 out of 140 pairs that were initially showed disconcordance for chronic fatigue were found to meet the rigorous inclusion criteria, regarding monozygosity, demographics, other medical conditions and other factors that could possibly influence the study (Byrnes et al., 2009). A discordant twin pair was defined as one twin meeting criteria for either CFS or idiopathic chronic fatigue (ICF) and the other twin was required never to have experienced impairing unusual fatigue or tiredness lasting more than one month.

Several rounds of PCR of XMRV from cellular DNA from white blood cells from patients and controls were performed using conditions described by Lombardi et al. (2009). The primers used were: X419F ATCAGTTAACCTACCCGAGTCGGAC; X1154R GCCGCCTCTTCTTCATTGTTCTC; XGAGOF CGCGTCTGATTTGTTTTGTT; and XGAGOR CCGCCTCTTCTTCATTGTTC. The first PCR was carried out in a mixture containing 1xPCR buffer II (Applied Biosystems) with 2.5 mM MgCl2 , 200 µM dNTPs, 0.4 µM of the forward and reverse primers, and 2.5 U AmpliTaq Gold DNA polymerase (Applied Biosystems) and 5 µl of the clinical sample in a total volume of 50 µl. The reactions were incubated 10 min at 94°C followed by (94°C for 1 min, 52°C for 1 min, 72°C for 2 min) x 35 cycles, followed by 72°C for ten minutes, and they were subsequently held at 4°C. The second PCR was carried out using 2 µl from the first PCR in a mixture containing 1xPCR buffer II (Applied Biosystems) with 2.5 mM MgCl2 , 200 µM dNTPs, 0.4 µM of the forward and reverse primers, and 2.5 U AmpliTaq Gold DNA polymerase (Applied Biosystems) and 5 µl of the clinical sample in a total volume of 50 µl. The reactions were incubated 10 min at 94°C followed by (94°C for one minute, 52°C for one minute, 72°C for two minutes) × 35 cycles, followed by 72°C for five minutes, and they were subsequently held at 4°C.

A synthesized and cloned XMRV fragment (Blue Heron Bio) was used as a positive control. The sequence of the synthesized fragment was:

GCCCATTCTGTATCAGTTAACCTACCCGAGTCGGACTTTTTGG
AGTGGCTTTGTTGGGGGACGAGAGACAGAGACACTTCCCGCC
CCCGTCTGAATTTTTGCTTTCGGTTTTACGCCGAAACCGCGCC
GCGCGTCTGATTTGTTTTGTTGTTCTTCTGTTCTTCGTTAGTTTT
CTTCTGTCTTTAAGTGTTCTCGAGATCATGGGACAGACCGTAA

CTACCCCTCTGAGTCTAACCTTGCAGCACTGGGGAGATGTCCA
GCGCATTGCATCCAACCAGTCTGTGGATGTCAAGAAGAGGCG
CTGGGTTACCTTCTGTTCCGCCGAATGGCCAACTTTCAATGTA
GGATGGCCTCAGGATGGTACTTTTAATTTAGGTGTTATCTCTCA
GGTCAAGTCTAGAGTGTTTTGTCCTGGTCCCCACGGACACCC
GGATCAGGTCCCATATATCGTCACCTGGGAGGCACTTGCCTAT
GACCCCCCTCCGTGGGTCAAACCGTTTGTCTCTCCTAAACCCC
CTCCTTTACCGACAGCTCCCGTCCTCCCGCCCGGTCCTTCTGC
GCAACCTCCGTCCCGATCTGCCCTTTACCCTGCCCTTACCCTC
TCTATAAAGTCCAAACCTCCTAAGCCCCAGGTTCTCCCTGATA
GCGGCGGACCTCTCATTGACCTTCTCACAGAGGATCCCCCGC
CGTACGGAGCACAACCTTCCTCCTCTGCCAGGGAGAACAATG
AAGAAGAGGCGGCCACCACCTCCGT.

This fragment was also used in a titration experiment using serial dilution. We found that the detection limit of the nested PCR was between 1 to 5 copies of the fragment/PCR reaction. The PCR products were analyzed by 1% agarose gel electrophoresis.

RESULTS

PCR for XMRV was carried out for each patient and control DNA sample using the conditions described in the foregoing. The same samples have previously been used for detection of other viruses via PCR, including hepatitis C virus and GB virus C, and they have been found to be suitable for PCR analysis (Sullivan et al., 2011). We carried out nested PCR to detect provirus in the human DNA samples. Despite a detection level of 1-5 copies per PCR, and repeated PCRs, none of the twins suffering from CFS or their healthy siblings were found to be positive for XMRV (Figure 1). A non-reproducible band close to the correct size was seen for one of the unaffected twins in one PCR reaction (Figure 1c), and the affected twins showed no such products. The PCR conditions were closely modeled on those used in the original XMRV study (Lombardi et al., 2009), where XMRV was first detected in CSF patients. The results thus conflict the initial reports by Lombardi et al. (2009) and Lo et al. (2010) and are consistent with the hypothesis that the positive findings were due to an artifact. These results are in agreement with other studies performed in Europe (McClure and Wessely, 2010).

DISCUSSION

We did not detect XMRV in our CFS cases and controls. The high detection level of the PCR assay and the quality of the DNA samples used in this study makes it highly unlikely that the negative result is caused by low-level infections. There are recent indications that there is some variation among the XMRV and XMRV-like viruses found in CFS (Lo et al., 2010). We therefore cannot completely rule out the possibility that a related virus could have been present in our samples but which was undetected. This possibility would require the presence of specific geographical sequence variation in XMRV and no data currently exists to address this issue. In addition, no such

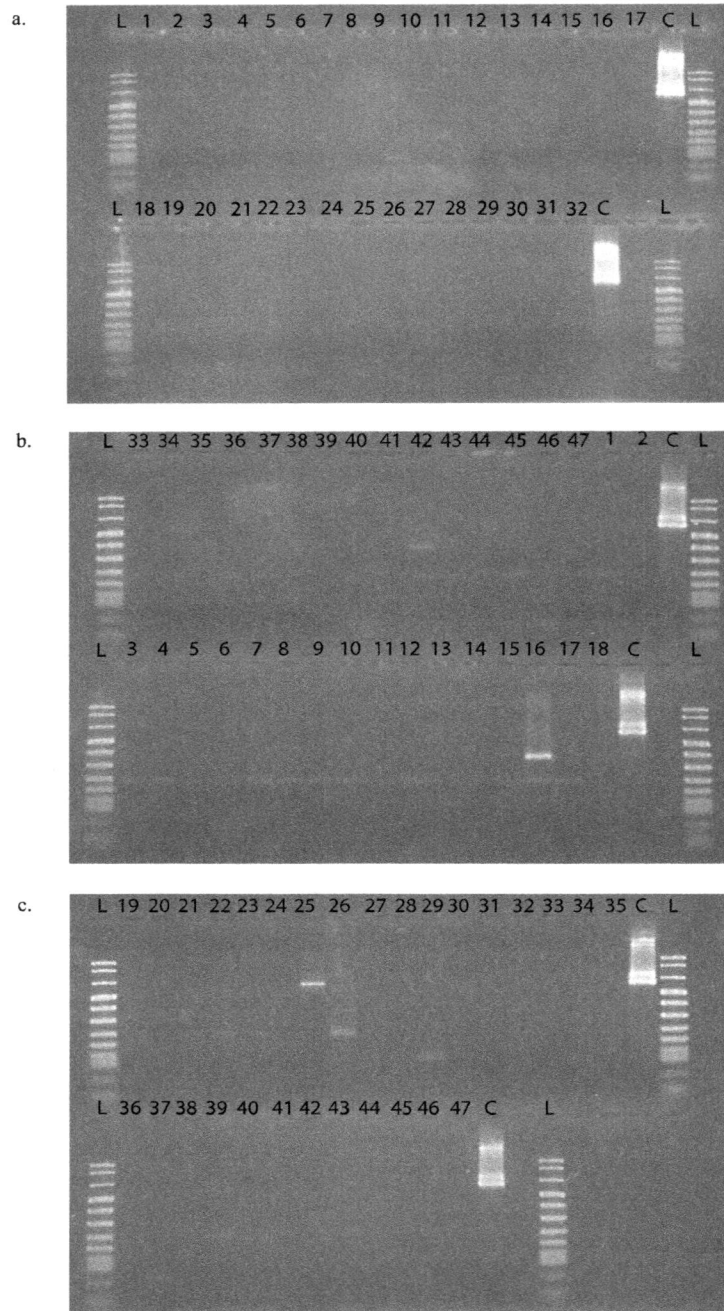

Figure 1. Gel images showing the results of the second PCR step in the nested PCR for XMRV for all patients and controls. Each gel row also contains a size marker and the PCR product from the positive control. The first 47 samples are from the twins affected with CSF and the remaining PCRs are from the healthy twins. No PCR products matching the XMRV control could be detected.

virus was found in the metagenomic search for viruses that was carried out in these samples (Sullivan et al., 2011).

The results of this study of XMRV are in concordance with other negative studies of European CFS cases and controls. As mentioned previously, multiple recent studies have indicated that positive XMRV findings were caused by laboratory contamination and that XMRV is not involved in human disease (Hue et al., 2010; Smith, 2010; Sato et al., 2010; Oakes et al., 2010; Robinson et al., 2010; Shin et al., 2011; Knox et al., 2011). Our results do not conflict with this possibility. Our results thus support the conclusion that XMRV is not involved in the etiology of CFS. Other explanations, such as that the pattern of XMRV infection is different depending on the geographic location, or that the virus variants present in Europe are significantly different from those present in the two American sample cohorts where it was detected, appear unlikely in the light of recent findings.

ACKNOWLEDGEMENTS

This project was funded by R01 AI056014 to PFS from the National Institute of Allergy and Infectious Diseases of the US National Institutes of Health. Additional funding was obtained from the Swedish Research Council.

REFERENCES

Byrnes A, Jacks A, Dahlman-Wright K, Evengard B, Wright FA, Pedersen NL, Sullivan PF (2009). Gene expression in peripheral blood leukocytes in monozygotic twins discordant for chronic fatigue: no evidence of a biomarker. PLoS One 4: e5805.

Devanur LD, Kerr JR (2006). Chronic fatigue syndrome. J. Clin. Virol., 37: 139-150.

Fukuda K, Strauss SE, Hickie I, Sharpe MC, Dobbins JG, Komaroff A (1994). The chronic fatigue syndrome: a comprehensive approach to its definition and study. Annals of Internal Med., 121: 953-959.

Hempel S, Chambers D, Bagnall AM, Forbes C (2008) Risk factors for chronic fatigue syndrome/myalgic encephalomyelitis: a systematic scoping review of multiple predictor studies. Psychol. Med., 38: 915-926.

Hue S (2010). Disease-associated XMRV sequences are consistent with laboratory contamination. By Stephane Hue and 12 others. Retrovirology, 7: 111.

Knox K, Carrigan D, Simmons G (2011). No evidence of murine-like gammaretroviruses in CFS patients previously identifi ed as XMRV-infected. Science: published online June 2. DOI:10.1126/science.1204963.

Komaroff AL, Buchwald DS (1998). Chronic fatigue syndrome: An update. Ann. Rev. Med., 49: 1-13.

Lo SC, Pripulova N, Li B, Komaroff AL, Hung G-C, Wang R, Alter HJ (2010). Detection of MLV-related virus gene sequences in blood of patients with chronic fatigue syndrome and healthy blood donors. Proc. Natl. Acad. Sci. USA, 107: 15874-15879.

Lombardi VC, Ruscetti FW, Das Gupta J, Pfost MA, Hagen KS, Peterson DL, Ruscetti SK, Bagni RK, Petrow-Sadowski C, Gold B, Dean M, Silverman RH, Mikovits JA (2009). Detection of an infectious retrovirus, XMRV, in blood cells of patients with chronic fatigue syndrome. Science, 326: 585-589.

Lorusso L, Mikhaylova SV, Capelli E, Ferrari D, Ngonga GK, Ricevuti G (2009). Immunological aspects of chronic fatigue syndrome. Autoimmun Rev., 8: 287-291.

McClure M, Wessely S (2010). Chronic fatigue syndrome and human retrovirus XMRV. BMJ, 340: c1099.

Mihrshahi R, Beirman R (2005). Aetiology and pathogenesis of chronic fatigue syndrome: a review. N Z Med. J., 118: U1780.

Oakes B, Tai AK, Cingöz O, Henefield MH, Levine S, Coffin JM, Huber BT (2010). Contamination of human DNA samples with mouse DNA can lead to false detection of XMRV-like sequences. Retrovirol., 7: 109.

Paprotka T, Delviks-Frankenberry KA, Cingöz O (2011). Recombinant origin of the retrovirus XMRV. Science 2011; published online June 2. DOI:10.1126/science.1205292.

Reeves WC, Lloyd A, Vernon SD, Klimas N, Jason LA, Bleijenberg G, Evengard B, White PD, Nisenbaum R, Unger ER (2003). Identification of ambiguities in the 1994 chronic fatigue syndrome research case definition and recommendations for resolution. BMC Health Serv. Res., 3: 25.

Robinson MJ, Erlwein OW, Kaye S, Weber J, Cingoz O, Patel A, Walker MM, Kim WJ, Uiprasertkul M, Coffin JM, McClure MO (2010). Mouse DNA contamination in human tissue tested for XMRV. Retrovirol., 7: 108.

Sato E, Furuta RA, Miyazawa T (2010) An endogenous murine leukemia viral genome contaminant in a commercial RT-PCR kit is amplified using standard primers for XMRV. Retrovirology, 7: 110.

Shin CH, Bateman L, Schlaberg R (2011). Absence of XMRV and other MLV-related viruses in patients with chronic fatigue syndrome. J. Virol., published online May 4. DOI:10.1128/JVI.00693-11.

Smith RA (2010). Contamination of clinical specimens with MLV-encoding nucleic acids: implications for XMRV and other candidate human retroviruses. Retrovirology, 7: 112.

Sullivan PF, Allander T, Lysholm F, Pedersen NL, Jacks A, Evengård B, Andersson B (2011). An Unbiased Search for Infectious Agents in Monozygotic Twins Discordant for Chronic Fatigue. BMC Microbiol., 11: 2.

6

Eggplant blister mottled virus (EBMV): A possible new potyvirus characterized from Iraq

Rakib A. Al-Ani, Mustafa A. Adhab* and Kareem A. H. Ismail

Department of Plant Protection, College of Agriculture, University of Baghdad, Iraq.

A possible new potyvirus infecting eggplants (*Solanum melongena* L.) was characterized from Iraq. The virus caused characteristic mottling, crinkling, blistering and stunting accompanied by severe abnormalities on new leaves and fruits. The virus was mechanically transmitted to *Gomphrena globosa* and *Zinnia elegans* producing necrotic local lesions (NLL) within 5 days of inoculation. The virus induced systemic latent infection on *Chenopodium amaranticolor* unlike any other known eggplant virus. The other test plants reacted systemically to the virus with different symptoms. No symptoms were observed on *Chenopodium quinoa*, *Chenopodium murale*, *Vigna unguiculata*, *Capsicum annum*, and *Phaseolus vulgaris*. Results of virus-vector relationship showed that, the virus transmitted by *Myzus persicae* in non-persistant manner. Purified preparation of the virus with absorption ratio 260/280 nm of 1.26, and yield of 4.45 mg/100 g virus infected was obtained. SDS-PAGE separation of purified viral particles indicated viral coat protein of 29 Kd. Electron microscopy of negatively stained crude extracts from symptomatic eggplants revealed flexuous particles of 720 nm. An antiserum at titer of 1/1024 against the purified virus was also prepared by four intramuscular injections of the virus. Based on the differences in symptoms on herbaceous host plants and similarity to potyvirus particles and coat protein (CP) size, we propose the name eggplant blister mottled virus (EBMV) for this possible new virus, member of potyvirus genus, characterized for the first time from Iraq.

Key words: Potyvirus, blister, mottling, eggplant, *Solanum melongena*.

INTRODUCTION

Eggplant (*Solanum melongena* L.) is one of the most important vegetable crops in Iraq. It is grown in open field during summer and in plastic and glasshouses during winter. It has been reported that eggplants are infected by several viruses causing significant damage due to plant stunting, crinkle, mottling accompanied by leaf and fruit abnormalities. Of these viruses, eggplant mottle dwarf virus (EMDV), a member of rhabdoviruses has been reported mainly from Mediterranean regions since 1969. It has also reported from northern Africa, southern

Europe and Middle East (Martelli and Russo, 1973; Martelli and Hamade, 1986; AL-Musa and Lockhart, 1990; Ghorbani, 1995; Ciffuo et al., 1999; Aramburu et al., 2006). The virus is known to be transmitted mechanically and by leafhoppers *Agalia vorobjevi* (Babaie and Izadpanah, 2003). Another virus, Eggplant mottle crinkle virus (EMCV), was first reported from Lebanon (Makkouk, 1981), then India (Raj et al., 1989), and recently from Iran (Rasoulpour and Izadpanah, 2008). The virus, having spherical particles of 37-40 nm in diameter, has been considered as a member of Tombusvirus that is transmitted mechanically (Lommel, 2000). Eggplant severe mottle virus (ESMV) was first identified in Nigeria and was reported to be transmitted by aphids *Myzus persicae* and *Aphis craccivora* in non-persistent manner (Ladepo, 1988; Cherif and Mateli, 1992; Brunt et al., 1996). Eggplant mosaic virus (EMV) in eggplants was first reported from west India by Briand et al. (1997). The virus is transmitted by sap, flea beetles *Epitrix Fuscula* (Gibbs and Harrison, 1971; Debrat et al.,

---
*Corresponding author. E-mail: maa_adhab@hotmail.com

Abbreviations: NLL , Necrotic local lesions; **CP**, coat protein; **EBMV**, eggplant blister mottled virus; **EMCV**, eggplant mottle crinkle virus; **ESMV**, eggplant severe mottle virus; **EMV**, eggplant mosaic virus; **PVY**, Potato virus Y; **AMV**, alfalfa mosaic virus; **CMV**, cucumber mosaic virus; **CLL**, chlorotic local lesions.

Figure 1. A, symptoms of mottle; B, associated with blister on the leaves of *S. melongena*; C, severe abnormalities on the new leaves.

1977; Dsonza et al., 1990; Rubeiro et al., 1996) and aphids *M. persicae* in non–persistent mode (Verma and Lai, 1967). Purified EMV consist of spherical particles of 28-30 nm in diameter. Potato virus Y (PVY), member of potyvirus is common in eggplants. It is transmitted by aphids *M. persicae* in non-persistent mode (Sastry et al., 1974). PVY consist of flexuous particles of 760 nm in long. Alfalfa mosaic virus (AMV) and Cucumber mosaic virus (CMV) are also reported to infect eggplants. The two viruses are transmitted mechanically to wide host plants and by several species of aphids in non-persistent manner (Kemp and Troup, 1977; Tanne and Sara, 1980; Brunt et al., 1996; Fleysh et al., 2001). Purified preparation of AMV contained bacilliform particles, while CMV consisted of spherical particles measuring 30 nm in diameter.

In recent years, an outbreak of a new disease was observed on eggplants in plastic and glasshouses cultivation in Iraq. The disease suspected to be of viral origin and was characterized by leaf mottling, crinkle, plant stunting, accompanied by severe malformation of upper new leaves and fruits. The present study was conducted to identify and characterize the causal virus by biological, serological and molecular means.

MATERIALS AND METHODS

Survey and sample collection

Symptomatic eggplants leaves (Figure 1) from plastic/glasshouse cultivation in Baghdad region of Iraq were collected.

Biological characterization

Host range study

The symptomatic leaves were homogenized in 0.1 M Potassium phosphate buffer PH 7.0 containing 0.2% Na-diethyl dithiocarbamate (Na-DIECA) at 1:4 (g/ml) (Al-Ani and Rathi, 1984). The homogenate was filtered through 2 layers of cheese cloth and the filtrate was centrifuged at 3000 rpm for 10 min. The supernatant obtained was mechanically inoculated on to carborandum dusted leaves of various test plants *viz.*; *S. melongena* L., *C. amaranticolor*

Coste and Reyn., *Datura stramonium* L., *Datura ennoxia*, *Dolicus Lablab* L., *G. globosa* L., *Lycopersicon esculentum* Mill., *Nicotiana clevelandii* Gray., *N. glutinosa* L., *Nicotiana tabacum* cv. Turkish, *N. tabacum* cv. White burly, *Petunia hybrida* Hord., *physalis floridana* Rydb., *Vigna sienensis* Endl., *Vinca rosea* = L., *Withania samnifera*, *Zinnia elegans* jacq., *Cucumis melo* L., *Cucumis sativus* L., *Cucurbita pepo* L., *Malva parviflora* L., *Solanum nigrum* L. at 3-4 leaves stage. The inoculated plants were maintained in insect–proof cages in a greenhouse for four weeks till symptom development.

Virus-vector relationship: Groups of aphids *M. persicae* and *Aphis fabae* reared on *Lactuca sativa* and *Beta vulgaris* respectively were placed on virus–infected eggplants in insect–proof cages for 0.5, 1, 5, 10, 30 and 60 min for acquisition/access periods. The insects were then transferred on healthy eggplants (5 aphids/plant) for 24 h. In other trial, viruliferous aphids were caged with healthy eggplants for 1/2, 1, 5, 10, 30 and 60 min for determination of minimum inoculation access period. The inoculated plants were maintained in growth room at 26 ± 2°C with 14 h photoperiods for symptoms development. Similarly, adults of whitefly *Bemisia tabaci* were collected from eggplants field by aspirator and caged with virus infected eggplants for 24 h. Then the whiteflies were transferred to healthy eggplants (5 whiteflies/plant) for 24 h for inoculation. The inoculated plants were maintained in growth room till symptoms development. The virus in the inoculated plants was confirmed by enzyme–linked immunosorobent assay (ELISA) using indigenously raised polyclonal antiserum.

Serological characterization

Virus purification: Pure isolate of virus was obtained by successive isolation of single lesion from *Z. elegans* grown in insect–proof cages in a greenhouse. The virus was then propagated in eggplants that serve as source for virus purification. The virus was purified according to the procedure described by Rowhani and stace-smith (1979) for Potato leaf roll virus.

Antiserum production and enzyme–linked immunosorobent assay (ELISA): Antiserum was obtained by 4 administrations of purified virus in to a New Zealand rabbit of 1 year age intramuscularly at 10 days intervals. For each administration 0.5 mg/ml of virus emulsified with an equal volume of Freund's incomplete adjuvant was injected. The blood was collected 15 days after the last injection through the ear marginal vein. The final bleed was allowed to clot for 1 h at room temperature and finally the antiserum was obtained by centrifugation at 3000 rpm for 10 min. The antiserum was maintained at -20°C until use. DAS-ELISA was done to detect the virus in the infected plants using polyclonal

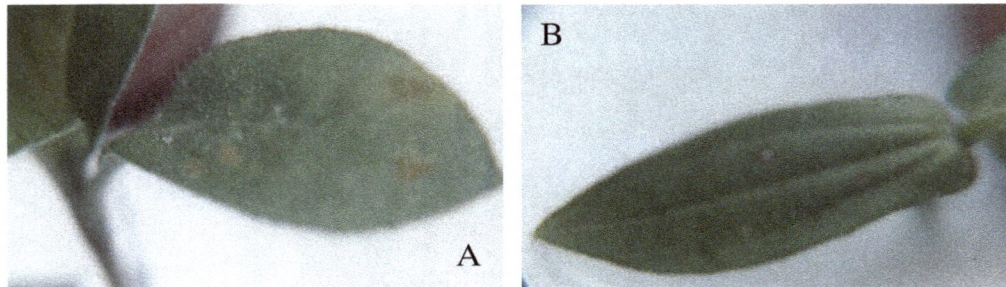

Figure 2. Necrotic local lesions on leaves of *G. globosa*. A, *Z. elegansl*; B, sap inoculated by EBMV.

Figure 3. Symptoms of vein clearing associated with mottling and faint chlorosis on the leaves of *D. ennoxia*. A, *D. stramonium*; B, *Nicotiana glutinosa*; C, *N. tabacum* var Samsun; D, *S. nigrum*; E, *W. samnifera*; F, sap inoculated by EBMV.

antiserum prepared against the virus and linked with alkaline phosphatase according to Clark and Adams (1977).

Characterization of viral protein: The molecular weight of viral coat protein was determined by sodium dodecyl sulphate (0.1% SDS)-polyacrylamide gel electrophoresis (10% PAGE) of purified virus in according to Laemmli (1970). The gel was stained for 30 min in 0.25% commassie brilliant blue dissolved in methanol/water/acetic acid (5:5:1) and destained in 10% acetic acid with 5% methanol solution to visualize protein bands.

Electron microscopy: Copper grids (300 mesh) coated with 0.4% formvar solution in chloroform were floated on a droplet of an extract from virus infected plants. The grids were blotted dry, floated on a droplet of water, and then transferred for 1 min to a droplet of 0.1% solution of potassium phosphotungstate in 0.1 M phosphate buffer. The grids were then observed by transmission electron microscopy.

RESULTS

Among mechanically inoculated test plants, *G. globosa* and *Z. elegans* reacted to the virus by producing necrotic local lesions on the inoculated leaves after 5 days of inoculation by extracts from leaves of virus infected eggplants (Figure 2). The virus induced systemic symptoms on *S. melongena* after 16 to 20 days of inoculation. The symptoms began as small faint chlorotic lesions on the leaves, developed to mottle or mosaic associated with blisters on the blades and finally to severe abnormalities of the new leaves associated with some degrees of necrosis on the veins and stunting of plants (Figure 1). Symptoms of vein clearing, mild mottling on the leaves developed to faint chlorosis on *D. stranonium*, *D. ennoxia*, *Lycopersicon esculentum*, *Nicotiana glutinosa*, *N. tabacum* cv. Samsun, *N. tabacum* cv. Turkish, *S. nigrum* and *Withania somnifera* after 15 days of sap inoculation from infected eggplants (Figure 3). Mottling associated with chlorosis and vein-banding on the leaves of *Malva parviflora*, *N. clevelandii*, *Physalis floridana*, *Petunia hybrida*, *Vigna sinensis* Endle., and *Vinca rosea* were observed after 10–12 days of sap inoculation. Symptom of mottling and mosaic on the leaves of *C. sativus, Cucumis melo,* and *Cucurbita pepo*

Figure 4. Symptoms of mottling, vein banding associated with chlorosis on the leaves of Nicotiana clevelandii. A, *P. floridana*; B, *P. hybrid*; C, *V. sinensis* Endle; D, *V. rosea*; E, inoculated by EBMV.

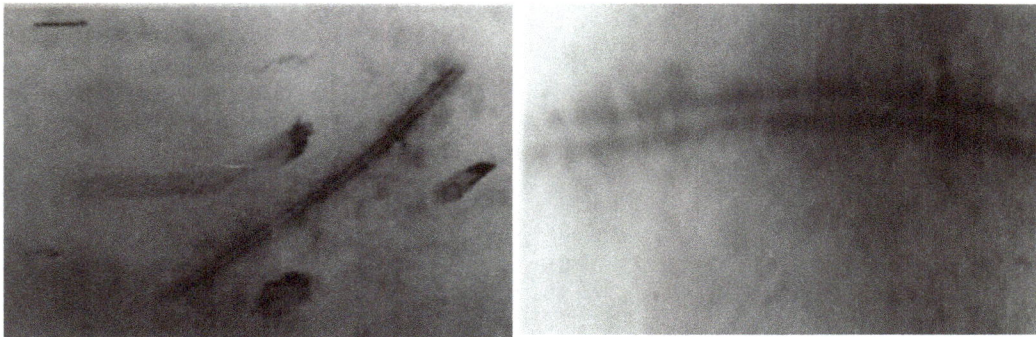

Figure 5. Electron micrographs of virus extract from infected eggplant leaves showing flexuous particle of about 720 nm long (The bar is 100 nm long).

with some degrees of stunting were observed upon mechanical inoculation by the virus (Figure 4). No symptoms were found on *Capsicum annum*, and *Solanum tuberosum*. The infection of *C. amaranticolor* by the virus as was detected by DAS-ELISA and was found to be latent mention type of symptoms observed on *C. amaranticolor*.

Virus–vector relationship: Symptoms characteristic of the virus appeared when healthy eggplants exposed to group of aphids *M. persicae* pre–caged on virus infected plants for 1/2 and 1 min. *A. fabae* and *B. tabaci* failed to transmit the virus. The same periods were required for

inoculation healthy plants. This indicated that *M. persicae* transmitted the virus in non–persistent manner.

Virus purification: Results showed that the procedure used for virus purification was suitable to obtain acceptable quantity of purified virus with absorption ratio 260/280 of 1.26. The virus yield was 4.45 mg/100 gm from infected eggplant leaves calculated using the PVY extinction coefficient = 2.8.

Electron microscopy: The observation of samples from infected eggplant leaves by transmission electron microscope revealed the presence of flexuous particles of

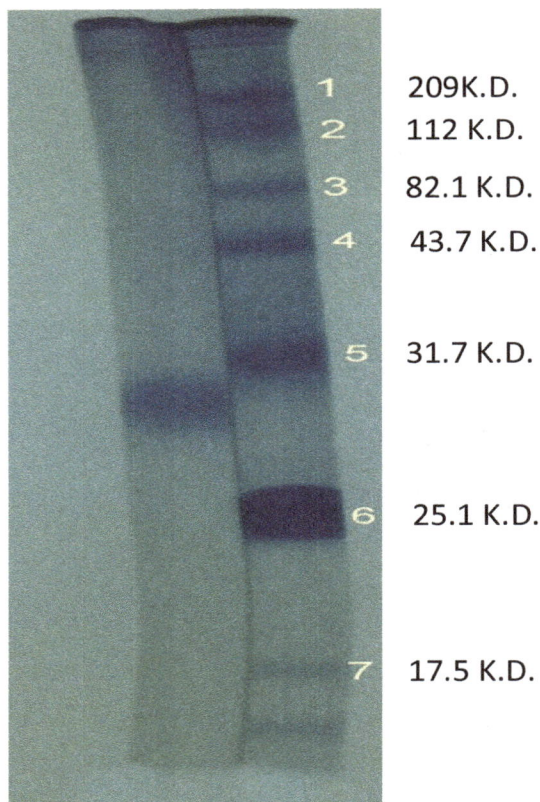

Figure 6. SDS-Polyacrylamide gel electrophoresis pattern of a dissociated sample of purified EBMV showing one band (left) represent the coat protein, of about 29 Kd estimated according to marker proteins of known molecular weight co-migrated with the viral protein (right).

about 720 nm in long (Figure 5). This indicated that the virus may be a member of potyvirus.

Serological analysis: An antiserum against the virus obtained after 4 intramuscularly administration of purified virus had a titer of 1/1024.

Viral coat protein characterization: The analysis of a sample of purified virus on SDS–PAGE revealed single protein subunit of 29 Kd determined by using proteins of known MWs co-migrating with the virus extract on the gel (Figure 6).

DISCUSSION

A possible new virus causing severe damage on eggplants was characterized by biological, serological, and molecular means. The symptoms induced on test plants upon inoculation by sap extracts from virus infected eggplants showed that, the virus seemed

different from AMV, CMV, EMCV, EMV, and PVY, these viruses induced chlorotic local lesions (CLL) on *C. amaranticolor* (Kemp and Troup, 1977; Makkouk et al., 1981, Raj et al., 1989; Brunt et al., 1996), while this plant is unsusceptible to EMDV and ESMV (Ladipo et al., 1988, Al–Musa and Lockhart, 1990, Brunt et al., 1996), but the infection of this plant by the virus in question found to be latent. Results showed that *D. stramonium* react systemically to AMV, ESMV, and EMV (Bos and Jsper, 1971; Gibbs and Harrisson, 1971) and by forming local lesions to CMV, and EMCV (Makkouk et al., 1981), immune to PVY (Sastry, 1974). To the virus in this study, *D. stramonium* reacted systemically. The virus infected *G. globosa* systemically, this host react to EMV and CMV by forming chlorotic local lesions (Gibbs and Harrisson, 1971), but not susceptible to AMV, EMV, PVY and ESMV. Unlike to all viruses infecting eggplant, discussed in this paper, the virus induced necrotic local lesions (NLL) on *Z. elegans*.

Like AMV, CMV, PVY, ESMV and EMCV, which are known to be transmitted by aphids *M. persicae* in non–persistent manner (Sastry et al., 1974; Kemp and Troup, 1977; Ladipo, 1988, Cherif and Marteli, 1992; Brunt et al., 1996; Fleysh et al., 2001), our virus was found to be transmitted by *M. persicae* in non-persistant manner.

Result of electron microscopy and coat protein analysis indicated that our virus was a member of potyvirus group like PVY and ESMV. It was reported that the MW of potyvirus genus is between 28–47 Kd (Lopez moya et al., 1994; Hull, 2002; Chang et al., 2002; Chen et al., 2006; Adhab, 2009), but this virus is different from these two viruses based on the differences in the symptoms induced on *D. stramonium*, *C. amaranticolor* and *Z. elegans*, which mean that our virus may be representing a new virus infecting eggplant in Iraq. The name proposed is EBMV due to characteristic blistering symptoms on eggplant.

ACKNOWLEDGEMENT

We would like to express our grateful to Dr. Aijaz A. Zaidi (Scientist and Head, Floriculture Division, Institute of Himalayan Bioresource Technology, Palampur, India), for his valuable help in putting the manuscript in this final version.

REFERENCES

Adhab MA (2009). Study of coat proteins for three plant viruses and the possibility of use it as markers in its detection. M.Sc. Thesis, University of Baghdad, College of Agriculture, Iraq. (in Arabic).
Al-Ani RA, Rathi YP (1984). Plant Viruses, Principles of Applied Experiments, University of Baghdad Press (in Arabic), P. 274.
AL-Musa AM, Lockhart B (1990). Occurrence of *Eggplant mottled dwarf virus* in Jordan. J. Phytopathal., 128: 283-287.
Aramburu J, Galipienso L, Tornos T, Matas M (2006). First report of *Eggplant mottled dwarf virus* in mainland Spain. Plant

Pathol., 55: 565.

Babaie GH, Izadpanah K (2003). Vector transmission of *Eggplant mottled* dwarf virus in Iran. J. Phytopathol., 151: 679-682.

Bos L, Jasper MJ (1971). Alfalfa mosaic virus. Description of plant viruses. CMI/AAB, No. 46.

Briand JP, Bouley JP, Witz J (1997). Self assembly of Eggplant mosaic virus protein. Virology, 76: 664-669.

Brunt AAK, Cabtree MJ, Dallwitz AJ, Gibbs JW (1996). Viruses of plant. Description and list from the (VIDE) data base (AL) international, 1484 p.

Chang CA, Chen CC, Hsu HT (2002). Partial characterization of two *potyviruses* associated with golden spider lily severe mosaic disease. Acta Horticulturae, 568: 127-134.

Chen CC, Hsu HT, Cheng YH, Huang CH, Liao JY, Tsai HT, Chang CA (2006). Molecular and serological characterization of a distinct *potyvirus* causing latent infection in *calla lilies*. Bot. Stud., 47: 369-378.

Cherif C, Mateli GP (1992). Outbreaks and new records, Tunisia mottled dwarf of eggplant. FAO PL. Prol. Bull., 33: 166-167.

Ciffuo M, Roggero P, Masenga V (1999). Natural infection of muskmelon be Eggplant mottled dwarf rhabdovirus in Italy. Plant Dis., 83: 78.

Clark MF, Adams AN (1977). Characteristic of the microplate methods of enzyme– linked immunosorbent assay for the detection of plant viruses. J. Gen. Virol., 34: 475-483.

Debrat EA, Lastra R, Deuzeategui RC (1977). *Solanum Seaforthianum*, A weed host of Eggplant mosaic virus in Venezuela. Plant Dis. Reporter, 61: 628-631.

Desonza VBVI, Gergerich RC, Kim KS, Langham MAC (1990). Properties and Cytopathology of a tymovirus isolated from eggplant. Phytopathology, 81: 1092-1098.

Fleysh N, Deka D, Darth M, Koprowiski H, Yusibov V (2001). Pathogenesis of *Alfalfa mosaic virus* in Soybean (*Glycin max*) and expression of chimeric Rabies peptide in virus infected soybean plants. Phytopathology, 91: 941-947.

Ghorbani S (1995). Identification of *Eggplant mottled* dwarf virus (EMDV) in Tehran province. Proc. 12th Plant Protection Cong. Iran, Karadj, Iran, p. 178.

Gibbs AJ, Harrisson BD (1971). Eggplant mosaic virus and its relationship to Indian Potato Latent virus. Ann. Appl. Biol., 64: 225-231.

Hull R (2002). Matthews' Plant Virology. Fourth edition. Academic Press, London, UK., 1001 p.

Kemp WJ, Troup PA (1977). *Alfalfa mosaic virus* occurring naturally on eggplant in Ontaria. Plant Dis. Reptr., 61: 393-396.

Laemmli UK (1970). Cleavage of structural proteins during the assembly of the head of bacteriophage T4. Nature, 227: 680-685.

Ladipo JI, Leseman DE, Koeing R (1988). Host ranges, serology and Cytopathology of eggplant and tomato strains of *Eggplant severe mottled virus*, a new potyvirus from Nigeria. J. Phytopathol., 122: 359-371.

Lommel SA, Marelli GP, Russo M (2000). Genus Tombusvirus. In: MHV VanRegenmortel, Fauquet CM, Bishop DHYL, Carstens E, Estes M (Eds.), Virus Taxonomy, seventh report of the international committee on taxonomy of viruses, NewYork, NY, pp. 819-825.

Lopez MJJ, Canto T, Lopez AD, Diaz RJR (1994). Differentiation of Mediterranean *Plum pox virus* isolates by coat protein analysis. Plant Pathol., 43: 164-171.

Makkouk KM, Koenig R, Lessemann DE (1981). Characterization of a Tombusvirus isolated from eggplant. Phytopathology, 71: 572-577.

Martelli GP, Hamadi A (1986). Occurrence of *Eggplant mottled virus* dwarf in Algeria. Plant Pathology, 35: 595-597.

Martelli GP, Russo M (1973). *Eggplant mottled dwarf virus*. CMI/AAB Descriptions of Plant Viruses, 115: 4.

Raj SK, Aslam M, Srivastava KM, Singh BP (1989). Occurrence and identification of Eggplant mottled crinkle virus in India. J. Phytopathol., 3: 283-288.

Rasoulpour R, Izadpanah K (2008). First report of *Eggplant mottled crinkle virus* in geranium in Iran. Plant Pathol., 57: 397.

Rowhani A, Stace-Smith R (1979). Purification and characterization of Potato leaf roll virus. Virology, 98: 45-54.

Ruberio S, Kitajima GEW, Oliveira CRB, Koeiny R (1996). A strain of Eggplant *mosaic virus* isolated from naturally infected tobacco plants in Brazil. Plant Dis., 80: 44-49.

Sastry KS, Sastry KSM, Singh J (1974). A mosaic disease of eggplant *Solanum melongena* L. Phytopathologia Medeterranea, 13: 176-178.

Tanne E, Sara ZG (1980). Cucumber mosaic virus on eggplant in Israel. Plant Dis., 64: 371-372.

Verma GS, Lai R (1967). Occurrence of mosaic disease on Brinjal. Indian Phytopathol., 20: 243-247.

7

The first isolation of chikungunya virus from non-human primates in Malaysia

Y. Apandi[1]*, W. A. Nazni[2], Z. A. Noor Azleen[2], I. Vythilingam[3], M. Y. Noorazian[3], A. H. Azahari[2], S. Zainah[1] and H. L. Lee[2]

[1]Virology Unit, Institute for Medical Research, Kuala Lumpur, Malaysia.
[2]Medical Entomology Unit, Institute for Medical Research, Kuala Lumpur, Malaysia.
[3]Parasitology Unit, Institute for Medical Research, Kuala Lumpur, Malaysia.

Chikungunya is a mosquito borne disease caused by chikungunya virus (CHIKV). The virus is transmitted to human by Aedes genus mosquitoes. Transmission cycles of CHIKV can be man - mosquito - man (urban cycle) or animal - mosquito - man (sylvatic cycle). Sylvatic transmission cycle of CHIKV has been described in Africa and may play a role in re-emergence of CHIKV infection. In Malaysia, CHIKV- neutralizing antibodies have been detected among wild monkeys in mid 1960s but so far CHIKV has never been isolated in monkeys.

Key words: Chikungunya virus, sylvatic cycle, non human primates, CHIKV genotypes

INTRODUCTION

Sera samples were collected from wild monkeys from 3 states of Peninsular Malaysia. All were inoculated into Vero and BHK cells; isolates were then identified and confirmed by polymerase chain reaction (PCR) and DNA sequencing techniques. These four isolates from monkeys were blast searched in GenBank found to be CHIKV with higher similarity to CHIKV isolated from klang outbreak in 1998 and Bagan Panchor outbreak in 2006 but were totally different from outbreak in Peninsular Malaysia in 2008.

This is the first to report CHIKV isolated from non human primates, thus confirming the existence of sylvatic transmission cycle in Malaysia. However, their roles as reservoir of infection need to be further investigated.

Text

Chikungunya virus (CHIKV) is an arbovirus from the genus Alpha virus in the family of *Togaviridae* was first isolated in Tanzania in 1953 (Ross, 1956). Between the 1960s and 1980s, these viruses had been reported circulating in many countries in the central, southern and Western Africa and recently a large epidemic occurred in Indian Ocean Islands and India (Vidya et al., 2007). The spread of CHIKV infections are normally sustained by human-mosquito-human transmission (urban cycle) but reports suggest primates-mosquito transmission (sylvatic cycle) maintains the virus in the wild in Africa (sylvatic cycle). Here we present the first evidence for a sylvatic cycle for CHIKV primates in Asia.

Currently CHIKV can be grouped into three distinct genotypes as Asian, Central/East African and West African genotypes based on the phylogenetic analysis of the E1 gene (Powers et al., 2000; Schuffenecker et al., 2006). Recent explosive epidemic of African genotype in Indian Ocean Islands and India since 2005 and other parts of Asia, Africa and Europe showed that international travelers have disseminated new strain of the virus, some into region from which CHIKV has been absent (Townson and Nathan, 2008). This scenario had changed the geographical origin of CHIKV worldwide.

Re-emergence of CHIKV infection has been reported in many parts of the world (Rao, 1971; Thuang et al., 1975; Thaikruea et al., 1997) and recently occurred in the India Ocean Islands and India (Schuffenecker et al., 2006; Vidya et al., 2007) with European countries such as France, Germany, Italy, Norway and Switzerland also experiencing imported cases of CHIKV infection from people returning from endemic areas (D'Ortenzio et al.,

*Corresponding author. E-mail: apandi@imr.gov.my

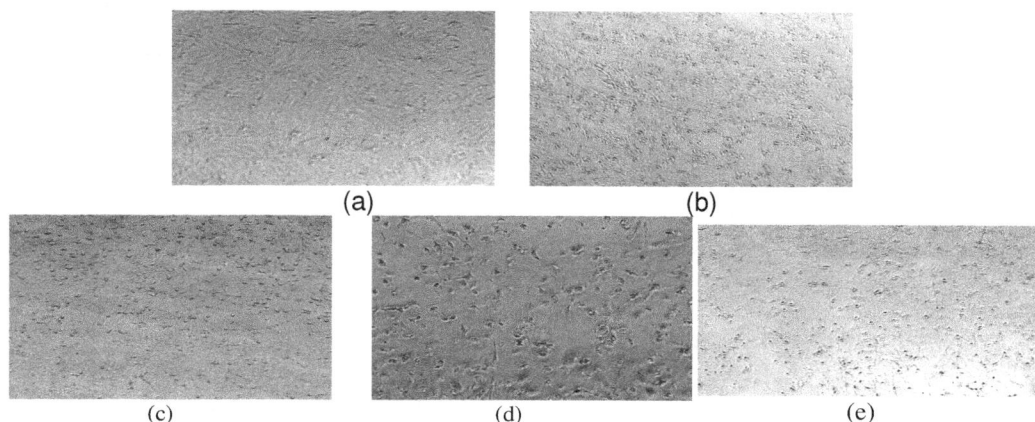

Figure 1. (a) Uninfected African Green Monkey Kidney (Vero) cell at 10X, (b) Sample M125 after 4 days of inoculation in Vero cell at 10X; (c) Sample M127 after 4 days of inoculation in Vero cell at 10X; (d) Sample M128 after 4 days of inoculation in Vero cell at 10X and (e) Sample M129 after 4 days of inoculation in Vero cell at 10X.

2009; Townson and Nathan, 2008; Pialoux et al., 2007; WHO, 2006). In Asia particularly, CHIKV activity was documented from its isolation in Bangkok in 1958 (Fields et al, 1996) and since then the transmission of this virus from Asian genotype continued and caused many outbreaks such as in Kolkata and Southern India in 1963 and 1964 respectively, in Indonesia during 2001 - 2003 (Laras et al., 2005) even though the African genotype was responsible for outbreaks India and Indian Ocean Islands during 2005 - 2006 (Yergolkar et al., 2006).

In Malaysia, CHIKV infection was first recorded in 1998 - 1999 in Port Klang affecting more than 51 people (Lam et al., 2001), which was followed by the outbreak in Bagan Panchor, Perak in 2006 (Kumarasamy et al., 2006). Both outbreaks were attributed to the Asian genotype of CHIKV. After a period of almost 2 years, the re emergence of CHIKV infection occurred in July 2008 which started in Tangkak, Johor then spread to other states in Peninsular Malaysia such as Negeri Sembilan and Malacca, with more than 2000 cases recorded (MOH, 2008). In March 2009, more than 1600 cases had been reported, with the states of Kuala Lumpur, Selangor and Kelantan, recording the highest number of cases (MOH, 2009). This outbreak is very important since it was the first chikungunya outbreak caused by CHIKV belongs to the Central/East African genotype in Malaysia (Maizatul, 2009). Clinically, the patients presented with fever, arthralgia, myalgia and most often the symptoms were indistinguishable from dengue infection. However, unlike dengue which caused morbidity and mortality, there has never been any mortality from CHIKV infection in Malaysia, although mortality has been reported in Reunion Island (Michault and Staikowsky, 2009) and suspected in India (Mavalankar et al., 2008).

In this study, the main objective was to detect the presence of viruses especially arboviruses in non human primates. A total of 105 sera were collected from wild long tailed macaques (*Macaca fascularis*) from 3 states of Peninsular Malaysia namely Pahang, Selangor and Wilayah Persekutuan Kuala Lumpur from August 2007 till January 2008. All sera were inoculated onto 3 different cell lines: C6/36, Vero and BHK cells. Cultures were observed daily for cytopathic effect (CPE) and samples which did not exhibit any CPE after 2 passages or after a total of 14 days incubation were regarded as negative. Four from 5 sera collected from Kuala Lipis in Pahang showed CPE only in Vero cells 3 - 4 days after inoculation. The virus showed characteristic of enterovirus CPE in which the cells became refractile and rounded cells in loose clusters as shown in Figures 1a - e.

Based on the CPE characteristics, simian enterovirus was thought to be the most likely virus. Therefore published primers 188 - 222, 012-011, OL68-EVP2 and 189 - 222, which were reported by Oberste et al. (2002) to detect simian enteroviruses in non human primates were used for PCR-based identification. Unforyunately all these primers failed to amplify any product as shown in Figures 2a, b, c and d.

Rudnick and Lim in 1986 reported the presence of arboviruses, especially dengue virus, in non human primates in Malaysia. Flavivirus universal primers and dengue universal primers (Lanciotti et al., 1992) were thus used in PCR analysis of these isolates; however, none were able to amplify any PCR product. Due to re emergence of CHIKV infection, that in Malaysia started in mid 2008, the isolates were then amplified using the primers NSP1-C and NSP1-S and E1-C and E1-S (Hasebe et al., 2002), which amplify 354 bp of the non-structural protein P1 and 294bp of the glycoprotein E1 gene of CHIKV, respectively. The amplifications of PCR products are shown in Figures 3a and b.

These products were sequenced in both directions using NSP-1 and E-1 primers. Blast searches in Gen-Bank revealed that these isolates were CHIKV and were

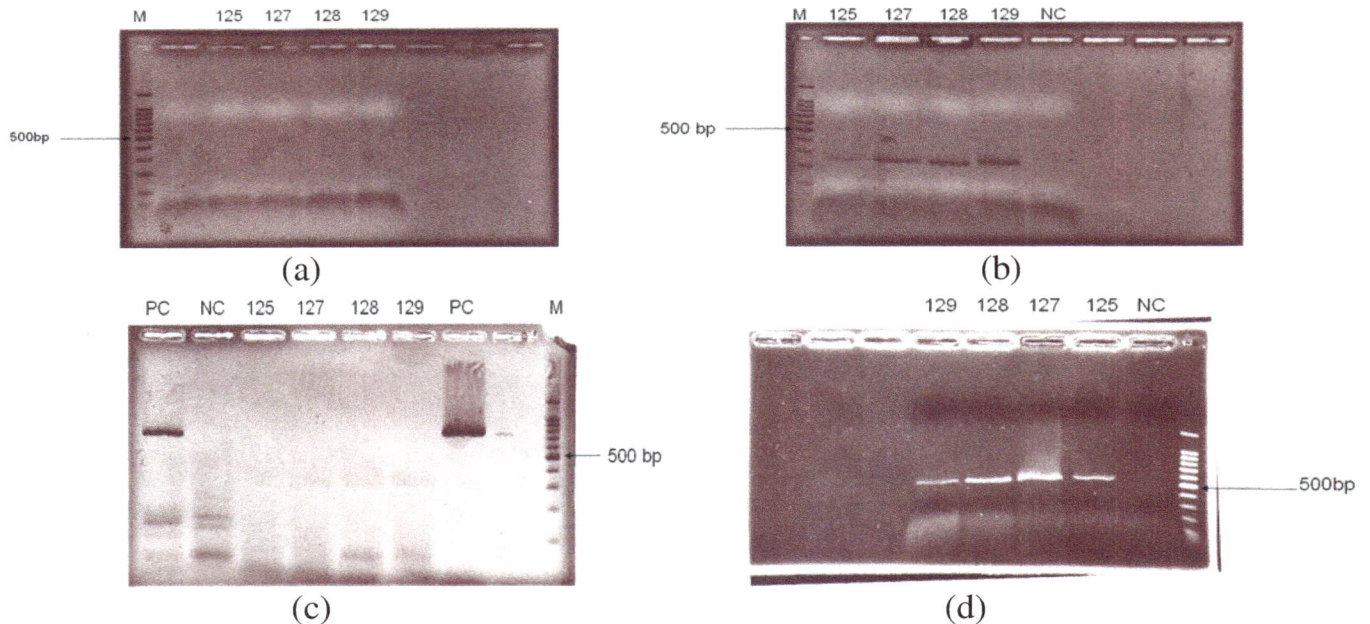

Figure 2. Amplification of four isolates by using (A): primer 188-222; (B) primer 012-011; (C) OL68-EVP2 and (D) primer 189-222. M: 1kb DNA ladder, BioLabs, New England.

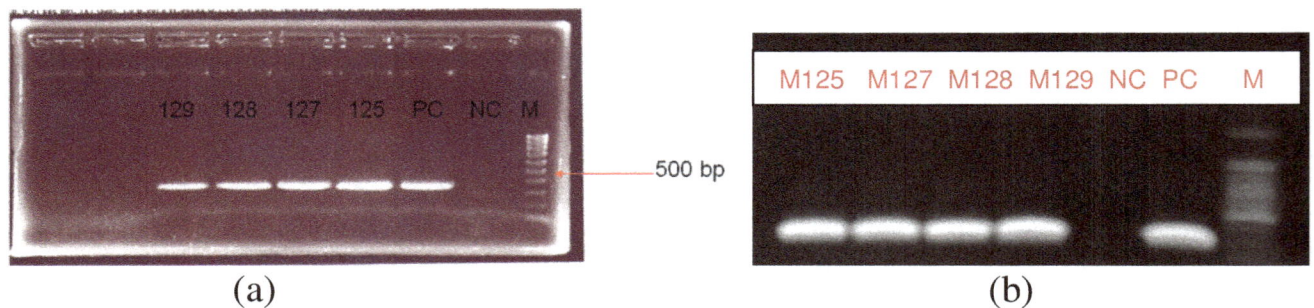

Figure 3. Amplification of four isolates by using (a): primer NSP1-C and NSP1-S; (b) primer E1-S and E1-C; M: 100bp DNA ladder, BioLabs, New England.

closely related to the human CHIKV isolated in Bagan Panchor in 2006 and Klang in 1998 outbreaks. They belonged to the Asian genotype clusters but distinct from the current CHIKV circulating in Malaysia in 2008, which has been identified as the Central/East African genotype (Maizatul, 2009). Phylogenetic analysis of these isolates together with CHIKV representatives from other regions is shown in Figure 4.

The human CHIKV isolates from the 2006 outbreak in Bagan Panchor, Malaysia were closely related to the isolates from the 1998 outbreak in Klang. Interestingly the four isolates from monkeys were closely related to these viruses, even though there were isolated from different hosts. The high sequence similarities suggests that cross transmission between human and non-human primates occurred in Malaysia and that the 2006 and 1998 out-

breaks may have arisen from virus circulating in monkeys, supporting the view that CHIKV is endemic in Malaysia (Abubakar et al., 2007).

Before 1998, CHIKV had not been isolated from humans or animals and no clinical disease caused by CHIKV had been reported in Malaysia. However, serologic survey of human serum samples collected during 1965 - 1969 in West Malaysia showed neutralizing antibodies to CHIKV among adults, especially those inhabiting the rural northern and eastern states bordering Thailand (Marchette et al., 1980). Earlier studies also found evidence of CHIKV-neutralizing antibodies in wild monkeys, pigs, and chickens (Marchette et al., 1980) suggested that a CHIKV sylvatic transmission cycle may exists in Malaysia even though CHIKV never been isolated from Asian monkeys (Chhabra et al., 2008). In

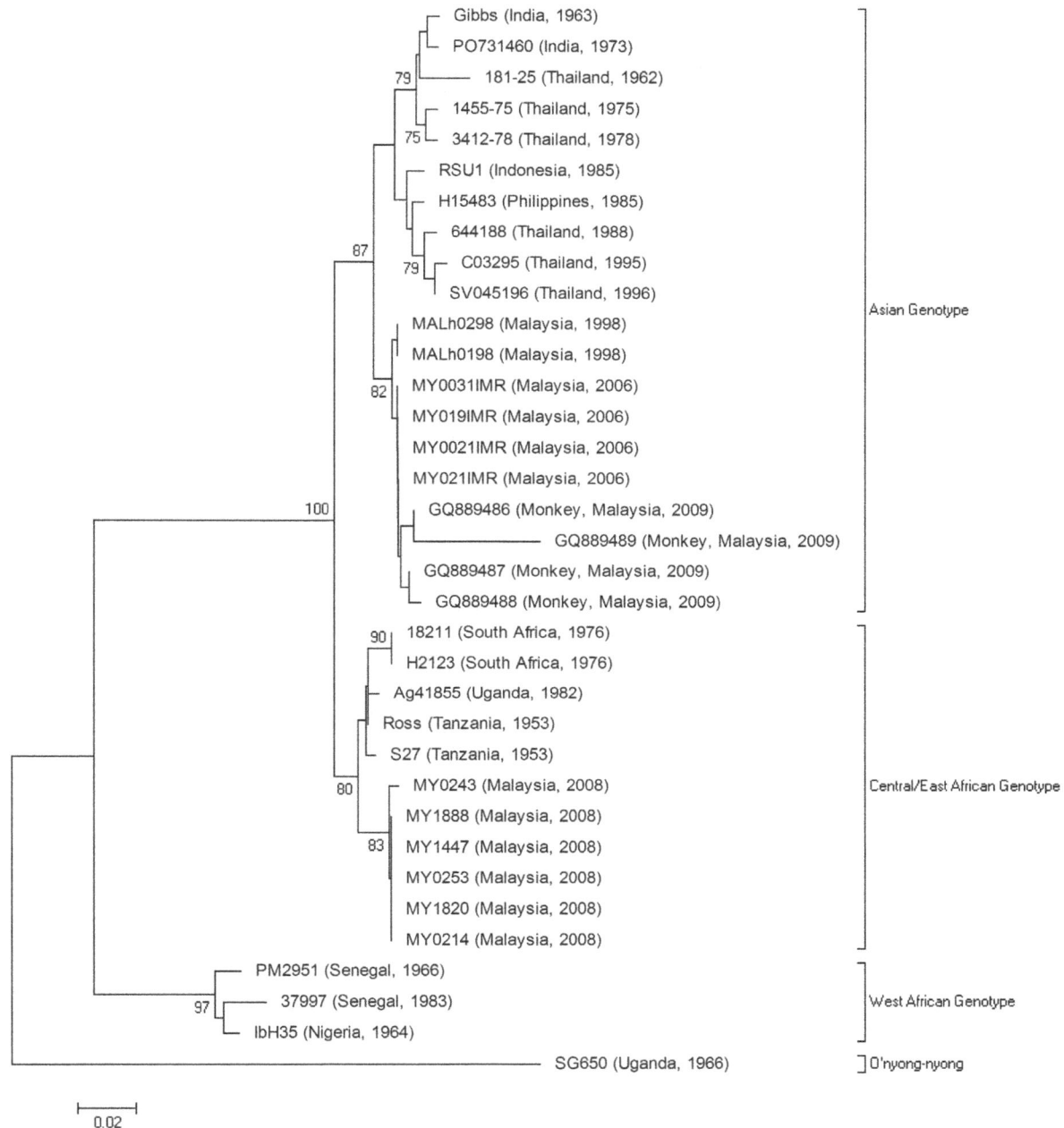

Figure 4. Phylogenetic tree of partial glycoprotein E1 sequences (257bp) of CHIKV inferred using the Neighbor-Joining method from the software MEGA 4. The evolutionary distances were computed using the Maximum Composite Likelihood method. Genotype Asian, Central/East African and West African are indicated by square brackets with O'nyong-nyong virus as an out-group. Four CHIKV isolated from monkeys labeled as GQ889486, GQ889487, GQ889488 and GQ889489. Representative strains of each genotype obtained from GenBank are labeled using the following format: 'isolate'-'Country of origin'- 'Year isolation'. Bootstrap values (> 75%) for 1,000 pseudo replicate dataset are indicated at branch nodes.

the absence of active surveillance since the 1965 study (Marchette et al., 1980), whether the apparent absence of CHIKV over the years and between the 2 recent outbreaks in Malaysia is due to an unidentified sylvatic transmission cycle or silent transmission among humans is difficult to establish.

Africa sylvatic transmission involving wild primates may play a role in the emergence and re-emergence of CHIKV infection (Diallo et al., 1999). The virus likely circulates among wild primates and many species of Aedes mosquitoes, with serological evidence demonstrating the presence of antibodies in humans and wild primates

(Adesina and Odelola, 1991; Jupp and McIntosh, 1998; Rodhain et al., 1989). This zootic cycle probably supports the virus with occasional human leading to epidemic (Lumsden, 1955). To date, a vertebrate reservoir has not been identified outside Africa to support the historical evidence by Carey (1971) that CHIKV originated in Africa. The evidence presented herein illustrates that viruses from the 2006 and 1998 CHIKV outbreaks in Malaysia are closely related to CHIKV isolates from monkeys suggesting the existence of sylvatic cycle for CHIKV in Malaysia

ACKNOWLEDGEMENTS

The authors wish to thank the Director of the Institute for Medical Research and Director-General of Health, Ministry of Health for permission to publish this paper.

REFERENCES

Abubakar S, Sam IC, Wong PF, MatRahim NA, Hooi PS Roslan N (2007). Reemergence of endemic chikungunya, Malaysia. Emer. Infect. Dis. 13(1): 147-149.

Adesina OA, Odelola HA (1991). Ecological distribution of Chikungunya haemagglutination inhibition antibodies in human and domestic animals in Nigeria. Trop. Geogr. Med. 43: 271-275.

Carey DE (1971). Chikungunya and dengue: a case of mistaken identity? J. Hist. Med. Allied Sci. 26: 243-262.

Chhabra M, Mittal V, Bhattacharya D, Rana UVS, Lal S (2008). Chikungunya fever: A re-emerging viral infection. Indian J. Med. Microbiol. 26(1): 5-12.

Diallo M, Thonnon J, Traore-Lamizana M, Fontenille D (1999). Vectors of chikungunya virus in Senegal: current data and transmission cycles. Am. J. Trop. Med. Hyg. 60: 281–6.

D'Ortenzio E, Grandadam M, Balleydier E, Dehecq JS, Jaffar-Bandjee MC, Michault A, Andriamandimby SF, Reynes JM, Filleul L (2009). Sporadic cases of chikungunya, Réunion Island, Eurosurveillance 14 (35): 1-2.

Fields BN, Knipe DM, Howley PM (1996). Alphavuruses. In: Fields of Virology Third Edition; V 1. Lippincott-Raven Publishers pp.858-898.

Hasebe F, Parquet MC, Pandey BD, Mathenge EG, Morita K, Balasubramaniam V, (2002). Combined detection and genotyping of Chikungunya virus by a specific reverse transcription polymerase chain reaction. J. Med Virol: 67: 370-374.

Jupp PG, McIntosh BM (1988). Chikungunya virus disease. In The Arboviruses : Epidemiol. Ecol. Edited by T. P. Monath. Boca Raton, FL: CRC Press. 2: 137-157.

Kumarasamy V, Prathapa S, Zuridah H, Chem YK, Norizah I, Chua KB (2006). Re-Emergence of Chikungunya Virus in Malaysia. Med. J. Malaysia 61(2): 221-225.

Lam SK, Chua KB, Hooi PS, Rahimah MA, Kumari S, Tharmaratnam M (2001). Chikungunya infection – an emerging disease in Malaysia. Southeast Asian J. Trop. Med. Public Health. 32: 447–51.

Lanciotti RS, Calisher CH, Gubler DJ, Chang GJ, Vorndam AV (1992). Rapid detection and typing of dengue viruses from clinical samples by using reverse transcriptase-polymerase chain reaction. J. Clin. Microbiol. 30(3): 545-551.

Laras K, Sukri NC, Larasati RP, Bangs MJ, Kosim R, Djauzi Wandra T, Master J, Kosasih H (2005). Tracking the re-emergence of epidemic chikungunya virus in Indonesia. Trans. R. Soc. Trop. Med. Hyg. 99: 128–141.

Lumsden WH (1955). An epidemic of virus disease in Southern Province, Tanganyika territory in 1952-53 II: General description and epidemiology. Trans. R. Soc. Trop. Med. Hyg. 49: 33-57.

Maizatul A (2009). Analisa jujukan gen glikoprotein envelop E1 virus chikungunya Malaysia bagi tahun 2008. Thesis B.Sc (Hons). Universiti Putra Malaysia.

Mavalankar D, Shastri P, Bandyopadhyay T (2008). Increased mortality rate associated with chikungunya epidemic, Ahmedabad, India. Emerg. Infect. Dis. Mar; 14(3): 412-5.

Marchette NJ, Rudnick A, Garcia R (1980). Alphaviruses in Peninsular Malaysia: II. Serological evidence of human infection. Southeast Asian J. Trop. Med. Public Health. 11: 14–23.

Michault A, Staikowsky F (2009). Chikungunya: First step toward specific treatment and prophylaxis. J. Infec. Dis. 200: 489-491.

MOH (2008). Daily Report on Chikungunya outbreak, Ministry of Health, Malaysia.

MOH (2009). Daily Report on Chikungunya outbreak, Ministry of Health, Malaysia.

Oberste MS, Maher J, Pallansch MA (2002). Molecular phylogeny and proposed classification of the simian picornaviuses. J. Virol. 76(3): 1244-1251.

Pialoux G, Gauzere BB, Jaureguiberry S, Strobel M (2007). Chikungunya, an epidemic arbovirosis. Lancet Infect. Dis. 7(5): 319-327.

Powers AM, Brault AC, Tesh RB, Weaver SC (2000). Re-emergence of chikungunya and o'nyong-nyong viruses: evidence for distinct geographical lineages and distant evolutionary relationships. J. Gen Virol. 81: 471–479.

Rao TR (1971). Immunological surveys of arbovirus infections in South East Asia, with special reference to dengue, chikungunya and Kyasanur Forest disease. Bull. WHO 44: 585-591.

Rodhain F, Gonzalez JP, Mercier E, Helynck B, Larouze B, Hannoun C (1989). Arbovirus infections and viral haemorrhagic fevers in Uganda: a serological survey in Karamoja district, 1984. Trans. R. Soc. Trop. Med. Hyg. 83: 851-854.

Ross RW (1956). A laboratory technique for studying the insect transmission of animal viruses, employing a bat-wing membrane, demonstrated with two African viruses. J. Hyg. (Lond) 54: 192–200.

Rudnick A, Lim TW (1986). Dengue fever studies in Malaysia. Ins. Med. Res Malaysia Bulletin 23: 127-147.

Schuffenecker I, Iteman I, Michault A, Murri S, Frangeul L, Vaney MC, Lavenir R, Pardigon N, Reynes JM (2006). Genome microevolution of chikungunya viruses causing the Indian Ocean outbreak. PLoS Med 3, e263.

Thaikruea L, Cheareansook O, Reanphumkarkit S (1977). Chikungunya in Thailand: a re-emerging disease?. SEA J. Trop. Med Pub. Health. 28: 358-364.

Thuang U, Ming CK, Swe T, Thein S (1975). Epidemiological features of dengue and chikungunya infections in Burma. SEA J. Trop. Med Pub Health 6: 276-283.

Townson H, Nathan MB (2008). Resurgence of chikungunya. Trans. R. Soc. Trop. Med. Hyg. 102(4): 308-309.

Vidya AA, Shubham S, Sarah C, Rashmi SG, Atul MW, Santosh MJ, Sudeep AB, Akhilesh CM (2007). Genetic divergence of chikungunya viruses in India (1963-2006) with special reference to the 2005-2006 explosive epidemic. J. Gen. Virol. 88: 1967-1976.

WHO (2006). Outbreak news. Wkly Epidemiol Rec. 81, 105–116.

Yergolkar PN, Tandale BV, Arankalle VA, Sathe PS, Sudeep A, Gandhe SS, Gokhle MD, Jacob GP, Hundekar SL, Mishra AC (2006). Chikungunya outbreaks caused by African genotype, India. Emerg. Infect. Dis. 12: 1580–1583.

8

Hepatitis C virus infection: A review of the current and future aspects and concerns in Pakistan

H. Akbar, M. Idrees*, S. Manzoor, I. ur Rehman, S. Butt, M. Z. Yousaf, S. Rafique, Z. Awan, B. Khubaib, M. Akram and M. Aftab

National Center of Excellence in Molecular Biology University of Punjab, Lahore-Pakistan.

Hepatitis C virus (HCV) is the major etiological agent of hepatitis. It infects 200 million people worldwide and 85% of them could develop chronic hepatitis, liver function failure or hepatocellular carcinoma. Hepatitis C is rapidly emerging as a major health problem in developing countries like Pakistan with prevalence rate of 10% and genotype 3a is the most prevalent. Here, approximately 80% of infections proceed to chronic infection and infected blood is the primary route of spread. In Pakistan, about 75% of patents do not receive standard anti HCV therapy (Interferon + Ribavirin) and of the 25% that do receive such treatment, the SVR rate is 60 - 70%. This review is designed to cover the information about the status of HCV in Pakistan with major focus on its prevalence, genotypes, current diagnostic assays, available therapies and treatment outcomes. The present review further emphasizes the need to uncover exact HCV prevalence rate in the country, to develop diagnostic assays based on local genotype, to understand the interaction between HCV genotype 3a genes and cell line genes responsible HCV pathogenesis. In addition, this review discusses the need for the generation of infectious pseudo particle of HCV as a potential vaccine, to investigate DNA base vaccine, or siRNA-based anti HCV approaches for our local genotypes.

Key words: Hepatitis C, prevalence, genotypes, treatment, Pakistan.

INTRODUCTION

Hepatitis C Virus was isolated in 1989; a member of Flaviviridae and is a major pathogen of hepatitis, liver cirrhosis and hepatocellular carcinoma (HCC) (Kato, 2001; Bhandari and Wright, 1995; Shepard et al., 2005; Giannini and Brechot, 2003). It is estimated that 3.3% of the population globally (lower in Europe 1.03% and highest in Africa 5.3%) and 10% of the Pakistani population is chronically infected with this viral pathogen (Hepatitis C, 2009, Raza et al., 2007; Farhana et al., 2009; Idrees and Riazuddin, 2009). In industrial countries the HCV accounts for 20% of acute and 70% of chronic cases of hepatitis (Farhana et al., 2009).

Major HCV infections lead to chronic hepatitis, which results in progressive fibrosis ultimately resulting in cirrhosis, liver failure and an increased risk of hepatocellular carcinoma (Shepard et al., 2005; Giannini and Brechot, 2003; Freeman et al., 2001; Jacobs et al., 2005). According to Farhana et al. (2009), 40% of HCV infections end stage cirrhosis, 60% hepatocellular carcinoma and 30% of liver transplantation (Farhana et al., 2009).

Pakistan is the sixth most populous country in the world with total estimated population of 170 million and 803,940 Km^2 land area (Idrees and Riazuddin, 2008). It is situated in the Western part of the Indian subcontinent, with Afghanistan and Iran on the west, India on the east and the Arabian Sea on the south. It is a federation of 4 provinces (Balochistan, North West Frontier Province, Punjab and Sindh), a capital territory and federally administered tribal areas. The present review summarizing the current information available about prevalence, most prevalent genotype, pathogenesis, modes of transmission, diagnostic assays, treatment response rate in Pakistan and also emphasizes for the further needs to

---

*Corresponding author. E-mail: idreeskhan@cemb.edu.pk.

(A)

(B)

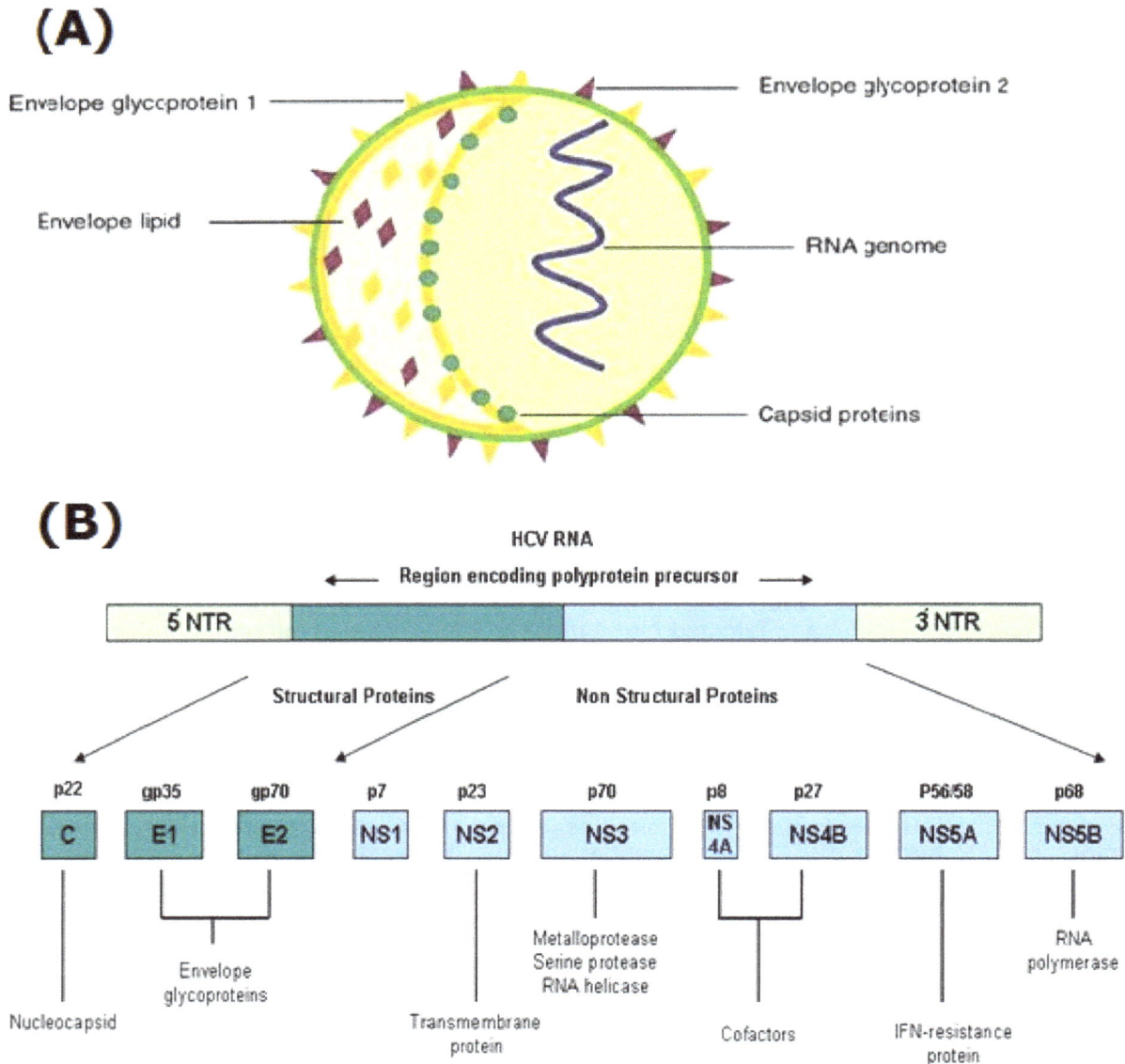

Figure 1. (A). Model structure of HCV: The left-hand side of the illustration shows the viral surface of envelope lipids and glycoproteins; the right-hand side shows the RNA genome encased by capsid proteins. (B). HCV genome organization: Proteins encoded by the HCV genome. HCV is formed by an enveloped particle harbouring a plus-strand RNA of 9.6 kb. The genome carries a long open-reading frame (ORF) encoding a polyprotein precursor of 3010 amino acids. Translation of the HCV ORF is directed via a 340 nucleotide long 5' nontranslated region (NTR) functioning as an internal ribosome entry site; it permits the direct binding of ribosomes in close proximity to the start codon of the ORF. The HCV polyprotein is cleaved co- and post-translationally by cellular and viral proteases to produce 10 mature proteins as indicated, including the four structural proteins: core, E1, E2, and p7 and the 6 non-structural proteins, NS2, NS3, NS4a, NS4b, N4a, NS5B. Putative functions of the cleavage products are shown (Adopted from NET).

eradicate this from the country.

Hepatitis C virus (HCV)

HCV is an enveloped virus consisting of a single positive-sense strand RNA genome of about 9.6 kb, which encodes a polyprotein of 3010 amino acids processed by cellular- and virally-encoded proteases into 4 structural proteins (core, p7 and the envelope glycoprotein, E1 and E2) and 6 nonstructural (NS) proteins (NS2, NS3, NS4A, NS4B, NS5A and NS5B) (Figure 1) (Dubuisson, 2007; Choo et al., 1991; Reed and Rice, 2000; De Francesco and Steinkuhler, 2000; Lin et al., 1994; Posta et al., 2008). Based on genome sequence similarity, international standardization of nomenclature recently classified HCV into 6 major genotypes (1 - 6) and more than 70 sub genotypes which differ 30 and 15% at the nucleotide level, respectively. Their prevalence and distribution are linked to geographical location and mode of transmission. (Simmonds, 2004; Roman et al., 2008; Simmonds et al., 1994).

Prevalence and genotypes of HCV in Pakistan

Hepatitis C is rapidly emerging as a major health problem in developing countries including Pakistan (Raza et al., 2007; Khan et al., 2008). It is estimated recently that 200 million individuals are infected with HCV worldwide including approximately 17 million in Pakistan (Narendra et al., 2004; Idrees and Riazuddin, 2008). Prevalence of HCV may be different in different regions and various groups of the same community (Idrees et al., 2008). Hospital-based studies revealed prevalence rates of 5.31% (Islamabad), 2.45% (Rawalpindi), 4.06% (Multan), 20.89% (Faisalabad), 4-6% (Karachi), 9% (Mardan), 5% (Buner, NWFP) and 25.7% (Northern Areas) (Farhana et al., 2009; Chaudhary et al., 2007; Jehangir et al., 2006; Hashmie et al., 1999; Kazmi et al., 1997; Khan et al., 2004; Muhammad and Jan, 2005; Tariq et al., 1999). Slightly higher prevalence of HCV was recorded in the earth quake effect areas of Pakistan in 2005 (Khan et al., 2008). A recent study showed that in Pakistan more than 90% of HCV positive subjects were unaware of HCV infection in general (Idrees et al., 2008).

Previously, 3351 serum samples with viral load > 500 IU/ml were successfully typed by genotype-specific genotyping assay. Out of which 2165 genotyped patients belonged to Punjab region, 823 patients to N.W.F.P., 239 to Sindh and 124 patients were from Balochistan (Idrees and Riazuddin, 2008). It has been observed that the prevalence of HCV increases with age which may be due to increasing exposure to risk factors (Idrees et al., 2008).

Recently an improved genotyping method for detection of HCV genotypes and subtypes in Pakistan was developed that is based on entire core region and a part from 5 non-coding region (5'NCR) with specific primer, the current system is reliable, sensitive, specific and economical for genotyping assay of for Pakistani HCV isolates (Idrees, 2008). According to a recent study, in Pakistan the observed genotypic distribution were 3a (49.05%), 3b (17.66%), 1a (8.35%), 2a (7.52%), 1b (3.01%), 4 (1.49%), 3c (0.75%), 2b (0.80%), 5a (0.18%), 1c (0.15%), 6a (0.12%), 2c (0.09%) and mixed infection (4.80%). During this study genotype 1c, 2c, 3c, 4, 5a and 6a were isolated for the first time from Pakistan (Idrees and Riazuddin, 2008). It has been confirmed by other studies that the most prevalent genotype in Pakistan is 3a with rate of 50% followed by genotype 3b and 1a, respectively (Idrees et al., 2008; Idrees and Riazuddin, 2008; Idrees, 2008; Sarwat et al., 2008).

Pathogenesis of HCV

Hepatitis C virus (HCV) causes acute and chronic heaptitis and approximately 85% of HCV infections progress to chronic infections, which often results in liver disease including variable degree of hepatic inflammation, oxidative stress, steatosis, fibrosis, cirrhosis, hepatocellular carcinoma and insulin resistance (Hoofnagle, 2002; Jacobs et al., 2005; Alter, 1997). In Pakistan approximately 80% of HCV cases develop to chronic infection which may increase in the coming decade (Idrees and Riazuddin, 2009; Khan et al., 2008). Hepatitis C virus RNA synthesis and protein expression affect cell homeostasis by modulation of a wide range of activities, including gene expression alteration, cell signaling, transcriptional modulation, transformation, apoptosis, steatosis, fibrosis, oxidative stress, membrane rearrangement, vesicular trafficking and immune response (Basu et al., 2006; Chang et al., 2008; Li et al., 2007). HCV is now viewed as a true metabolic syndrome associated with type II diabetes, hypertension, dyslipidemia, cardiovascular disease and atherosclerosis (Sheikh et al., 2008).

Mode of transmissions of HCV

HCV is blood born pathogen and high-risk have been observed in the recipients of multiple or repeated blood transfusions or blood products, intravenous drug abusers, prisoners, hemodialysis patients, healthcare workers exposed to needle stick and sharps injuries. In about 50% of infected patients (so-called 'sporadic' cases) have no obvious risk factor (Roman et al., 2008; Farhana et al., 2009).

In Pakistan approximately 70% of the cases were acquired in the hospitals via reuse of syringes and major/minor surgery that is very common in Pakistan (Idrees and Riazuddin, 2008). The overall observed mode of transmission in Pakistan were: multiple use of needles/syringes (61.45%), major/minor surgery/dental procedures (10.62%), blood transfusion and blood products (4.26%), sharing razors during shaving or circumcision by barbers (3.90%), piercing instruments, nail clippers, tooth brushes, siwaks, in less than 1% due to needle stick, from infected mother to baby and sexual transmission. For about 20.35% subjects the mode of transmission is unclear in this country (Farhana et al., 2009; Idrees et al., 2008; Idrees and Riazuddin, 2008) that is very dangerous situation. Injecting vitamins and antibiotics are very common in cities, towns and villages of Pakistan that play a major role in the HCV infection spread. Several studies from Pakistan showed that the average number of injections per person per year is more than 9 injunctions in this country that is 1 of the highest frequencies of injections anywhere in the world (Jafri et al. 2006; Khan et al. 2000). In addition, at a time of previous mass vaccination about 25 - 30 years back there was no concept of safe injection practices was largely received at schools and villages that might be a source of HCV contamination. Other studies also reported that several times vaccinations at the public health-care facilities included sharing of syringes (Jafri et al., 2006; Khan et al., 2000; Ministry of Health, 2002). It has been established from Pakistan that subjects who had received more injections were more likely to be infected with HCV and these non-sterile syringes or needles may be the source of HCV infections (Idrees et al., 2008; Jafri et al., 2006; Khan et al., 2000). Presently the major source of hepatitis spread in this country is the use of previously used re-pack syringes. Even the Ministry of Health Government of Pakistan has published a survey in year 2002 that shows that in Pakistan more than 72% therapeutic injections and 50% immunization injections in public health-care facilities are unsafe and potentially dangerous for the spread of infections including HCV. Several other studies shows that even still the use of multiple-dose vials is common in many government and private sector hospitals that may also be an important risk factor in the transmission of HCV infection ((Jafri et al., 2006; Khan et al., 2000; Ministry of Health, 2002; Siddiqi et al., 2002). In many areas of Pakistan still sharing razors during shaving by barbers are common. Majority of the males, 20 years and older generally use barber shops for their shaving needs. According to a recent study higher anti-HCV prevalence in males compared to females could be due to this additional exposure to used and non-sterile shaving blades (Idrees et al., 2008). The study further describes that in 9% female subjects the probable transmission mode was the sharing of piercing needle/instrument. In females, piercing of nose and ears in group settings using non sterilized needles is a common practice in the area. According to this recent study, some of minor risk factors included sharing nail clippers, tooth brushes, Siwak, needle stick, from infected mother to baby and sexual transmission. The authors of that study have mentioned that in about ¼th of the anti-HCV positive subjects it could not ascertain any known risk factor and the transmission of infection is sporadic in this country.

Diagnosis and treatment

Diagnosis of hepatitis C is based on serological assays and HCV RNA. For screening and epidemiological surveillance enzyme-linked Immunosorbant assay (ELISA) and a confirmatory recombinant Immunoblot assay are initially used which detect HCV-specific antibodies (anti-HCV). Qualitative Polymerase chain reaction (PCR) is used to find out the presence of the viral genome in order to confirm active infection. Quantitative PCR is also helpful to monitor disease activity and response to treatment (Farhana et al., 2009).

Apart from quantitative PCR, genotyping has become increasingly important for routine laboratory diagnosis and the genotype should be taken into consideration when prescribing therapy. Both the duration and sustained response to current standard therapy regimens are strongly associated with the HCV genotypes (Idrees, 2008; Davis and Lau, 1997; Heathcote et al., 2000; Zeuzem et al., 2000). Evidence suggested that the patient with type 2 and type 3 HCV infections are more likely to have a sustained response to therapy than patient with type 1 (Idrees, 2008; Dusheiko et al., 1996).

In Pakistan, about 60 to 70% of the patients having a sustained virological response to therapy but at least 75% of patients have no therapeutic benefit (Mujeeb et al., 1997). In Pakistan more than 75% of Pakistanis are living below the line of poverty. Though more than 80 different brands of interferons are available in the market of this country, however they are very costly and are out of the reach of poor HCV infected patients. From 2006 - 2008, about 20,000 patients received interferon treatment free of charge from "Prime Minister Program for the prevention and Control of Hepatitis" but that is only 0.01% of the total cases and even after 12 years 3 out of 4 patients are still far away from treatment. Currently, the combination of interferon alpha and nucleoside analogue ribavirin is recommended for Pakistani patients with high reported sustained viral response rates (Idrees and Riazuddin, 2009; Farhana et al., 2009). Our previous observations showed that end of treatment response rates to IFN plus ribavirin therapy is very high (67%) in Pakistan. And significantly higher sustained virological response (SVR) rates were observed in Pashtoon (69.2%) as compared to, Punjabi (45.5%), Sindhi (45.5%) and Balochi (50%) for the patients received IFN-alpha plus ribavirin. In the same study we have seen highest SVR in patients with HCV genotype 2 (69.7%) followed by genotype 3 (57.3%) and lowest SVR in genotype 1 infection (24.3%) (Idrees and Riazuddin, 2009).

Recently, significantly increased response has been observed for polyethylenglycol (PEG)-conjugated interferon alpha as compared to conventional interferon alpha in the patients of genotype 1 (42 - 46%), genotypes 2 and 3 (76 - 82%) (Farhana et al., 2009; Foster and Mathurin, 2008). Side effects of interferon alpha are numerous and severe and require discontinuation of therapy in 2 - 10% of patients. The early side effects involve the inconvenience of subcutaneous administration of the medicine three times (or once) weekly for 6 - 12 months (Farhana et al., 2009; Foster and Mathurin, 2008).

Future need of studies

Currently no authentic country wide data is available on the actual prevalence rate of HCV in Pakistan (Idrees et al., 2008). In order to delineate the risk groups and risk factors of HCV infection it is important to conduct epidemiological studies that depict an accurate prevalence of the disease in the general population. A few epidemiological studies addressing the issue HCV prevalence conducted in various parts of Pakistan used blood donors as the study population; it does not reflect an accurate prevalence of an infection in the general population of a country like Pakistan. Recently, Pakistan Medical research council (PMRC) has done a survey to find out the actual prevalence rate of hepatitis in Pakistan, which is important but the survey have low international credibility or none at all as the rapid test method was used for the screening of sera samples for the detection of anti-HCV. Here we suggest that the first and most important task is to find out the true prevalence of hepatitis C infection in the whole country at least using third generation ELISA for effective screening. In addition HCV RNA PCR should be performed for the anti-HCV positive sera to find out the true active HCV infection in Pakistan. In Pakistan about 90% of the HCV positive subjects are unaware of their HCV infection in general. So if we really want to defeat HCV, we must need to educate people and ensure that people know about that they are infected, so that they may receive anti-viral treatment and take steps to protect their colleagues, life partners, children and other family members. On government level Prime Minister of Islamic Republic of Pakistan had launched a program for the control and eradication of hepatitis from the whole country in year 2005. The program was very active in years 2006 - 2008 however; due to political changes in the country it is not functional as it was planned. Therefore it is suggested to resume the program and allocate more funds for at least 5 years. Additionally more efforts are also need on all levels to eradicate the problem from the country.

Several studies have already confirmed that genotyping is very important and useful test that should be done before starting anti-viral therapy (Idrees, 2008; Davis and Lau, 1997; Heathcote et al., 2000; Zeuzem et al., 2000). Therefore rapid, economic and reliable genotyping methods are required to be developed that may be used in routine in all clinical laboratories of the country. These locally developed assays will have more specificity and sensitivity as compared to other commercially available assays in the market.

Oxidative stress, fibrosis and steatosis generally seem to appear together in the liver with HCV infection and contributes in the development of Hepatocellular carcinoma (HCC) (Emerit et al., 2000; DeMaria et al., 1996; Rubbia-Brandt et al., 2000).In human HCV genotype 3, which is most common in Pakistan, is more commonly associated with steatosis (Idrees et al., 2008; Rubbia-Brandt et al., 2000; Roingeard and Hourioux, 2008). Moreover, association has been reported of genotype 3a core protein with lipid droplets (Rubbia-Brandt et al., 2000; Roingeard and Hourioux, 2008). Therefore the study of interaction between HCV protein and various genes involve in oxidative stress, fibrosis, steatosis and HCC development will very important and helpful for controlling HCV pathogenesis and effective drug designing for our local genotypes.

The current treatment is neither economical nor fully effective in all patients and carries significant side effects. Clearly, novel therapeutic strategies are urgently required in order to prevent the infection due to HCV. Several experimental anti HCV approaches have been presented recently and are discussed here at potential approaches to be implemented in Pakistan. Where therapy, is not widely available, vaccination is an advantageous option. Currently there is no approved anti HCV vaccine. It has been reported that infectious pseudo particle of HCV containing functional E1-E2 envelope protein complex mimic the early steps of parental HCV and may be suitable for the development of much needed antiviral therapies (Bartosch et al., 2003). Thus to prevent the infection due to HCV we will need to generate therapeutic vaccines by using pseudo particles of HCV 3a local genotype. This knowledge will also be helpful to study the early stages of viral life cycle like attachment and entry into host cells for our local setup. An additional experimental approach towards anti HCV vaccination could be nucleic acid immunization, which is the most recent approach in vaccine development. A DNA-based vaccine usually consists of purified plasmid DNA carrying sequences. Nucleic acid immunization is the most recent approach in vaccine development. A DNA-based vaccine usually consists of purified plasmid DNA carrying sequences encoding for an antigen of interest under the control of eukaryotic promoter. Many studies have been published on the development of DNA-based vaccines against HCV. It was reported that the injection into muscle cells of plasmids constructs expressing HCV core protein generate strong cytotoxic T-lymphocyte (CTL) activity, as assessed both *in vivo* and *in vitro*, and are promising candidates as antiviral agents (Tokushige et al., 1996; Gehring et al., 2004). The development of DNA base vaccine for our local setup will be valuable work and will be helpful in eradicating the whole nation from the blood born pathogen.

Finally, it has been suggested that RNA interference (RNAi) induced by small interfering RNA (siRNA), has potential as effective therapeutic agents for HCV infections (Trejo-Ávila et al., 2007). Synthetic siRNAs targeted against sequences in the protein-coding regions of core, NS3 and NS5B resulted in profound, up to 100-fold inhibition (Trejo-Ávila et al., 2007; Kapadia et al., 2003; Randall et al., 2003). siRNA is a rapid, inexpensive and sequence-specific gene silencing method. We suggest that such a study, concentrated on the 3a genotype which is common in Pakistan, will be helpful to catch potential therapeutic agents for our local genotypes and new therapeutic targets.

Conclusion

HCV infections represent a major threat to Pakistani population. There is a dreadful need to screen the whole nation for this "Silent Killer". Presently no country wide data is available about the actual prevalence rate of HCV in Pakistan. The development of diagnostic assays based on local genotype will be more sensitive, specific and economical. Due to the vital importance, HCV genotyping should be available in all clinical laboratories as a routine laboratory diagnosis and should be carried out before the start of anti-viral treatment. As the current treatment is neither economical nor fully effective in all patients and carries significant side effects, the need of the hour is to find out new therapeutic strategies and targets for local genotypes of HCV, using of siRNA technology and generation of protein and DNA base vaccines for our local setup.

REFERENCES

Alter MJ (1997). Epidemiology of hepatitis C. Hepatology 26: 62S–65S.

Bartosch B, Dubuisson J, Cosset FL (2003). Infectious hepatitis C virus pseudo-particles containing functional E1–E2 envelope protein complexes. J. Exp. Med. 197: 633-642.

Basu A, Meyer K, Lai KK, Saito K, Di Bisceglie AM, Grosso LE, Ray RB, Ray R (2006). Microarray analyses and molecular profiling of Stat3 signaling pathway induced by hepatitis C virus core protein in human hepatocytes. Virology 349: 347-58.

Bhandari BN, Wright TL (1995). Hepatitis C: an overview. Annu. Rev. Med. 46: 309-317.

Chang ML, Yeh CT, Chen JC, Huang CC, Lin SM, Sheen IS, Tai DI, Chu CM, Lin WP, Chang MY, Liang CK, Chiu CT, Lin DY (2008). Altered expression patterns of lipid metabolism genes in an animal model of HCV core-related, non-obese, modest hepatic steatosis. BMC. Genomics 9: 109-117.

Chaudhary IA, Samiullah U, Khan SS, Masood R, Sardar MA and Mallhi AA (2007). Seroprevalence of hepatitis B and C among the healthy blood donors at Fauji Foundation Hospital, Rawalpindi. Pak. J. Med. Sci. 23: 64-7.

Choo QL, Richman KH, Han JH, Berger K, Lee C, Dong C, Gallegos C, Coit D, Medina-Selby R, Barr PJ, Weiner AJ, Bradley DW, Kuo G, Houghton M (1991). Genetic organization and diversity of the hepatitis C virus. Proc. Natl. Acad. Sci. 88: 2451-2455.

Davis GL, Lau JN (1997). Factors presitive of a beneficial response to therapy of hepatitis C virus. Hepatology 26: 122.s-127.s.

De Francesco R, Steinkuhler C (2000). Structure and function of the hepatitis C virus NS3–NS4A serine proteinase. Curr. Top. Microbiol. Immunol. 242: 149-169.

DeMaria N, Calantroni A, Faginoli S, Guang-Jun L, Rogers BK, Farinati F, Van Thiel DH, Floyd RA (1996). Association between reactive oxygen species and disease activity in chronic hepatitis C. Free Radic. Biol. Med. 21: 291-295.

Dubuisson J (2007). Hepatitis C virus proteins. World J. Gastroenterol. 13: 2406-2415.

Dusheiko G, Schmilovitz H, Brown D, Mcomish F, Yap PL, Simmonds P (1996). Hepatitis C virus genotypes: an investigation of type specific differences in geographic origin and disease. Hepatology 26: 122.s-127.s.

Emerit I, Serejo F, Filipe P, Alaoui Youseefi A, Fernandez A, Costa A, Freitas J, Ramalho F, Bapista A, Carneiro de Mura A (2000). Clastogenic factors as biomarkers of oxidative stress in chronic hepatitis C. Digestion. Int. J. Gast. 62: 200-207.

Farhana M, Hussain I, Haroon TS (2009). Hepatitis C: the dermatologic profile J. Pak. Assoc. Derm. 18: 171-181.

Foster G, Mathurin P (2008). Hepatitis C virus therapy to date. Antivir. Ther. 1: 3-8.

Freeman AJ, Dore GJ, Law MG, Thorpe M, Von Overbeck J, Lloyd AR, Marinos G, Kaldor JM (2001). Estimating progression to cirrhosis in chronic hepatitis C virus infection. Hepatology 34: 809 - 816.

Gehring S, Gregory SH, Kuzushita N, Wands JR (2004). Type 1 interferon augments DNA-based vaccination against hepatitis C virus

core protein J. Med. Virol. 75: 249-257.

Giannini C, Brechot C (2003). Hepatitis C virus biology Cell Death and Differentiation. Cell. Death. Differ. 10(Suppl 1): S27-38.

Hashmie ZY, Chaudhary AH, Ahmad M, Ashraf M (1999). Incidence of healthy voluntary blood donors at Faisalabad. Professional 6: 551-5.

Heathcote EJ, Shiffman ML, Cooksley WGE, Dusheiko GM, Lee SS, Balart L, Reindollar R, Reddy RK, Wright TL, Lin A, Hoffma J, De Pamphilis J (2000). Peginterferon alfa-2a in patients with chronic hepatitis C and cirrhosis. New Engl. J. Med. 343: 1673-80.

Hepatitis C: The Facts: The Epidemic - Worldwide Prevalence. (2009). http://www.epidemic.org/theFacts/theEpidemic/worldPrevalence/

Hoofnagle JH (2002). Course and outcome of hepatitis C. Hepatology. 36: S21–S29.

Idrees M, Riazuddin S (2008). Frequency distribution of hepatitis C virus genotypes in different geographical regions of Pakistan and their possible routes of transmission. BMC. Infect. Dis. 8-69.

Idrees M, Riazuddin S (2009). A study of best positive predictors for sustained virological response to interferon alpha plus ribavirin therapy in naïve hepatitis C patients. BMC Gast. pp.9-5.

Idrees M, Lal A, Naseem M, Khalid M (2008). High prevalence of hepatitis C virus infection in the largest province of Pakistan. J. Dig. Dis. 9: 95-103.

Idrees M (2008). Development of an improved genotyping assay for the detection of hepatitis C virus genotype and subtypes in Pakistan. J. Virol Method. 150: 50-56.

Jacobs JM, Diamond DL, Chan EY, Gristenko MA, Qian W, Stastna M, Bass T, CampII DG, Carithers RL, Smith RD, Katze MG (2005). Proteome Analysis of full length hepatitis C Virus (HCV) Replication and Specimens of Posttransplantation Liver from HCV infected patients. JVI. 79(12): 7558-7569.

Jafri W, Jafri N, Yakoob J, Islam M, Tirmizi SFA, Jafar T (2006). Hepatitis B and C: prevalence and risk factors associated with seropositivity among children in Karachi, Pakistan. BMC Infect. Dis. 6:101.

Jehangir W, Ali F, Shahnawaz U, Iqbal T, Qureshi HJ (2006). Prevalence of hepatitis B, C and HIV in blood donors of South Punjab. Esculapio. 2: 6-7.

Kapadia SB, Brideau-Andersen A, Chisari FV (2003). Interference of hepatitis C virus RNA replication by short interfering RNAs. Proc. Natl. Acad. Sci. 100: 2014-2018.

Kato N (2001). Molecular virology of hepatitis C virus. Acta. Med. Okayama. 55(3):133-159.

Kazmi K, Sadaruddin A, Dil AS, Zuberi SJ (1997). Prevalence of HCV in blood donors. Pak. J. Med. Res. 36: 61-2.

Khan AJ (2000). Unsafe injections and the transmission of hepatitis B and C in a Periurban community in Pakistan. Bull World Health Organ. 78: 956–963.

Khan MSA, Khalid M, Ayub N, Javed M (2004). Seroprevalence and risk factors of Hepatitis C virus (HCV) in Mardan, N.W.F.P. Rawal. Med. J. 29: 57-60.

Khan S, Rai MA, Khan A, Farooq A, Kazmi SU, Ali SH (2008). Prevalence of HCV and HIV infections in 2005-Earthquak areas of Pakistan. BMC Infect. Dis. 8-147.

Khan UR, Janjua NZ, Akhtar S, Hatcher J (2008). Case-control study of risk factors associated with hepatitis C virus infection among pregnant women in hospitals of Karachi-Pakistan. Trop. Med. Int. Health 13(6): 754-61.

Li Y, Boehning DF, Qian T, Popov VL, Weinman SA (2007). Hepatitis C virus core protein increases mitochondrial ROS production by stimulation of Ca2+ uniporter activity. FASEB J. 21: 2474-2485.

Lin C, Lindenbach BD, Pragai BM, McCourt DW, Rice CM (1994). Processing in the hepatitis C virus E2– NS2 region: identification of p7 and two distinct E2-specific products with different C termini. J. Virol. 68: 5063–5073.

Ministry of Health (2002). Annual report, Director General Health, 2001–2002. Government of Pakistan.

Mujeeb SA, Jamal Q, Khannam R (1997). Prevalence of hepatitis B surface antigen and HCV antibodies in hepatocellular carcinoma cases in Karachi Pakistan. Trop Doct. 27(1): 45–6.

Muhammad N, Jan A (2005). Frequency of hepatitis C in Bunir, NWFP. J. Coll. Phys. Surg. Pak. 15: 11-14.

Narendra M, Dixit Jennifer E, Layden-Almer, Layden TJ, Perelson AS

(2004). Modelling how ribavirin improves interferon response rates in hepatitis C virus infection. Nature. 432: 922-924.

Posta J, Ratnarajahc S, Lloyd AR (2008). Immunological determinants of the outcomes from primary hepatitis C infection. Cell. Mol. Life Sci. 66(5): 733-56.

Randall G, Grakoui A, Rice CM (2003). Clearance of replicating hepatitis C virus replicon RNAs in cell culture by small interfering RNAs. Proc. Natl. Acad. Sci. 100: 235-240.

Raza SA, Clifford GM, Franceschi B (2007). Worldwide variation in the relative importance of hepatitis B andhepatitis C viruses in hepatocellular carcinoma: a systematic review. Br. J. Cancer. 96(7): 1127 – 1134.

Reed KE, Rice CM (2000). Overview of hepatitis C virus genome structure, polyprotein processing, and protein properties. Microbiol. Immunol. 242: 55–84.

Roingeard P, Hourioux C (2008). Hepatitis C virus core protein, lipid droplets and steatosis. J. Viral. Hepat. 15(3): 157-164.

Roman F, Hawotte K, Struck D, Ternes AM, Servaiss JY, Arendt V, Hoffman P, Hemmer R, Steub T, Seguin-Devaux C, Schmit JC (2008). Hepatatis C virus genotype distribution and transmission risk factors in luxembourg from 1991 to 2006. WJG; ISSN 1007-9327.

Rubbia-Brandt L, Quadri R, Abid K, Giostra E, Male PJ, Mentha G, Spahr L, Zarski JP, Borisch B, Hadengue A, Negro F (2000). Hepatocyte effect of hepatitis virus genotype 3a. Hepatology 33: 106-115.

Sarwat A, Naeem M, Hussain A, Kakar N, Babar ME, Ahmad J (2008). Prevalence of hepatitis C virus (HCV) genotypes in Balochistan. Mol. Biol. Rep. DOI 10.1007/s11033-008-9342-0.

Sheikh MY, Qadari I, Friendman JE, Sanyal AJ (2008). HCV infection: Pathways to metabolic syndrome. Hepatology 47(6): 2127-33.

Shepard CW, Finelli L, Alter MJ (2005). Global epidemiology of hepatitis C virus infection. Lancet. Infect. Dis. 5: 558 – 567.

Siddiqi S, Hamid S, Rafique G, Chaudhry SA, Ali N, Shahab S, Sauerborn R (2002). Prescription practices of public and private health care providers in Attock District of Pakistan. Int. J. Health Plan Manag. 17: 23-40.

Simmonds P (2004). Genetic diversity and evolution of hepatitis C virus–15 years on. J. Gen. Virol. 85: 3173 - 3188.

Simmonds P, Alberti A, Alter HJ, Bonino F, Bradley DW, Brechot C, Brouwer JT, Chan SW, Chayama K, Chen DS (1994) A proposed system for the nomenclature of hepatitis C viral genotypes. Hepatology 19: 1321-1324.

Tariq WU, Hussain AB, Karamat KA, Ghani E, Hussain T, Hussain S (1999). Demographic aspects of hepatitis C in Northern Pakistan. J. Pak. Med. Assoc. 49: 198-201.

Tokushige K, Wakita T, Pachuk C, Moradpour D, Weiner DB, Zurawski VR Jr, Wands JR (1996). Expression and immune response to hepatitis C virus core DNA-based vaccine constructs. Hepatology 24(1): 14-20.

Trejo-Ávila L, Elizondo-González R, Trujillo-Murillo KC, Zapata-Benavides P, Rodríguez-Padilla C, Rivas-Estilla AM (2007). Antiviral therapy: Inhibition of Hepatitis C Virus expression by RNA interference directed against the NS5B region of the Viral Genome. Ann. Hepatol. 6(3): 174-180.

Zeuzem S, Feinman SV, Rasenack J, Heathcote EJ, Lai MY, Gane E, O'Grady J, Reichen J, Diago M, Lin A, Hoffman J, Brunda MJ (2000). Peginterferon alfa-2a in patients with chronic hepatitis C. N Engl. J. Med. 343(23): 1666-72.

9

Feasibility and factors affecting global elimination and possible eradication of rabies in the world

A. A. Ogun[1], I. O. Okonko[2]*, A. O. Udeze[3], I. Shittu[4], K. N. Garba[2], A. Fowotade[5], O. G. Adewale[6], E. A. Fajobi[7], B. A. Onoja[2], E. T. Babalola[8] and A. O. Adedeji[8]

[1]Department of Epidemiology, Medical Statistics and Environmental Health, Faculty of Public Health, College of Medicine, University of Ibadan, Ibadan, Nigeria.
[2]Department of Virology, Faculty of Basic Medical Sciences, College of Medicine, University of Ibadan, University College Hospital (UCH) Ibadan, Nigeria.
[3]Virology Unit, Department of Microbiology, Faculty of Sciences, University of Ilorin, Ilorin, Nigeria.
[4]Department of Viral Research, National Veterinary Research Institute, P.M.B. 01, Vom, Plateau State, Nigeria.
[5]Department of Medical Microbiology and Parasitology, University of Ilorin Teaching Hospital, Ilorin, Kwara State, Nigeria.
[6]Department of Biochemistry, Olabisi Onabanjo University, Ago-Iwoye, Ogun State, Nigeria.
[7]Department of Basic Sciences, Federal College of Wildlife Management, New Bussa, Niger State, Nigeria.
[8]Department of Veterinary Microbiology and Parasitology, Faculty of Veterinary Medicine, University of Ibadan, Ibadan, Nigeria.

This article reviews the feasibility of global eradication of rabies and factors affecting eradication of rabies in the world. Effective vaccines are now available against many viruses making eradication a viable proposition. As in the case of smallpox, the following questions should be addressed when the feasibility of eradication of a particular human virus disease is considered. Is the disease worth eradicating? Is there any animal reservoir? Is there a carrier state? Is effective vaccination available? How communicable is the rabies? What level of coverage is required for eradication? What are the possibilities for rabies control in reservoir hosts? Can rabies be controlled in wildlife reservoirs? Can the population density of reservoir hosts be reduced? Can contact between wild dogs and domestic dogs be minimized? Whether a virus disease can be eradicated or not depends on many factors, not least on the will power to implement such a policy. These factors include human (increased human activities and international travel; lack of adequate public awareness, proper surveillance, emergency preparedness planning, solid commitment and resourced initiatives among others); socioeconomic (major ecologic changes, agricultural practices, poverty, increasing demands for meat etc.); animal factors (illegal importation, population increase, migration of dogs, stray animals etc.); and vaccines and vaccination (low vaccination coverage and potent vaccines, vaccine failure, inferior vaccine quality, vaccine shortage, high cost, existence of multiple hosts, reservoir and healthy carriers etc.). Rabies eradication is not feasible because of the extensive factors and the inability to eliminate reservoirs with existing technology. However, elimination of human rabies in urban areas may be possible through different strategies. Vaccination of stray dogs could lead to the eradication of rabies in countries where dog rabies is the sole source of human exposure. Research to design strategies for rabies control globally, is urgently needed. Additional genetic work will help to set priorities for the conservation of populations which may be genetically unique for spread of rabies and other related diseases.

Key words: Animal factor, continual endemicity, effective vaccines, eradicability, feasibility, human, socioeconomic, vaccination.

HISTORICAL BACKGROUND OF DISEASE ERADICATION

Viruses account for the bulk of infectious diseases. Rabies infection in humans is still a major public health problem, causing upwards of 20,000 deaths per year. Most human deaths from rabies occur in tropical resource-limited countries (Zinsstag et al., 2007). In Africa and Asia, an estimated 24,000 - 70,000 persons die of rabies each year (Knobel et al., 2005). The domes-tic dog is the main source of exposure and vector for human rabies (Zinsstag et al., 2007). Rabies is acute, progressive, fatal encephalitis caused by viruses in the Family Rhabdoviridae, Genus Lyssavirus. Rabies virus is the representative member of the group. Warm-blooded vertebrates are susceptible to experimental infection, but major primary hosts for disease perpetuation encompass bats and mammalian carnivores. The dog is the global reservoir, and important wild carnivores include foxes, raccoons, skunks and mongoose, among others (Rupprecht et al., 2004). Rabies is one of the oldest known diseases of mankind, yet it has been only slightly more than 100 years since Pasteur developed the first vaccine for post-exposure treatment. Since this first crude nerve tissue vaccine, numerous other rabies vaccines for human use have been developed and used with varying degrees of effectiveness and safety.

Attempts to control human rabies and other infectious diseases have a long history: animal and human vaccines provide efficient weapons for prevention. Contagious pleuropneumonia of cattle, a disease that had been imported into the United States in 1843, was declared eradicated from the country in 1892, following a 5-year, $2-million campaign to identify and slaughter infected animals (Cockburn, 1961 reviewed by Ginsberg and Woodroffe, 1997). The eventual eradication of smallpox as a result of the use of Jennerian vaccination was predicted by Edward Jenner, as well as by Thomas Jefferson, in the early 19th century. Following the emergence of the germ theory and more systematic approaches to disease control in the mid-19th century, the concept of eradication of a disease first became popular briefly around the turn of the century (MMWR, 1993). The Rockefeller Foundation began campaigns to eradicate hookworm in 1907 and yellow fever in 1915 (MMWR, 1993). Both these campaigns against diseases of humans failed: the hookworm campaign because mass treatment of affected populations with anthelmintic therapy reduced the severity of individual infections but rarely eliminated them and thus did not prevent rapid reinfection and the campaign against yellow fever because of the previously unknown, inaccessible cycle of disease among nonhuman primates living in forests (Andrews and Langmuir, 1963; Soper, 1965 reviewed by Ginsberg and Woodroffe, 1997).

Acceptance of the concept of eradication declined during the late 1920s and early 1930s, after the futility of the eradication of hookworm and yellow fever was recognized (MMWR, 1993). The concept became popular again in the late 1940s, following the elimination of Anopheles gambiae mosquitoes from Brazil and Egypt,

the elimination of malaria from Sardinia, reductions in the prevalence of yaws in Haiti, and the introduction of a stable freeze-dried vaccine against smallpox (Cockburn, 1961; Soper, 1965 reviewed by Ginsberg and Woodroffe, 1997). By 1955, WHO had declared goals of global eradication of yaws and malaria, and in 1958 it adopted the goal of smallpox eradication as well (MMWR, 1993). The yaws campaign failed, partly because persons with inapparent latent cases were not adequately treated, in addition to persons with clinical disease. Many such latent infections relapsed to produce infectious lesions soon after mass treatment teams visited a community. Later, disease-specific control measures were withdrawn prematurely, allowing the infection to reappear in several areas (Hopkins, 1985a reviewed by Ginsberg and Woodroffe, 1997).

The achievement of global smallpox eradication in 1977 and its official certification by WHO in 1980 did not at first bring about the acceptance of the concept of eradication (MMWR, 1993). Concerns were raised that a new eradication effort might detract from efforts to focus attention on the need for developing comprehensive primary health services, rather than focusing on one or two diseases. However, several diseases (e.g., schisto-somiasis, rotavirus diarrhea, brucellosis, and leprosy) that were then being considered as possible targets for global eradication did not have potential for success given the current technology. Several reports and conferences have considered the potential for eradicating other diseases, of which poliomyelitis, mumps, and rubella were among those most frequently cited (Ginsberg and Woodroffe, 1997). Reports in 1980 and 1985 both concluded that no other major disease was then a potential candidate to be targeted by a global eradication campaign (Evans, 1985 reviewed by Ginsberg and Woodroffe, 1997). After the concept of eradication was accepted again in the late 1980s, some observers considered a disease to be unsuitable for eradication to the extent that it differed from smallpox or that the intervene-tion against it differed from smallpox vaccine (Hopkins, 1985b reviewed by Ginsberg and Woodroffe, 1997). In this third period of acceptance, WHO has targeted dracunculiasis and poliomyelitis for eradication (MMWR, 1993).

The eradication of smallpox from the world in 1977 proved the feasibility of infectious disease eradication. The International Task Force for Disease Eradication (ITFDE) is assessing the potential for global eradication of other infectious diseases. An important part of the work done by International Task Force for Disease Eradication (ITFDE) was to help identify key impediments to improved prevention and control of the diseases under discussion such as rabies, even if the disease was not considered to have potential as a candidate for eradication. One such "noneradication outcome" was the impetus that the members of the ITFDE gave to initiating a demonstration project to control intestinal parasites among school children in Ghana (MMWR, 1993). The criteria that the International Task Force for Disease Eradication (ITFDE) developed and their conclusions after reviewing more

*Corresponding author. E-mail: mac2finney@yahoo.com.

than 90 diseases has been previously reported (CDC, 1992a, b; MMWR, 1993).

However, rabies was eradicated from England in 1896 (MMWR, 1993). No indigenous cases of human rabies have occurred in the United Kingdom since 1902, and only 20 cases have been imported into England and Wales since 1946. Person-to-person transmission of rabies is very rare. A theoretic risk of transmission through infected body fluids exists, but the only documentted cases of person-to-person transmission occurred in people who received corneal transplants from donors who died of undiagnosed rabies. The diagnosis can be confirmed during life by detecting rabies virus antigens in corneal impressions or skin biopsies (WHO, 1997). Traditionally, reliance upon long-term, widespread, government-supported programmes aimed at population reduction of animals at risk has been unsuccessful as the sole means of rabies control, based in part upon economical, ecological and ethical grounds. In contrast, immunization of domestic dogs with traditional veterinary vaccines by the parenteral route led to the virtual extinction of canine-transmitted rabies in developed countries. Taken from this basic concept of applied herd immunity, the idea of wildlife vaccination was conceived during the 1960s, and modified-live rabies viruses were used for the experimental oral vaccination of carnivores by the 1970s (Rupprecht et al., 2004).

Effective vaccines are now available against many viruses making eradication a viable proposition. Whether a virus disease can be eradicated or not depends on many factors, not least on the will power to implement such a policy. Many new, emerging and re-emerging diseases of humans are caused by pathogens which originate from animals or products of animal origin. A wide variety of animal species, both domestic and wild, act as reservoirs for these pathogens, which may be viruses, bacteria or parasites. Given the extensive distribution of the animal species affected, the effective surveillance, prevention and control of zoonotic diseases pose a significant challenge. In Africa, such high levels of population immunity are rarely achieved due to a number of reasons. Oral immunization has been shown to be an effective means of inducing high levels of immunity in fox populations in several European countries, and this technique has been mooted as a means of overcoming the logistical problems of delivering injectable rabies vaccines to dogs (Perry and Wandeler, 1993). Dog rabies control relies principally on the mass immunization of dogs in order to achieve population immunity levels sufficient to inhibit rabies transmission (Perry and Wandeler, 1993). Devising an appropriate technology for human rabies immunization includes new regimens of administration, one of which, a revised intramuscular regimen requiring only four doses and three clinic visits, proved highly efficient for postexposure treatment. This article reviews the feasibility of global eradication of rabies and factors affecting eradication of rabies in the

world.

CRITERIA AND CONDITIONS FOR ASSESSING ERADICABILITY OF A DISEASE

The criteria and conditions for assessing eradicability of a disease according to MMWR (1993) include: 1). Scientific Feasibility, 2) Epidemiologic vulnerability (e.g., existence of nonhuman reservoir; ease of spread; natural cyclical decline in prevalence; naturally induced immunity; ease of diagnosis; and duration of any relapse potential), 3) Availability of effective, practical intervention (e.g., vaccine or other primary preventive, curative treatment, and means of eliminating vector). Ideally, intervention should be effective, safe, inexpensive, long-lasting, and easily deployed. 4) Demonstrated feasibility of elimination (e.g., documented elimination from island or other geographic unit), 5) Political Will/Popular Support, 6) Perceived burden of the disease (e.g., extent, deaths, other effects; true burden may not be perceived; the reverse of benefits expected to accrue from eradication; relevance to rich and poor countries), 7) Expected cost of eradication (especially in relation to perceived burden from the disease), 8) Synergy of eradication efforts with other interventions (e.g., potential for added benefits or savings or spin-off effects) and 9) Necessity for eradication rather than control.

Smallpox eradication as a model for global cooperation

The eradication of smallpox serves as a model for global cooperation. It is unlikely that every virus disease is eradicable. Every friend of humanity must look with pleasure on this discovery (smallpox vaccination), by which one evil more is withdrawn from the condition of man; and must contemplate the possibility that future improvements and discoveries may still more and more lessen the catalogue of evils (MMWR, 1993). Smallpox was the first and only virus disease to be completely eradicated. It was eradicated through mass vaccination, and more importantly, the tracing and isolating of known cases and contacts. There were certain features of smallpox which made it a relatively easy target for eradication and this include: 1) Smallpox was a severe disease with significant morbidity and mortality. It had already been eradicated from many developed countries before the WHO campaign began, thus demonstrating the feasibility of global eradication. Eradication would result in significant savings in terms of the cost of vaccination to non-endemic countries. Therefore, the will power was there for eradication. 2) The disease was characteristic and thus easily diagnosed; therefore, there were no cultural or social barriers to case tracing and control. 3) There were no animal reservoirs for smallpox.

4) The incubation period for smallpox was long and infected individuals were infectious after the incubation period. The communicability of smallpox was low. Therefore, people living in the area surrounding a known case of smallpox can be readily protected by vaccination. 5) There was no carrier state and 6) there was only one serotype of virus and an effective vaccine was available which conferred lifelong immunity.

Four factors/conditions enabled the eradication of smallpox: 1) no reservoir of the virus existed except in humans; 2) nearly all persons infected with smallpox had an obvious, characteristic rash and were infectious for a relatively short period; 3) the natural infection conferred lifelong immunity; and 4) a safe, effective (even in newborns), and inexpensive vaccine was available that was also highly stable in tropical environments (Hopkins et al., 1985 reviewed in Ginsberg and Woodroffe, 1997). Any program for the eradication of a particular virus would involve universal vaccination of children, preferably at as young an age as possible. As in the case of smallpox, the following questions should be addressed when the feasibility of eradication of a particular human virus disease is considered.

Is the disease (rabies) worth eradicating?

In the case of severe viral diseases like smallpox and poliomyelitis, the case for eradication is straightforward. However, for milder virus diseases such as chickenpox and mumps, the case for eradication is less straightforward. The priorities would differ between developed and developing countries. The cost-benefit ratio of such an eradication program must be taken into account (Ginsberg and Woodroffe, 1997).

Is there any animal reservoir?

Viruses with a known animal reservoir such as rabies will be very difficult to eradicate. Eradication of the virus would involve eradication of the virus from the animal reservoir as well. This would involve the vaccination of the reservoir. In the case of rabies, eradication from some countries has been achieved by strict quarantine procedures for imported animals and vaccination of pets. However, eradication of rabies from the wildlife would prove to be extremely difficult (Ginsberg and Woodroffe, 1997).

Is there a carrier state?

Viruses which could produce a carrier state or persistent infection in humans such as hepatitis B and the herpes viruses would be very difficult to eradicate unless the carriers could be treated. The possibility that some of these viruses may be vertically transmitted from mother to child by transplacental transmission would confer extra difficulty in terms of eradication (Ginsberg and Woodroffe, 1997).

Is effective vaccination available?

Effective vaccines are now available against some viruses such as rubella, measles and poliomyelitis. These viruses are antigenically stable and are restricted to one or a few serotypes (Ginsberg and Woodroffe, 1997). Other viruses such as influenza A and B are antigenically unstable and the formulation of the vaccine has to be modified annually. These viruses would be almost impossible to eradicate by universal vaccination. Vaccines are still not available against the majority of viruses such as HIV and hepatitis C. In some of the cases, the development of a vaccine is very difficult not because of the multiplicity of serotypes as in the case of rhinoviruses and HIV. In other cases such as in the case of RSV, the development of a vaccine had proved to be very difficult because of other practical difficulties. In the case of RSV, the vaccine would have to be more immunogenic than the natural virus because natural infection does not appear to confer long term immunity.

How communicable is the rabies? What level of coverage is required for eradication?

A 100% coverage rate for vaccine uptake is unlikely to be attainable. However, eradication does not require 100% coverage as herd immunity will impede the transmission of the virus. The coverage rate required for eradication depends mainly on the transmissibility of the virus. Example; smallpox had a low rate of transmission whereas measles had a high rate of transmission. Thus measles is proving difficult to eradicate in countries such as the US even though the vaccine coverage is very good (Ginsberg and Woodroffe, 1997).

IMPETUS FOR ERADICATION OR ELIMINATION OF OTHER DISEASES

Smallpox Eradication Program (SEP)

The 31-year-old success of the Smallpox Eradication Program (SEP) provides an impetus for eradication or elimination of other diseases. A symposium sponsored by the Fogarty International Center of the National Institutes of Health to consider post-SEP possibilities in 1980 identified yaws, measles, and polio as the most likely candidates for eradication (Duffy et al., 1990). In 1986, the the World Health Assembly resolved to "eliminate" Guinea worm disease (Resolution WHA 39.21), the first such resolution since the smallpox campaign; in 1989, the Assembly added the deadline for eradicating Guinea worm disease in "the 1990s" (Resolution 42.29). According to Duffy et al. (1990), Global 2000 and the African Regional Office of WHO have set the informal goal of eradicating Guinea worm disease by 1995. In 1988, the World Health Assembly officially established the goal of eradicating polio by the year 2000 (Resolution WHA 41.28).

Other eradication model for global cooperation

In their discussions, two diseases (Guinea worm disease-dracunculiasis and Poliomyelitis) were judged to be eradicable and three to be candidates (Rabies, yaws and endemic syphilis) for elimination of transmission or of clinical symptoms; three were not considered candidates (measles, tuberculosis and Leprosy-Hansen disease) for eradication at this time. Guinea worm eradication was considered feasiable if the necessary commitment and resources can be mobilized (Duffy et al., 1990). The ITFDE will help publicize efforts and funding needs, and worldwide polio eradication is deemed technically possible by the year 2000; an improved vaccine would facilitate eradication of polio. The ITFDE agreed to write to the heads of state of several nations in the Americas to solicit their support for this hemisphere's goal of eliminating polio by the end of 1990 (Duffy et al., 1990).

Global eradication of measles

Global eradication of measles is not currently feasible because of the high communicability of measles and the suboptimal serologic responses to vaccines administered to young infants. After the ITFDE conference, WHO recommended use of high-titered Edmonsten-Zagreb vaccine beginning at 6 months of age developing countries; however, an improved vaccine is still needed. Global eradication of tuberculosis is not now feasible. Better tools for diagnosis, case-finding, prevention, and treatment need to be developed, and the application of current short-course therapy in developing countries needs to be greatly increased   and leprosy (Hansen disease) eradication worldwide is not feasible now (Duffy et al., 1990).

Elimination of blindness

Elimination of blindness caused by onchocerciasis appears feasible through vector control and treatment with ivermectin. Because of the cost, duration, and difficulty of effective larviciding and the absence of a drug to kill the adult worms, eradication of the infection altogether is not now feasible (Duffy et al., 1990).

Eradication of yaws and endemic syphilis

Eradication of yaws and endemic syphilis is not feasible under present conditions. However, elimination of the transmission of these diseases in certain areas appears feasible (Duffy et al., 1990). Tests need to be developed that can reliably distinguish the organisms that cause yaws, endemic syphilis, and pinta from those that cause venereal syphilis (Burke et al., 1985 reviewed by Duffy et al., 1990).

Regional goals of eliminating polio, measles, or neonatal tetanus

Different WHO regions have also established regional goals of eliminating polio, measles, or neonatal tetanus over the next decade, starting with the elimination of polio from the Americas by the end of this year. India and China aim to eliminate leprosy transmission within their borders by the year 2000, and the United States has set a national goal of eliminating tuberculosis by 2010 (defined as an annual case rate of less than one per million population (CDC, 1992a, b).

FACTORS CONTRIBUTING TO DIFFICULTY IN GLOBAL ERADICATION OF RABIES AND CONTINUAL ENDEMITY IN DEVELOPING COUNTRIES OF THE WORLD

Rabies in humans can be prevented by appropriate postexposure prophylaxis and through vaccination of the animal vector, which is not, however, always available and affordable in resource-limited countries (Zinsstag et al., 2007). Infectious pathogens that originate in wild animals have become increasingly important throughout the world in recent decades, as they have had substantial impacts on human health, agricultural production, wildlife-based economies and wildlife conservation (Bengis et al., 2004). Globally, dog-transmitted rabies represents the largest threat to human health. In order to prevent the transmission of rabies in a dog population, it is theoretically necessary to vaccinate a minimum of 60 to 70% of the dogs. Even countries with potentially sufficient resources, however, do not often meet and sustain these rates. One reason for such failure might be that individual dog owners might feel that it is too expensive to vaccinate their pets (Meltzer and Rupprecht, 1998 reviewed by Tamashiro et al., 2007). Risk factors for Rabies including risk behaviours, associated conditions, protective factors, and unrelated factors. However, eradication of rabies from the wildlife would prove to be extremely difficult. Transmission would confer extra difficulty in terms of eradication. Whether a virus disease can be eradicated or not depends on many factors (Tamashiro et al., 2007).

Factors that can increase the risk of rabies include: travelling or living in developing countries where rabies is more common, including countries in Africa (Tamashiro et al., 2007). Pathogens can emerge either through introduction into a new population or when the interaction with the vector changes; emergence is also influenced by microbiological adaptation and change, global travel patterns, domestic and wild animal contact and other variants in human ecology and behaviour (Tapper, 2006). Although the discovery of such zoonoses is often related to better diagnostic tools, the leading causes of their emergence are human behavior and modifications to natural habitats (expansion of human populations and their encroachment on wildlife habitat), changes in agricultural practices, and globalization of trade. However, other factors include wildlife trade and translocation, live animal and bush meat markets, consumption of exotic foods, development of ecotourism, access to petting zoos, and ownership of exotic pets (Chomel et al., 2007). Factors such as the size and intensity of infectious foci, the rapidity of spread, difficulty of eradication, etc., will influence these risk factors. However, the main factors that contribute to the increase in the number of cases of rabies and making eradication of rabies in the world difficult include:

Human factors

Meslin et al. (2000) described the direct and indirect implications for public health of emerging zoonoses. Direct implications are defined as the consequences for human health in terms of morbidity and mortality. Indirect implications are defined as the effect of the influence of emerging zoonotic disease on two groups of people, namely: health professionals and the general public (Meslin et al., 2000). Professional assessment of the importance of these diseases influences public health practices and structures, the identification of themes for research and allocation of resources at both national and international levels (Meslin et al., 2000). The perception of the general public regarding the risks involved considerably influences policy-making in the health field. However, other human factors affecting eradication of rabies are:

Sharp and exponential rise of global human activity

Human activities may also be a source of wildlife infection, which could create new reservoirs of human pathogens (Chomel et al., 2007). The emergence of these pathogens (rabies, human immuno deficiency virus/acquired immune deficiency syndrome, influenza A, Ebola virus and severe acute respiratory syndrome) as significant health issues is associated with a range of causal factors, most of them linked to the sharp and exponential rise of global human activity (Bengis et al.,

2004). Jebara (2004) described the new challenges that this brings for individual countries and the international community.

Deforestation, development of human habitat, and mining activities

Deforestation, development of human habitat, and mining activities has been suggested as risk factors associated with the reemergence of vampire bat rabies in humans in the Amazon Basin (Chomel et al., 2007). In 2004, 46 persons died of rabies transmitted by vampire bats, mainly in Brazil (22 cases) and Colombia (14 cases); only 20 human cases of rabies were transmitted by dogs in all Latin America (Schneider et al., 2005). A similar trend was again observed for 2005 (Chomel et al., 2007). Such a zoonosis is a good example of deforestation and agricultural development leading to human habitat expansion into natural foci of a viral infection.

Human encroachment on wildlife habitat

The exponential growth of the human population, from ≈1 billion in 1900 to 6.5 billion in 2006, has led to major ecologic changes and drastic wildlife habitat reduction. Many examples of the emergence or reemergence of zoonoses related to human encroachment on wildlife habitats exist (Chomel et al., 2007). Leading factors for emergence of zoonoses are unbalanced and selective forest exploitation and aggressive agricultural development associated with an exponential increase in the bushmeat trade (Wolfe et al., 2005a; Chomel et al., 2007).

Increasing international travel

Rabies is a fatal disease, and increased international travel is one of the important factors affecting eradication of rabies globally (Shaw et al., 2009). Importation of infectious diseases to new countries is likely to increase among both travellers and immigrants (Fenner et al., 2007). Approximately 80 million people from resource-rich areas worldwide travel to resource-poor countries every year (WTO, 2006) and are exposed to many infections that are no longer prevalent in the countries where they live (Fenner et al., 2007). Moreover, visitors to developing countries are sometimes unaware of the rabies risk posed by dog bites and thus may not seek appropriate medical attention for such bites (Nadin-Davis et al., 2007). The occasional cases of rabies reported in industrialized countries, such as the United Kingdom, are often the result of exposure while travelling in developing countries such as India (Solomon et al., 2005). Arguin et al. (2000) reported that of the 36 cases of human rabies that have occurred in the United States since 1980, 12 (33%) were presumed to have been acquired abroad. Twenty people died of rabies in France between 1970 and 2003 (compared to 55,000 yearly worldwide), 80% on returning from Africa. Dogs were the contaminating animals in 90% of the cases and children were the most common victims (Peigue-Lafeuille et al., 2004). In France, people travelling abroad, particularly to Africa, are warned not to approach unknown animals (especially dogs) or to try to import them, and are advised to comply with vaccinal recommendations for travellers, particularly for toddlers (Peigue-Lafeuille et al., 2004). We should recognizing that humans and animals are part of a global community with frequent travel and translocation, the risks of disease introduction, particularly with sub-clinical or incubating animals, are real and present (Castrodale et al., 2008). In the United States, it is recommended that international travellers likely to come in contact with animals in canine rabies-enzootic areas that lack immediate access to appropriate medical care, including vaccine and rabies immune globulin, should be considered for preexposure prophylaxis (Arguin et al., 2000). However, travelers going to endemic areas need to take precautions (Shaw et al., 2009).

Lack of emergency preparedness planning

Emergency preparedness planning for animal diseases is a relatively new concept that is only now being applied in Africa. Information can be drawn from numerous recent disease epidemics involving rinderpest, contagious bovine pleuropneumonia (CBPP) and Rift Valley fever. These examples clearly demonstrate the shortcomings and value of effective early warning with ensured early reaction in the control of transboundary animal disease events. On a global level, the human health sector lags behind the animal health sector in the assessment of potential threats, although substantive differences exist among countries in the state of national preparedness planning for emerging diseases (Merianos, 2007).

Lack of good laboratory and proper surveillance

Isolation of LBV from terrestrial wildlife in study reported by Markotter et al. (2006a) serves as further confirmation of our lack of understanding of the incidence and host range of lyssaviruses in Africa. Poor surveillance of rabies-related viruses and poor diagnostic capability in most of Africa are large contributors to our lack of information and the obscurity of the African lyssa viruses (Markotter et al., 2006a). The lack of surveillance data on emerging zoonoses from many developing countries means that the burden of human, livestock and wildlife disease is underestimated and opportunities for control interventions thereby limited (Merianos, 2007). Many countries in postcommunist transition face a sharp increase in zoonotic diseases resulting from the breakdown of government-run disease surveillance and control and weak private health and veterinary services (Zinsstag et al., 2007). However, laboratories continue to be sadly behind the times in terms of equipment and skills for diagnosing the emerging

pathogens as can be readily observed in developing nation such as Nigeria (Okonko et al., 2008).

Lack of appropriate control methods

Lethal methods of dog population control are even more expensive, and attempting to control rabies by reducing dog populations has not worked for any extended period. Despite the availability of techniques to improve the global rabies situation, limitations in surveillance and epidemiologic investigations impede the institution of such measures (Knobel et al., 2005). In industrialized countries, diagnosis of rabies in animals is achieved by using rabies-specific fluorescein-conjugated antibody to detect viral antigen in brain smears; however, ante-mortem diagnosis in humans must rely on less-invasive methods (Nadin-Davis et al., 2007). The utility of PCR-based methods to detect rabies virus sequences in saliva and other body fluids has been reported, and PCR is being used in many industrialized countries. An additional component of rabies control in such countries is the application of viral typing methods to identify viral variants that circulate in specific host reservoirs; which are lacking in most developing countries (Nadin-Davis et al., 2007).

Lack of solid commitment and internationally resourced initiatives

Little public domain information is available on international coordinated responses to the deliberate introduction of biological pathogens. Terrorist events in the early 21st Century have increased awareness of the risks, but solid commitment and internationally resourced initiatives are still lacking (Lubroth, 2006). However, in many countries, little is known about the real cost of mass vaccination of dogs, and quantitative data are urgently needed to evaluate the cost-effectiveness of different rabies control strategies in resource-limited countries; rabies control strategies in developing countries are currently under review by WHO (Zinsstag et al., 2007).

Inappropriate management

The clinical signs and symptoms of the initial rabies case-patients in Africa may have been altered due to use of traditional medicines. The use of traditional medicines is common in rural settings in Africa and may result in toxicities, including abdominal and psychiatric symptoms and abnormal liver function test results (Luyckx et al., 2004; Cohen et al., 2007). These medicines could have contributed to the atypical manifestations in some cases. In addition, clinicians may have attributed some of the neurologic symptoms to herbal intoxication (Cohen et al., 2007). "In areas where there is a high prevalence of rabies, such as Africa and Asia, "the Alliance for Rabies Control added, "the need for vaccination has often been overlooked, despite the fact this would cost less than other health care programs," including administering post-exposure rabies immunization to save dog bite victims (Clifton, 2007). In the context of emerging zoonoses, comprehensive risk assessments are needed to identify the animal-human and animal-animal interfaces where transmission of infectious agents occurs and the feasibility of risk reduction interventions (Merianos, 2007). Shaw et al. (2009) in a study that examined for those who reported postexposure management to animals while abroad, reported that only 25% of the 459 post-travel records from October 1998 until February 2006 at two travel medicine clinics, in Auckland and Hamilton, received postexposure treatment consistent with WHO guidelines, reflecting inappropriate management abroad (Shaw et al., 2009).

Gross under-diagnosis

The incidence of rabies in many parts of Africa is unknown, but rabies is probably underdiagnosed (Mallewa et al., 2007). Cases of rabies may be incorrectly attributed to other causes of pyrexia and confusion common to rural Africa, including cerebral malaria, bacterial infections, and infection with HIV (Bleck and Rupprecht, 2005 reviewed by Mallewa et al., 2007; Cohen et al., 2007). In a study by Hayman et al. (2008), there were differences in seroprevalence between *E. Gambian us* and *E. helvum* with respect to LBV infection. The underlying cause of the difference in seroprevalence is unclear. Possible explanations include differential susceptibilities to infection; virus–host adaptation; different contact with the virus, including a recent epidemic in the *E. helvum* colony; or different population ecology (Hayman et al., 2008). No investigations into infections of humans were made during these investigations, but lyssavirus infections in humans in Africa are underdiagnosed (Mallewa et al., 2007; Hayman et al., 2008).

Gross under reporting

Rabies is no doubt underreported and probably misdiagnosed in Nigeria and elsewhere in Africa (Asselbergs, 2007; Fagbo, 2009; Woolf, 2009). According to Ogundipe et al. (1989), in their study the development and efficiency of the animal health information system in Nigeria as well as the completeness and immediacy of data supply by the system for the period between January, 1977 and December, 1984 revealed that the system was found to be characterized by: late, inaccurate and gross under reporting. And these constraints in reporting include inadequate personnel, poor diagnostic and reporting facilities (Okonko et al., 2008).

Poor coordination and cooperation

Because animal disease control, e.g. rabies control and ORV, is the responsibility of each federal state, insuffi-

cient cooperation in the planning of vaccination campaigns between neighbouring federal states has also been an important shortcoming (Müller et al., 2005). Preparations for international cooperation in response to disease disasters at the regional or continental levels are poorly coordinated and cooperation is limited, although intergovernmental and international organizations have been advocating for years that emergency responses to infectious disease outbreaks should be planned for and prepared at the national level (Lubroth, 2006).

Lack of knowledge and information

Many countries, especially those with resource constraints and those in sub-Saharan Africa, lack information on the distribution of zoonotic diseases (Zinsstag et al., 2007). Risks for zoonoses are considered negligible compared with those for diseases of higher consequence because the societal consequences of zoonoses are not recognized by the individual sectors. For example, outbreaks of Rift Valley fever in persons in Mauritania were mistakenly identified as yellow fever. However, transmission of zoonoses to humans can already be greatly reduced by health information and behavior. Authorities in Kyrgyzstan, for example, have started an information campaign to reduce brucellosis transmission to small-ruminant herders by encouraging them to wear gloves for lambing and to boil milk before consuming. Interventions in livestock should always be accompanied by mass information, education, and communication programs (Zinsstag et al., 2007).

Lack of risk communication

Lack of risk communication is one of the factors affecting polio eradication especially in Nigeria. Although, the boycott of immunization is no longer in effect, low participation during vaccination may persist reflecting a failure to implement risk communication (Agbeyegbe, 2007). The silence of the government over the alleged report by JNI and widely in the media that the government acknowledged the use of contaminated vaccines but claimed that the contaminated batch had been completely used, could be interpreted as indicative of the accurateness of the report. To address such situations, risk communication is increasingly becoming important in public health (Rudd et al., 2003 reviewed by Agbeyegbe, 2007). Risk communication offers a two-way communication process that presents the expert opinions based on scientific facts to the public, and acknowledges the fears and concerns of the public, seeking to rectify knowledge gaps that foster misrepresentation of risk (Leiss, 2004; Aakko, 2004 reviewed by Agbeyegbe, 2007; Okonko et al., 2009).

Lack of adequate public awareness

Lack of public education campaigns in the developing

world (Clifton, 2007) is another factor affecting rabies eradication. Though, public awareness of the human health risks of zoonotic infections has grown in recent years (Heeney, 2006), reliable data on rabies are scarce in many areas of the globe, making it difficult to assess its full impact on human and animal health (Awoyomi et al., 2007). Some steps are being taken. The worldwide communications networks have made inroads with SatelLife and ProMED online services. Physicians in developing nations can consult colleagues, libraries, or data banks for help with puzzling cases. Promising as they are, though, these "electronic conference tables" represent only a small piece of the solution. At present, the networks do not address the underlying causes of new and re-emerging infectious diseases. And they probably can do little to get to the heart of the real issues: educating the public, immunizing the children, improving sanitation, cleaning up the water, housing the homeless, feeding the starving (Hoel and Williams, 1997 reviewed by Okonko et al., 2008).

Lack of motivation

Several studies have documented the costs associated with wildlife-rabies epizootics (Shwiff et al., 2007; Sterner et al., 2009). According to the Alliance for Rabies Control, "the tools for effective rabies control are available. What is lacking are the motivation, commitment and resources to tackle the disease effectively (Clifton, 2007)."

Political instability

In several countries particularly Afganistan and Somalia, part of Parkistan, Sierra Leone, Sudan, Liberia, Congo, and Ethiopia political instability or armed conflicts make vaccination logistically difficult and unpredictable (Okonko et al., 2009). In addition to this internal politics in the 2003 immunization boycott, were ramifications from the international political arena. Anti-western sentiments have increased among Northern Muslims fundamentalist following the September 11, 2001 attacks and America's war on terrorism. Given the distrust and growing antagonism towards America, the involvement of the West in a program that benefits Muslims was viewed with suspicion (Agbeyegbe, 2007; Okonko et al., 2009).

Conflict-threat of armed militias and forced migration

Other challenges facing vaccination teams included the threat of armed militias that roam the area in search of opportunities to seize control over the local oil resources (Njoku, 2006). With the overwhelming increase in high-intensity local conflicts among political, ethnic, and religious rivals, government-based disease-surveillance systems have little or no chance of success (Okonko et al., 2008). The reintroduction of canine rabies into northern KwaZulu-Natal Province in 1976 followed an influx of refugees from Mozambique (Cohen et al., 2007). The possible contribution of increased immigration into Limpopo Province from Zimbabwe in recent years is difficult to quantify (Oucho, 2006; Cohen et al., 2007).

Cultural and religious beliefs

Cultural and religious beliefs will also contribute to the underreporting of human rabies that may arise from the consumption of infected apparently healthy dogs and cats. The [rabies-related lyssaviruses] Lagos bat and Mokola viruses still remain under-diagnosed in the human populace (Fagbo, 2009; Woolf, 2009). In case of measles and polio viruses, many Nigerians are blaming the outbreak on vaccination efforts; an attitude expert's fear may ruin previous gains in eradicating vaccine preventable disease in the country (Okonko et al., 2009). Most of the anti-immunization campaigns in Nigeria have been predominantly Muslim north of Nigeria, and a number of Muslim clerics have been quoted in the Nigerian media as claiming that vaccines are dangerous and cause sterility or illness (Adeija, 2007). Cultural and religious objections under vaccinations efforts, resulting in persistently low immunity in the population and consequently, a high incidence of emerging vaccine-derived viruses and reemergence of wild viruses (Okonko et al., 2009). Whereas the undercurrents between Muslims and Christians in Nigeria as well as Western donors may have been sufficient to begin the controversy on the vaccines e.g. polio vaccine, other factors helped to sustain it (Agbeyegbe, 2007).

Ignorance

Factors responsible for the continued endemicity of rabies in Nigeria have been determined in a study by Opaleye et al. (2006). In their study, the knowledge, attitude and practice study among residents of Osun State, Nigeria was reported. In their study among 679 individuals, only 33.4% of the respondents knew rabies could be prevented by vaccination, while 38.7% believed that the infection could be treated with herbs. Of the 387 victims of dog bite, 240 (62%) never sought prophylactic postexposure treatment. Of the 10 people who received postexposure treatment, only one received the appropriate treatment consisting of washing, disinfection of wounds, tetanus toxoid and complete antirabies immunization.

Government negligence

No vaccine is fully safe, perfectly potent and without risk of administering error (Clements et al., 1999 reviewed by Okonko et al., 2009). Deficiencies in national veterinary services have contributed to failures in early detection and response; in many regions investigation and diagnosis services have deteriorated (Rweyemamu et al.,

2000). Also, the inability of the Nigerian government to acknowledge the risk involved in vaccination however negligible raises doubt about the sincerity of the government, and positions the boycotts of polio vaccination proponents as a more reliable source of information. The government does not appear to have positioned itself as a credible authority to implement immunization programs (Olugbode, 2007; Okonko et al., 2009).

Government policy

In resource-limited and transitioning countries, many zoonoses are not controlled effectively because adequate policies and funding are lacking (Zinsstag et al., 2007). In Chang et al. (2002) analysis on expenditure reports to estimate the cost of rabies prevention activities. An estimated $13.9 million was spent in New York State in United States to prevent rabies from 1993 to 1998; yet Nigerian government has not seen the need to make such tremendous dedication and policy in the health sector.

Deception

According to Clifton (2007), policymakers in the developing world often seek for their cities the superficially animal-free appearance of a "modern" city that they see in Europe and the U.S., equating this with ridding themselves of rabies. But casual outdoor observation of European and U.S. cities by daylight is deeply deceptive. European and American cities support even more dogs, cats, and wild animals per thousand humans than the cities of the developing world. They have merely achieved a transition from hosting outdoor animals, seen in daytime, to hosting mostly indoor pets and nocturnal wildlife.

Socioeconomic factors

Poverty

Poverty is one other factor. Rabies in humans can be prevented by appropriate postexposure prophylaxis, which is not, however, always available and affordable in resource-limited countries (Zinsstag et al., 2007). Tissue-culture vaccines are expensive and they are not always used in all parts of the world. Industrialized countries have responded rapidly to recent zoonosis outbreaks and contained them well, but many resource-limited and transitioning countries have not been able to respond adequately because they lack human and financial resources and have not sufficiently adapted public health surveillance. In industrialized countries, an important part of successful zoonosis control has been compensating farmers for culled livestock. However, many resource-

limited countries would not be able to conduct such programs (Zinsstag et al., 2007).

Hunting with dogs

Hunting activities for the 10 million hunters in Europe generate a financial flux of almost 10 billion euros and ≈100,000 jobs. Europe is also the world's largest importer of venison (>50,000 tons/year). Similarly, in the United States, hunting activities generate >700,000 jobs (Chomel et al., 2007). The combination of urban demand for bushmeat (a multibillion-dollar business) and greater access to primate habitats provided by logging roads has increased the amount of hunting in Africa, which has increased the frequency of human exposure to primate retroviruses and other disease-causing agents (Chomel et al., 2007). However, urban (dog-mediated) rabies has almost been completely eradicated, except from cases from failed vaccination, increased mortality due to hunting and rabies, dogs and foxes and along their dispersal routes to maximize the effect of vaccination (Eisinger and Thulke, 2008). The viral ecotype that previously had been confined to urban dogs in Texas, variant's occurring in a hunting dog in Alabama in 1993. The last case of rabies in Italy was diagnosed in a fox, sheep and goats kept outdoors; thereby leading to prohibition of hunting with dogs (Eisinger and Thulke, 2008).

Increasing demands for bush meat

Another risk factor related to the emergence of zoonotic diseases from wildlife has been the considerable increase in consumption of bushmeat in many parts of the world, especially Central Africa and the Amazon Basin, where 1 - 3.4 million tons and 67 - 164 million kilograms, respectively, are consumed each year (Karesh et al., 2005 reviewed by Chomel et al., 2007). Similarly, several outbreaks of Ebola virus and other disease-causing agents in western Africa have been associated with consumption of bushmeat, mainly chimpanzees that were found dead (Chomel et al., 2007). Human butchering and consumption of animals potentially infected with rabies and other zoonotic viruses are not limited to Asia (Durosinloun, 2009; Fagbo, 2009; Woolf, 2009). In Africa, the bushmeat trade is generating hundreds of millions of dollars (Karesh et al., 2005 reviewed by Chomel et al., 2007). In the Congo Basin, trade and regional consumption of wild animal meat could reach 4.5 million tons annually; the demand for bushmeat in western and Central Africa could reach up to 4× the demand for bushmeat in the Amazon Basin (Wolfe et al., 2005b; Chomel et al., 2007). However, the fast-growing demand for meat in urban centers in resource-limited countries is leading to the intensification of livestock production systems, especially in periurban areas of these countries (Zinsstag et al., 2007). Because efficient zoonosis surveillance and food safety are lacking, the risk for zoonosis transmission is increasing, particularly in rapidly growing urban centers of resource-limited countries (Steinmann et al., 2005 reviewed in Zinsstag et al., 2007).

Dogs, cats and bats eating

Dogs purchased by restaurants are soon killed for consumption. With the exception of butchers, there would be insufficient time to transmit RABV to other dogs and humans. Until now, persons in China who eat dog meat have not been not considered at risk for rabies because no related infections have been reported (Tao et al., 2009). In Nigeria, dog eating is very common in states such as Plateau, Akwa Ibom, Cross River, Kaduna, Kebbi and Ondo. In fact, dog suya (barbequed dog meat) is sold publicly in the dog eating areas. In some areas such as Jos, only local and seasoned connoisseurs may easily distinguish restaurants where dog and other conventional meats are sold. Cat eating, though not as common as dog eating, can also be encountered, even in cosmopolitan places such as Lagos. While human consumption of bats is also common, there seems to have been little or no local effort (as per the limited information available) to evaluate the risk of rabies transmission (Durosinloun, 2009; Fagbo, 2009; Woolf, 2009).

Major ecologic, environmental and anthropogenic changes of the biosphere

Climatic changes in both East and West Africa were associated with an upsurge of emerging infections such as Rift Valley fever (Rweyemamu et al., 2000). Climatic changes are likely to have a direct impact on the presence and abundance of various pathogens and their vectors, so that with a warming climate exotic diseases may play a role in future UK livestock and wildlife disease management (Böhm et al., 2007). However, the world is experiencing an increase in emergent infections as a result of anthropogenic changes of the biosphere and globalization (Cabello and Cabello, 2008). Global warming unrestricted exploitation of natural resources such as forests and fisheries, urbanization, human migration, and industrialization of animal husbandry cause environmental destruction and fragmentation. These changes of the biosphere favour local emergence of zoonoses from their natural biotopes and their interaction with domestic animals and human populations, favour the dissemination of these zoonotic pathogens worldwide (Cabello and Cabello, 2008).

Globalization of animal trade

Globalization is leading to a rise in the emergence of diseases. The magnitude of the global movement of animals is staggering. In terms of sheer numbers, 37,858,179 individually counted live amphibians, birds, mammals, and reptiles were legally imported to the United States from 163 countries in 2000 - 2004. These imports included Asian macaques, South American rodents, and African great cats (Jenkins et al., 2007; Marano et al., 2007). Abazeed and Cinti (2007) and Swanepoel et al. (2007) reports on rabies and Marburg virus respectively illustrated zoonotic diseases with serious health implications for humans, and both have a common reservoir, the bat. Tumpey (2007) reminded us how globalization has had an impact on the worldwide animal trade. This worldwide movement of animals has increased the potential for the translocation of zoonotic diseases, which pose serious risks to human and animal health (Tumpey, 2007; Marano et al., 2007).

Liberalization of world trade

In the past decades, public health authorities within industrialized countries have been faced with an increasing number of food safety issues. The situation is equally serious in developing countries. The globalization of food (and feed) trade, facilitated by the liberalization of world trade, while offering many benefits and opportunities, also represents new risks (Domenech et al., 2006).

Wildlife trade and translocation

However, other factors include wildlife trade and translocation (Chomel et al., 2007). Wildlife trade provides mechanisms for disease transmission at levels that not only cause human disease outbreaks but also threaten livestock, international trade, rural livelihoods, native wildlife populations, and ecosystem health (Chomel et al., 2007). Worldwide, an estimated 40,000 primates, 4 million birds, 640,000 reptiles, and 350 million tropical fish are traded live each year (Karesh et al., 2005 reviewed by Chomel et al., 2007). International wildlife trade is estimated to be a US $6-billion industry (Check, 2004 reviewed in Chomel et al., 2007). Translocation of wild animals is associated with the spread of several zoonoses (Chomel et al., 2007). Rabies was introduced in the Mid-Atlantic States in the 1970s when hunting pens were repopulated with raccoons trapped in rabies-endemic zones of the southern United States (Chomel et al., 2007). Wild life such as wild deer can feature in the epidemiology of a wide range of livestock and human diseases in the United Kingdom by representing a source of disease via various transmission routes (Böhm et al., 2007). In Eastern Europe, raccoon dogs (Nyctereutes procyonoides) are becoming a new reservoir for rabies, in addition to the established red fox reservoir, as raccoon dogs have spread into new habitats from accidental release of animals raised for fur trade (Chomel et al., 2007). Infectious pathogens of wildlife affect not only human health and agricultural production but also wildlife-based economies and wildlife conservation. Zoonotic pathogens that infect domestic animals and wildlife hosts are more likely to emerge (Chomel et al., 2007).

Changes in agricultural practices

The emergence of rabies and its expansion has been directly linked to development of agricultural activities. In the late 1970s and early 1980s, a rabies epidemic occurred in free-ranging greater kudus (*Tragelaphus strepsiceros*) in Namibia (Hubschle, 1988 reviewed by Chomel et al., 2007). The kudu population had increased considerably in response to favourable conditions and human-made environmental changes (Chomel et al., 2007). Suitable conditions for transmission in the kudu population after initial infection by rabid carnivores are provided by the social behavior of kudus, such as browsing on thorny acacia trees and resultant lesions in the kudus' oral cavity, and excretion of relatively high titers of virus in the saliva of infected animals (Chomel et al., 2007). However, reemergence of zoonotic diseases that had been controlled from their domestic animal reservoirs is also of major concern. Wildlife may become new reservoirs of infection and may recontaminate domestic animals (Chomel et al., 2007). Most animal pathogens for which surveillance programs exist relate to farm animals, and few or no programs are specifically aimed at wildlife. Two different but complementary approaches are 1) to monitor the presence of specifically Identified pathogens that have emerged as human pathogens and 2) to investigate in a given wildlife species the presence of known or unknown infectious agents. Furthermore, conservation of habitat biodiversity is critical for preventing emergence of new reservoirs or amplifier species (Chomel et al., 2007).

Topography

This is another important factor affecting eradication of rabies. There are huge numbers of wild animals distributed throughout the world and the diversity of wildlife species is immense (Bengis et al., 2004). According to Bengis et al. (2004), each landscape and habitat has a kaleidoscope of niches supporting an enormous variety of vertebrate and invertebrate species, and each species or taxon supports an even more impressive array of macro- and micro-parasites.

Paving streets

Paving streets tends to eliminate feral pigs, since pigs need mud to wallow in. That tends to leave more habitats to monkeys, if free-roaming dogs disappears mostly macaques in Asia, baboons in Africa. Macaques and baboons do not run from feral cats, bite more often and more dangerously than dogs, are capable of transmitting more deadly diseases to humans than any other animals even though they rarely carry rabies, can outclimb cats, and are often smarter than the public policymakers whose misguided ideas about animal control invite their presence (Clifton, 2007).

Lack of good road network

Of the targeted 29 million children, 4 million reside in impoverished and hard-to-reach settlements across the Niger Delta Region of Nigeria (Njoku, 2006).

Capacity carrying

In effect, mechanization of transport and improvements in urban sanitation reallocated the carrying capacity of the human environment. Instead of supporting dogs and cats who lived directly off of refuse and rodents, the human environment evolved to support dogs and cats who lived on refuse that was processed into pet food, fed to them in human homes. This same reallocation of carrying capacity has occurred in Western Europe, and is occurring now in Eastern Europe, India, China, Ethiopia, and wherever else economic development is transforming former hubs of agrarian commerce into technologically developed modern cities (Clifton, 2007).

Animal factors

Migration of dogs

Also, outbreak of rabies in humans followed an outbreak in domestic dogs. Increasing numbers of human rabies cases in Africa have been attributed to the mobility of human and animal populations (Cohen et al., 2007). Other human-related activities, such as persons migrating with their dogs may also contribute to long-distance spread of rabies. While the majority of reported potential rabies exposures are associated with dog and cat incidents in most places, most rabies exposures have been derived from rabid wildlife (Roseveare et al., 2009). Can contact between wild dogs and domestic dogs be minimized? According to Ginsberg and Woodroffe (1997), this would depend upon local peoples' need for domestic dogs. More research is needed to determine whether domestic dogs' movements could be restricted by, for example, requiring that owned dogs be collared, that dogs be tied up at night, and shooting unaccompanied dogs (Wu et al., 2009). However, the animal population itself does not pose a rabies threat (Wu et al., 2009). Subsequently, international commerce, human and animal migration and travel, favour the dissemination of these zoonotic pathogens worldwide (Cabello and Cabello, 2008).

Migration of stray animals

According to Roseveare et al. (2009), stray cats were most frequently rabid among domestic animals. This underscores the need for improvement of wildlife rabies control and the reduction of interactions of domestic animals, including cats, with wildlife. Can contact between wild life and stray animals be minimized?

Migration of wildlife

Most emerging infectious diseases are zoonotic; wildlife constitutes a large and often unknown reservoir. Wildlife reservoirs of classical and emerging zoonoses persist in many countries and substantially slow control efforts for livestock (Smith et al., 2006). Though, wildlife is a major source of income, either directly for consumptive or productive use value or indirectly for touristic and scientific values. For instance, worldwide, deer farming has been developing dramatically. In New Zealand, ≈2 million farmed deer, half of the world's farmed deer population, generate an annual income of NZ $200 million (Chomel et al., 2007). Even in industrialized countries, wildlife-related activities can generate major income. In the United States, the total expenditure for wildlife-related activities was $101 billion in 1996, ≈1.4% of the national economy; it (wildlife) is now recognized as an important source of emerging human pathogens, including parasites (Polley, 2005). Wildlife can also be a source for reemergence of previously controlled zoonoses. Although the discovery of such zoonoses is often related to better diagnostic tools, the leading causes of their emergence are human behavior and modifications to natural habitats (expansion of human populations and their encroachment on wildlife habitat), changes in agricultural practices, and globalization of trade (Chomel et al., 2007).

Animal importation

Animals imported for commercial trade represent a substantial risk to human health (Marano et al., 2007). Animals are legally imported into the United States for many reasons. They are used for exhibitions at zoos; scientific education, research, and conservation programs; food and products; and in the case of companion animals, tourism and immigration. Increasingly, however, animals are being imported for a thriving commercial pet trade (Marano et al., 2007). In many cases the animals that are imported and traded are of species that are considered exotic (here defined as non-native species, animals not traditionally kept as pets, or both). This can be a risky business, as many shipments include a high volume of wild-caught versus captive raised animals. For most of these animals, there are no requirements for zoonotic disease screening either before or after arrival into the United States. There have been anecdotal reports of high rates of death among animals in these shipments (Marano et al., 2007). As a scientist, one might suggest solutions that employ familiar tools, such as post arrival screening of animals with reliable laboratory tests, empirical treatment for known

diseases (if such tests and treatments already existed), or quarantine of the animals for an appropriate length of time (Marano et al., 2007). Many of these solutions are not feasible or practical to use on the large volume of animals that are being imported and cannot be employed to prevent new or emerging pathogens or infections. Ultimately, import restrictions may be the only means of preventing introduction of exotic infections (Marano et al., 2007).

Illegal importation of animals

Illegal trade can also be a possible source of human infection (Chomel et al., 2007). Illegal importation of animals still poses a risk of rabies world wide. In March 2007, a puppy reportedly to have been recently imported from India into the United States was found to be positive for rabies by the Alaska Department of Health and Social Services. According to Castrodale et al (2008), this case report highlights several important public health issues. Animal-importation regulations, policies and practices are intended to minimize these risks and should be routinely evaluated and updated as needed in response to occurrences such as detailed in the Castrodale et al (2008) communication. And the illegal trade of exotic wildlife, with promises of considerable financial return in the underground markets, has disastrous implications for many endangered or threatened species (Marano et al., 2007).

Mating activities and dispersal

This is one of the animal factors that contribute to the difficulty in the eradication of rabies. According to Eisinger and Thulke (2008), different transmission patterns of rabies re-emerged out of the set of their model rules, for example the seasonal rabies peaks caused by mating activities and dispersal and the focal spatial pattern of the advancing rabies epidemic. In the study by Eisinger and Thulke (2008), any neighbouring cells of foxes within a distance of up to 3 cells may be infected, with a probability of 0.16^1, 0.16^2 and 0.16^3, respectively, because of mating contacts. There were hardly any infections during dispersal but juvenile foxes dispersing during their incubation period will cause standard transmission after settlement (Eisinger and Thulke, 2008).

Host population increase

As in a classic situation, outbreak of rabies in humans followed an outbreak in domestic dogs of the Africa region (Cohen et al., 2007). Increasing numbers of human rabies cases in Africa have been attributed to increasing numbers in animals (Cohen et al., 2007). Descriptive epidemiologic analysis showed that the increase in domestic dog populations has contributed to rabies epidemics in the 3 provinces in southern China (Wang, 2006). Increases in deer abundance as well as range expansion are likely to exacerbate the potential for disease persistence due to the formation of multi-species deer assemblages, which may act as disease reservoirs (Böhm et al., 2007). In a study by a dog and cat population ecologist John Marbanks in 1947 - 1950 reviewed by Clifton (2007), who estimated that there were only 600,000 street dogs in the already heavily motorized Northeast, but were 3.5 million in the South and 2.3 million in the Midwest, the two most agrarian parts of the U.S. More than 20 years passed before the U.S. dog and cat populations were again studied in depth. By then, in the early 1970s, the U.S. street dog population had disappeared. The feral cat population rose in the absence of street dogs to a peak of about 40 million circa 1990, and then fell with the advent of neuter/return to today's levels of about six million in winter, 12 million in summer (Clifton, 2007). In the interim, the number of cars and miles driven in the U.S. had tripled.

The pet dog and cat populations rose proportionate to the human population. The pet dog population had increased by just about exactly as much as the street dog population declined. The biomass of dogs and cats relative to human population remained almost the same (Clifton, 2007). A high seroprevalence to LBV in *E. helvum* population may pose a substantial public health risk because *E. helvum* is widely distributed in Africa and a food source in West Africa (Hayman et al., 2008). *E. helvum* resides in high-density populations (hundreds of thousands) and migrates annually, compared with *E. gambianus*, which resides in less dense colonies of tens or hundreds. *E. helvum* commonly forms large colonies in African cities in close proximity to humans and domestic animals and is a food source in West Africa (Hayman et al., 2008).

Neighbourhood contacts

Furthermore, our quest for close contact with wild animals puts us at risk for exposure to zoonoses (Chomel et al., 2007). In the study by Eisinger and Thulke (2008), eight neighbouring cells of foxes have a probability of 16% of getting infected in neighbourhood contacts (adjusted to hunting bag pattern). Infected fox incubates and gets infectious after a time period that is drawn randomly from a negative exponential distribution, with a minimum of 2 weeks and an effective mean of 3·5 weeks. During the following infectious period of 1 week, a fox can transmit the disease using the approach of infection communities or group infection rate. If there is at least one infectious fox in a cell, all other susceptible foxes within the cell become infected in intragroup contacts

(Eisinger and Thulke, 2008).

Frequency of consumption of brain (carnivorous animals)

Oral transmission of rabies could be produced in laboratory animals like mice, guinea pigs and hamsters using challenge virus strain (CVS) and 2 strains of street virus. Study of virus pathway following ingestion suggested predominant neural spread to brain and centrifugal spread to non neural organs like heart and kidneys. However it was found that virus dose required for oral infection was relatively very high (Madhusudana and Tripathi, 1990). The role of such a transmission in nature needs to be further evaluated, keeping in view the high dose of virus required for oral infectivity and the frequency of consumption of brain by carnivorous animals.

Consumption of exotic foods

Other factor affecting eradication of rabies includes consumption of exotic foods, development of ecotourism, access to petting zoos, and ownership of exotic pets. Furthermore, our quest for close contact with exotic pets puts us at risk for exposure to zoonoses (Chomel et al., 2007). Industrialized nations' new taste for exotic food has also been linked with various zoonotic pathogens or parasites (Chomel et al., 2007).

Development of ecotourism

One other factor affecting eradication of rabies includes development of ecotourism. For instance, wildlife tourism is among the top exporting activities of Tanzania and Kenya and generates an annual income of approximately half a billion US dollars (Chomel et al., 2007). Adventure travel is the largest growing segment of the leisure travel industry; growth rate has been 10% per year since 1985. This type of travel increases the risk that tourists participating in activities such as safaris, tours, adventure sports, and extreme travel will contact pathogens uncommon in industrialized countries (Chomel et al., 2007). A recent review of a global surveillance network's data set showed different demographic characteristics and different types of travel-related illnesses among immigrant-VFR, traveller-VFR, and tourist travellers (Leder et al., 2006; Fenner et al., 2007). More than 800 million tourist arrivals were registered worldwide in 2005, and an estimated 2% of the world's population lives outside the country of birth. Moreover, because ecotourism is becoming increasingly popular with international travellers, more cases of imported rabies are likely to occur in Europe, North America, and elsewhere in years to come (Fenner et al., 2007). Similarly, the increase of ecotourism, often in primitive settings with limited hygiene, can be associated with the acquisition of zoonotic agents (Chomel et al., 2007). In countries with large migrant populations, improved public health strategies are needed to reach visiting friends and relatives (VFR) travellers (Fenner et al., 2007). Therefore, development of appropriate programs for surveillance and for monitoring emerging diseases in their wildlife reservoirs is essential (Chomel et al., 2007).

Petting zoos

Some other factor affecting eradication of rabies includes access to petting zoos. Petting zoos, where children are allowed to approach and feed captive wildlife and domestic animals, have been linked to several zoonotic outbreaks, including infections caused by rabies (Bender and Shulman, 2004 reviewed in Chomel et al., 2007). More than 25 outbreaks of human infectious diseases associated with visitors to animal exhibits were identified during 1990–2000 (Chomel et al., 2007). Exposure to captive wild animals at circuses or zoos can also be a source of zoonotic infection.

Ownership of exotic pets

Also, one other factor affecting eradication of rabies includes ownership of exotic pets. Exotic pets are also a source of several human infections that vary from severe monkeypox related to pet prairie dogs or lyssaviruses in pet bats to less severe but more common ringworm infections acquired from African pygmy hedgehogs or chinchillas (Chomel et al., 2007). Eight cases of rabies caused by a new rabies virus variant were reported in the state of Ceará, Brazil, from 1991 through 1998. Marmosets (*Callithrix jacchus jacchus*) were determined to be the source of exposure (Chomel et al., 2007). These primates are common pets; most cases occurred in persons who had tried to capture them, and 1 case was transmitted by a pet marmoset. In 1999, encephalitis was diagnosed in an Egyptian rousette bat (*Rousettus egyptiacus*) that had been imported from Belgium and sold in a pet shop in southwestern France. The pet bat was infected with a Lagos bat lyssavirus and resulted in the treatment of 120 exposed persons (Chomel et al., 2007).

Traditional and local food and live animal markets

Traditional and local food markets in many parts of the world can be associated with emergence of new zoonotic diseases. Live animal markets, also known as wet markets, have always been the principal mode of commercialization of poultry and many other animal species. Such markets, quite uncommon in the United States and, until recently, in California, are emerging as a new mode of commercialization within specific ethnic groups for whom this type of trade assures freshness of the product but raises major public health concerns.

However, recent data suggest that civets may be only amplifiers of a natural cycle involving trade and consumption of bats (Chomel et al., 2007).

Persistence of rabies through transplantation

This is another factor that has persisted for so long. On July 1, 2004, CDC reported laboratory confirmation of rabies as the cause of encephalitis in an organ donor and three organ recipients at Baylor University Medical Center (BUMC) in Dallas, Texas. Hospital and public health officials in Alabama, Arkansas, Oklahoma, and Texas initiated public health investigations to identify donor and recipient contacts, assess exposure risks, and provide rabies postexposure prophylaxis (PEP) (CDC, 2004). As of July 9, 2004 PEP had been initiated in approximately 174 (19%) of 916 persons who had been assessed for exposures to the organ recipients or the donor. As a result of its public health investigation, the Arkansas Department of Health determined that the donor had reported being bitten by a bat (CDC, 2004). Srinivasan et al. (2005) reports documented the transmission of rabies virus from an organ donor to multiple recipients and this underscores the challenges of preventing and detecting transmission of unusual pathogens through transplantation (Srinivasan et al., 2005). In Germany, a recent case of rabies in a person who had visited India remained unidentified until after the patient's death; soft tissue transplantation from this patient resulted in rabies transmission to several organ recipients (Nadin-Davis et al., 2007).

Persistence of rabies in areas with extremely high density of settlements

In Germany, the local increase in the number of rabies cases and the resulting spread of rabies in recent years has been mainly due to the persistence of rabies in areas with an extremely high density of settlements in which ORV is severely hindered. This is a phenomenon that no other country in Europe has been confronted with (Müller et al., 2005). The local increase in the number of rabies cases and the resulting spread of rabies in Germany in recent years are mainly due to increased fox densities (Müller et al., 2005). Rabies reemerges periodically in China because of high dog population density (Wu et al., 2009).

Vaccines and vaccination factors

Vaccines are the basis of the medical and veterinary medical future, the belief being that, if a vaccine can be made to every disease, then all disease can be prevented. This presupposes that 1) disease attacks from outside and has nothing whatsoever to do with the person or animal themselves; 2) that vaccines actually always protect one hundred per cent; and 3) that vaccines themselves are only beneficial and cannot cause harm. None of these are true. There is a growing list of research into and information on the problems that can be caused by vaccines. There has been much debate on the subject of annual pet vaccination, chiefly in response to concerns voiced by pet owners. The veterinary profession is largely unaware of the range of side-effects vaccines can stimulate, and consequently they go unreported. A radical rethink of the vaccination programme is necessary - immunization programmes need not be abandoned, but reassessed.

Human attitudes and errors toward vaccination

In rural areas of Nigeria, nearly every family owns several dogs, most of which are free-roaming, without special diets, and unvaccinated against rabies (to save costs). Sale of these dogs to restaurants can increase a farmer's income (average 12–15 US dollars/dog). Free-roaming and unvaccinated dog populations may increase the likelihood of transprovincial spread of RABVs.

Poor attitude of vaccination record keeping

Quick, decisive action to detect and control novel pathogens, and thereby contain outbreaks and prevent further transmission, is frequently hampered by incomplete or inadequate data about a new or re-emerging pathogen (Tapper, 2006). There is a poor attitude of record keeping among Nigerians as it was practically impossible to obtain the vaccination records. Vaccinees should be educated in all immunization programs to keep their vaccination cards for future reference. Virology Laboratories also should be equipped with adequate test facilities to monitor post vaccination seroconversion among subjects (Ogunjimi, 2008).

Low vaccination coverage (by increasing nonimmune population), including factors contributing to low coverage

The single most important factor affecting eradication of rabies is the failure to immunize domestic dogs, which transmit rabies to humans (Fu, 2008). Descriptive epidemiologic analysis has also shown that low vaccination coverage has contributed to rabies epidemics (Wang, 2006). Mass vaccination of the domestic dog provides the most cost-effective and efficient strategy for controlling canine rabies and hence transmission from dogs to humans (Clifton, 2007). In Zimbabwe, dog rabies cases increased after 1990, after declining vaccination coverage associated with decreased resources and diversion of resources (Cohen et al., 2007). Low vaccination coverage in domestic dogs in Limpopo, South Africa over several years may have led to an accumulation of susceptible animals, which led to the reestablishment of transmission (Cohen et al., 2007). Rabies also reemerges periodically in China because of low vaccination coverage in dogs. Mass vaccination campaigns rather than depopulation of dogs should be a long-term goal for rabies control (Wu et al., 2009).

Lack of ideal cross-reactivity with modern biological

Significantly, less than ideal cross-reactivity with modern biological used for veterinary and public health interventions is a major cause for concern among these emerging viral agents [Duvenhage and Lagos bat viruses] (Nel and Rupprecht, 2007).

High cost of postexposure prophylaxis

Although rabies is preventable, the high cost of postexposure prophylaxis, compounded by the lack of education and awareness about rabies, limits use of postexposure prophylaxis in many developing countries (Nadin-Davis et al., 2007).

Lack of vaccine delivery systems and resources

Outbreak of rabies in humans followed an outbreak in domestic dogs (Cohen et al., 2007). Increasing numbers of human rabies cases in Africa have also been attributed to deteriorating infrastructure and resources for rabies control. Although, the tools for effective rabies control are available, lacking are the delivery systems, public education campaigns and resources to apply these technologies in the developing world (Clifton, 2007). To attempt control, and possibly elimination, of zoonoses, benefits to public health and society need to be demonstrated, particularly in countries with scarce resources (Zinsstag et al., 2007). Zinsstag et al. (2007) in their study presented examples from their work on brucellosis and rabies and demonstrated the circumstances for which zoonosis control would save money for resource-limited countries and likely reduce the occurrence of zoonoses worldwide.

Challenge in testing of vaccine efficacy

It has been suggested that handling stress could have compromised wild dogs' cell-mediated immune response to rabies infection - does vaccination induce a cell-mediated immune response? Cell-mediated immunity can be assayed in the laboratory from blood samples. The ultimate test of vaccine efficacy is challenge with a dose and strain of rabies virus known to be lethal to unvaccinated animals (Ginsberg and Woodroffe, 1997). However, establishing the necessary challenge conditions, followed by carrying out the challenge experiments themselves, would necessitate killing at least 20 - 30 captive wild dogs. The consensus of vets and biologists involved in research on rabies in wild dogs and other carnivores is that challenges would be both unnecessary and unethical - for this reason, applications for government licenses to carry out such experiments would probably be unobtainable (Ginsberg and Woodroffe, 1997). Nevertheless, the experiments suggested above would answer most of the questions that have been raised concerning the efficacy of inactivated rabies vaccines, without the need for carrying out challenge experiments (Ginsberg and Woodroffe, 1997).

Existence of multiple hosts

Sixty-six (66) viruses have been isolated from bats (Calisher et al., 2006). Bats (order Chiroptera, suborders Megachiroptera ["flying foxes"] and Microchiroptera) are abundant, diverse, and geographically widespread. These mammals provide us with resources, but their importance is minimized and many of their populations and species are at risk, even threatened or endangered. Some of their characteristics (food choices, colonial or solitary nature, population structure, ability to fly, seasonal migration and daily movement patterns, torpor and hibernation, life span, roosting behaviours, ability to echolocate, virus susceptibility) make them exquisitely suitable hosts of viruses and other disease agents (Calisher et al., 2006). Bats of certain species are well recognized as being capable of transmitting rabies virus, but recent observations of outbreaks and epidemics of newly recognized human and livestock diseases caused by viruses transmitted by various megachiropteran and microchiropteran bats have drawn attention anew to these remarkable mammals. It is clear that we do not know enough about bat biology; we are doing too little in terms of bat conservation; and there remain a multitude of questions regarding the role of bats in disease emergence (Calisher et al., 2006).

Existence of reservoir host

Most zoonoses are maintained in the animal reservoir but can cross over to humans as a result of different risk factors and behavioural traits. Hence, elimination of zoonoses such as rabies is possible only by interventions that vigorously target animal reservoirs (Zinsstag et al., 2007). Control of most zoonoses usually requires interventions outside the public health sector (Zinsstag et al., 2007). When one considers health from a point of view independent of species, including humans, domestic animals, and wildlife, zoonoses are part of a broader ecologic concept of health systems (Zinsstag et al., 2007). Many new, emerging and re-emerging diseases of humans are caused by pathogens which originate from animals or products of animal origin. A wide variety of animal species, both domestic and wild, act as reservoirs for these pathogens, which may be viruses, bacteria or parasites (Meslin et al., 2000). According to Meslin et al. (2000), given the extensive distribution of the animal species affected, the effective surveillance, prevention and control of zoonotic diseases pose a significant

challenge. Bats are listed as protected species across Europe (Stantic-Pavlinic, 2005). The role that bats have played in the emergence of several new infectious diseases has been under review. Bats have been identified as the reservoir hosts of newly emergent viruses such as Nipah virus, Hendra virus, and severe acute respiratory syndrome-like coronaviruses as well as rabies (Halpin et al., 2007).

Infection and disease in reservoir and spillover hosts determine patterns of infectious agent availability and opportunities for infection, which then govern the process of transmission between susceptible species (Daniels et al., 2007). In the United States, extensive reservoirs of the rabies virus exist in many diverse wild animal species, which continue to pose a serious risk of lethal infection of humans and cause an economic burden exceeding $1 billion annually (Dietzschold and Schnell, 2002). Previous experience with rabies control in foxes in Europe has clearly demonstrated that oral immunization with live vaccines is the only practical approach to eradicate rabies in free-ranging animals. However, unlike Europe where vulpine rabies was the only major reservoir, the Americas harbor a variety of species including raccoons, skunks, coyotes, and bats that serve as the primary reservoirs of rabies. Each of these animal reservoirs carries an antigenically distinct virus variant (Dietzschold and Schnell, 2002).

Existence of healthy carriers

Other authors have suggested that this trend may be caused by a carrier state in healthy dogs that remains undetected (Zhang et al., 2008). Tao et al. (2009) observed an infection rate of 2.3% in apparently healthy dogs from 15 cities in 3 provinces of China. Previous surveys in regions of high incidence of rabies showed different rates, ranging from 3.9 - 17.9% in dogs. The paper by Ajayi et al., (2006) also indicates a disturbing possibility of transmission of rabies by apparently healthy (free of overt rabies signs) stray dogs. If their observations are confirmed, this, in their words, "signifies a new dimension in the epidemiology of the disease in this environment where the high-risk practices are prevalent (Durosinloun, 2009; Fagbo, 2009; Woolf, 2009)." What's more intriguing epidemiologically and culturally is that the research by Ajayi et al., (2006) was carried out in Maiduguri; the overwhelming Muslim population in the city provides zero economic incentives for dog meat restaurants (Durosinloun, 2009; Fagbo, 2009; Woolf, 2009). However, the dogs were slaughtered in restaurants associated with 2 military barracks in the city (Durosinloun, 2009; Fagbo, 2009; Woolf, 2009).

Long incubation period

During most of the long incubation period of rabies, the virus likely remains close the site of viral entry.

Centripetal spread to the central nervous system and spread within the central nervous system occur by fast axonal transport. Neuronal dysfunction, rather than neuronal death, is responsible for the clinical features and fatal outcome in natural rabies (Jackson, 2003).

Existence of new and other emerging diseases

Polley (2005) discussed the linkages between wildlife, people, zoonotic parasites and the ecosystems in which they co-exist, and revisited definitions for 'emerging' and 're-emerging', and lists zoonotic parasites that can be acquired from wildlife including, for some, estimates of the associated global human health burdens and introduced the concepts of 'parasite webs' and 'parasite flow', provides a context for parasites, relative to other infectious agents, as causes of emerging human disease, and pointing out the drivers of disease emergence and re-emergence, especially changes in biodiversity and climate. *Angiostrongylus cantonensis* in the Caribbean and the southern United States, *Baylisascaris procyonis* in California and Georgia, *Plasmodium knowlesi* in Sarawak, Malaysia, *Human African Trypanosomiasis*, *Sarcoptes scabiei* in carnivores, and Cryptosporidium, Giardia and Toxoplasma in marine ecosystems are presented as examples of wildlife-derived zoonotic parasites of particular recent interest.

New emerging infectious diseases include "severe acute respiratory distress syndrome" (SARS) and avian influenza A H5N1 (Spicuzza et al., 2007). First cases of SARS, induced by a new strain of coronavirus, were described in China in 2002 and by May 2003 8360 cases and 764 deaths were reported by the WHO (Spicuzza et al., 2007). Avian influenza A H5N1 is another emerging infectious disease transmitted from avian species to humans, without clear evidence of transmission from human to human. The widespread outbreaks of H5N1 avian influenza in 2003-2004 have caused major problems for the poultry industry in many Asian countries. On January 2004 the disease crossed over to humans, for the first time in Vietnam, causing 74 deaths to date (mortality rate of 50%) in southeastern countries (Spicuzza et al., 2007). Unlike SARS, the avian flu occurs in rural areas, where people live in intimate contact with birds, and many of the victims are children < 5 years of age. The World Health Organization has adopted a global action plan to control avian influenza among chickens and ducks and at the same time to limit the threat of a human flu pandemic (Spicuzza et al., 2007).

Inconsistent vaccination and wrong vaccination regime

The local increase in the number of rabies cases and the resulting spread of rabies in recent years has also been attributed to inconsistent vaccination, e.g. missing complementary distribution of baits per hand in non-flying zones (Müller et al., 2005).

Inferior vaccine quality

The inferior quality of the domestically manufactured dog vaccine in China has been documented (Hu et al., 2008). According to Wu et al. (2009), vaccine quality control and mass production, rather than matching, are urgently needed and most important for addressing the current rabies problem in China. Any potent rabies vaccine will protect against rabies.

Vaccine shortage

Vaccine shortages can result from higher-than-expected demand, interruptions in production/supply, or a lack of resources to purchase vaccines (Hinman et al., 2006). Each of these factors has played a role in vaccine shortages especially in the United States during the past 20 years. Since 2000, the United States has experienced an unprecedented series of shortages of vaccines recommended for widespread use against 9 diseases, after more than 15 years without vaccine supply problems. In developing countries, the major cause of vaccine shortages is lack of resources to purchase them. Although there are several steps that could reduce the likelihood of future vaccine shortages, many would take several years to implement. Consequently, we will probably continue to see occasional shortages of vaccines in the United States in the next few years (Hinman et al., 2006).

Vaccine failure

Okoh (1982) reported 10 cases of apparent vaccine failure in Nigeria involving modified live (low egg passage chick embryo) vaccine in use during the study period. In their study, 4 of these cases of infection may actually have been induced by the vaccine. In 1989, Six cases of apparent vaccination failures in rural dogs given modified live virus, chicken embryo origin, low egg passage, Flury-type vaccine, was reported (Okolo,1989). Wiktor et al. (1984) reported that immunization of mice with a rabies vaccine (antigenic value, 10 international units) at a concentration 30-fold high than that necessary for complete protection against homologous challenge with rabies virus was not protective against Mokola infection and that no cross-reactivity between Mokola and rabies viruses was seen with cytotoxic T lymphocytes.

Unavailability of WHO recommended RIG

After severe exposure to suspected rabid animal, WHO recommends a complete vaccine series using a potent effective vaccine that meets WHO criteria, and administration of rabies immunoglobulin (RIG). RIG is not available globally (Yanagisawa et al., 2008), and is not marketed in Nigeria.

Use of low potency vaccines

When post-exposure prophylaxis are most health sector do not take time to know the sort of rabies vaccine being injected. It is important to know the sort of rabies vaccine injected abroad, because brain-tissue vaccines are less effective in inducing antibody than tissue-culture vaccines.

Safety problem

The currently available modified-live rabies virus vaccines have either safety problems or do not induce sufficient protective immunity in particular wildlife species. Therefore, there is a need for the development of new live rabies virus vaccines that are very safe and highly effective in particular wildlife species (Dietzschold and Schnell, 2002).

Failure to show an adequate antibody titre

In line with study carried out in other countries, Nigeria faces with problem of some vaccinated animals fail to show an antibody titre adequate to meet the requirements of the 0.5 IU/ml minimal threshold level accepted by WHO/OIE. In a study by Jakel et al. (2008) to identify specific risk factors in dogs and cats for post-vaccination rabies antibody titres below 0.5 IU/ml by FAVN test. Data on around 1,200 animals was analyzed. Most animals older than one year had already received more than one rabies vaccination. The influence of breed and sex on antibody titre seems to be insignificant. Young dogs have a high risk of results below 0.5 IU/ml after their first vaccination. This risk can be minimized by the application of a second vaccination and blood sampling according to the manufacturer's recommendations. An important factor for the test outcome might be the virus strain used in the vaccine (Jakel et al., 2008).

In another study in UK by Mansfield et al. (2004), after being vaccinated against rabies some cats and dogs fail to show an antibody titre adequate to meet the requirements of the UK Pet Travel Scheme. To investigate this problem, Mansfield et al. (2004) derived data from 16,073 serum samples submitted to the Veterinary Laboratories Agency for serological testing between 1999 and 2002, 1002 samples submitted to BioBest during March and April 2001, and 1264 samples associated with one make of vaccine submitted to BioBest between June, 2001 and January, 2003, were analyzed. The probability of antibody titre failing to reach at least 0.5 iu/ml was analyzed by logistic regression as a function of the choice of vaccine, the interval between vaccination and sampling, the sex and age of the animal, and its country of origin (Mansfield et al., 2004). In dogs, all these factors, except sex, had highly significant (P < 0.001) effects on the test failure rate, and in cats all the factors had a significant effect (P < 0.05).

Transboundary and transprovincial spread of RABVs

Transboundary animal diseases pose a serious risk to the world animal agriculture and food security and jeopardize international trade. The world has been facing devastating economic losses from major outbreaks of transboundary animal diseases (TADs) such as foot-and-mouth disease, classical swine fever, rinderpest, peste des petits ruminants (PPR), and Rift Valley fever. Lately the highly pathogenic avian influenza (HPAI) due to H5N1 virus has become an international crisis as all regions around the world can be considered at risk (Domenech et al., 2006). The geographic nature of these rabies foci or groups suggests that dogs are not moving, per se, but that human-related activities may account for these phenomena. Spread of RABVs from high-incidence regions, particularly by the long-distance migration or transprovincial movement of dogs caused by human-related activities, may be one of the causes of recent massive human rabies epidemics (Tao et al., 2009).

Limited number of related vaccine strains

Attenuated tissue culture-adapted and natural street rabies virus (RV) strains differ greatly in their neuroinvasiveness (Faber et al., 2004). A study by Tomori (1980) on wild caught shrews infected with rabies virus strains by the intramuscular, subcutaneous and oral routes suggested a mechanical role in the transmission of rabies. In his study, virus was isolated only from shrews infected with street or wild strains of rabies, but not with vaccine or fixed rabies strains. To identify the elements responsible for the ability of an RV to enter the CNS from a peripheral site and to cause lethal neurological disease, Faber et al. (2004) constructed a full-length cDNA clone of silver-haired bat-associated RV (SHBRV) strain 18 and exchanged the genes encoding RV proteins and genomic sequences of this highly neuroinvasive RV strain with those of a highly attenuated nonneuroinvasive RV vaccine strain (SN0). In their analysis of the recombinant RV (SB0), which was recovered from SHBRV-18 cDNA, they indicated that this RV is phenotypically indistinguishable from WT SHBRV-18. However, characterization of the chimeric viruses revealed that in addition to the RV glycoprotein, which plays a predominant role in the ability of an RV to invade the CNS from a peripheral site, viral elements such as the trailer sequence, the RV polymerase, and the pseudogene contribute to RV neuroinvasiveness (Faber et al., 2004). Faber et al. (2004) analyses also revealed that neuroinvasiveness of an RV correlates inversely with the time necessary for internalization of RV virions and with the capacity of the virus to grow in neuroblastoma cells.

Approximately 57 years ago, Johnson (1952 reviewed by Tao et al., 2009) speculated that RABV strains from Europe were transmitted into China through Hong Kong and Shanghai. The attenuated 3aG strain, which was isolated in Beijing in 1931, and the DRV strain, which was isolated in Jilin Province in 2002, is closely related to group III in a study by Tao et al. (2009). This finding by Tao et al. (2009) implies that a group of viruses that originated in Europe is present in China and is still circulating. The hosts of this group include not only domestic dogs but also other mammals likely infected by rabid dogs (DRV strain) as pointed out by Tao et al. (2009). Alternatively, the similarity among some RABVs circulating in dogs in China and the rest of the world as well as the international vaccine strains (Hu et al., 2008) should motivate health authorities in all over the world to revisit quality standards and adequacy for use of attenuated rabies vaccines to ensure that vaccine-related cases do not occur.

Role of rabies related viruses

According to Shope (1982), five viruses related to rabies occur in Africa. Two of these, Obodhiang from Sudan and kotonkan from Nigeria, were found in insects and are only distantly related to rabies virus. The other three are antigenically more closely related to rabies. Mokola virus was isolated from shrews in Nigeria, Lagos bat virus from fruit bats in Nigeria, and Duvenhage virus from brain of a man bitten by a bat in South Africa. The public health significance of the rabies-related viruses was emphasized in Zimbabwe where in 1981 a rabies-related virus became epizootic in the dog and cat population. It is postulated that the ancestral origin of rabies virus was Africa where the greatest antigenic diversity occurs and that the ancestor may have been an insect virus (Shope, 1982). Identification of the first case of LBV in a mongoose by Markotter et al. (2006a) underscores the need for surveillance of rabies-related viruses and the need for accurate identification of lyssavirus genotypes even if the host involved is normally only associated with RABV.

Spill over

With respect to LBV, Markotter et al. (2006a, b) have recently reported the likely persistence of this virus in pteropid bats in South Africa, which implicates continuous opportunity for spillover into terrestrial species. In determining the extent of risk to human and veterinary public health, it is important to establish the prevalence of LBV not only in bats but also in potential terrestrial animal vectors, to which mongoose species should be added, based on the finding of Markotter et al. (2006a).

Other vaccine related problem

Subsidized post-exposure vaccination is the standard response to rabies in India, China, and much of Africa. Post-exposure vaccination saves thousands of lives annually, despite many failures when dog bite victims fail to seek treatment soon enough, do not complete the full course of injections, or receive fake, expired, or obsoles-cent vaccines, a problem particularly prevalent in parts of India and China, where post-exposure vaccines are often made by local suppliers, using formulas elsewhere long abandoned (Clifton, 2007). In summary, however, the most likely cause of the vaccine failures lies in the vaccination protocols used. Each wild dog was given only a single dose of vaccine. However, administration of single doses of inactivated rabies vaccine to wild dogs held in captivity in Tanzania failed to bring about seroconversion, and preliminary vaccine trials in South Africa suggest that two doses must be given in order to achieve and maintain protective antibody levels. Further vaccine trials are urgently needed to determine the best protocol (Ginsberg and Woodroffe, 1997). Suzuki et al. (2008a, b) reported unsatisfactory titre level in their study in comparison with the results from other field investigations with inactivated tissue culture vaccines.

RABIES ERADICATION: HOW FEASIBLE?

Achievement of some or all of these interim milestones will increase support for global eradication of selected diseases (Duffy et al., 1990). Research to design strategies for disease control in wild dogs is also urgently needed. In particular: 1) can vaccines against rabies and canine distemper be delivered to wild dogs in a manner that is safe and effective? 2) Can these diseases be eradicated from their reservoir hosts, protecting wild dogs without vaccinating them directly? Rabies eradication is not feasible because of the extensive and animal and animal reservoirs of the virus and the inability to eliminate those reservoirs with existing technology. However, elimination of human rabies in urban areas may be possible through different strategies. Additional genetic work will help to set priorities for the conservation of populations which may be genetically unique (Ginsberg and Woodroffe, 1997) for spread of rabies and other related diseases.

Possibilities for control in reservoir hosts

In some circumstances, controlling disease in its reservoir hosts could be a better long-term solution than vaccinating wild dogs themselves. For example, rabies control in domestic dogs would protect people and their livestock as well as wild dogs. In other cases, however, it is not always clear that attempts to control disease in other species will provide effective protection for wild dogs (Ginsberg and Woodroffe, 1997). This highlights the need for more research, to address the following questions:

Does the administration of live canine distemper virus and rabies vaccines bring about seroconversion?

According to Ginsberg and Woodroffe (1997), one study,

of three litters of pups, found no evidence of seroconversion, while another found that adults given booster vaccinations did seroconvert. These results provide circumstantial evidence that, as suspected for rabies vaccination, more than one dose of vaccine might be needed to achieve and maintain protective antibody levels. In zoos that vaccinate wild dogs against CDV routinely, more studies could be carried out to assess the efficacy of different protocols. As for rabies, it would be useful to know whether multiple doses of vaccine are more effective than a single dose, whether dart-vaccination is as effective as vaccination by hand, and how often boosters must be given (Ginsberg and Woodroffe, 1997).

Can rabies be controlled in wildlife reservoirs?

Domestic dogs are important rabies reservoirs in East Africa, but in southern Africa wild species such as bat-eared foxes and jackals may be more important (Ginsberg and Woodroffe, 1997). Achieving anything approaching adequate vaccination cover in these species would be impossible if vaccines had to be delivered by hand, but oral vaccination is a possible alternative (Ginsberg and Woodroffe, 1997). This method of vaccine delivery has successfully eradicated rabies from red foxes in some parts of Europe and North America. However, although experimental administration of live oral vaccines to black-backed and side-stripe jackals has been shown to confer protection from rabies, the strain used proved highly pathogenic to baboons (Bingham et al. 1995 reviewed in Ginsberg and Woodroffe, 1997). Thus, more (ongoing) research, using other strains, is needed to perfect a method for vaccinating wild canids safely and effectively (Ginsberg and Woodroffe, 1997).

Can the population density of reservoir hosts be reduced?

In principal, reducing the density of reservoir hosts could lead to lower transmission rates and prevent disease from persisting in the population. The practical possibilities of doing this depend upon a number of factors. If the reservoir host was a wildlife species, controlling population size would rarely be possible. For domestic dogs, the possibilities would depend upon local peoples' requirement for those dogs (Ginsberg and Woodroffe, 1997).

Can contact between wild dogs and domestic dogs be minimized?

Again, according to Ginsberg and Woodroffe (1997), this would depend upon local peoples' need for domestic dogs. More research is needed to determine whether domestic dogs' movements could be restricted by, for example, requiring that owned dogs be collared, that

dogs be tied up at night, and shooting unaccompanied dogs.

Public health strategy of disease eradication

The public health strategy of disease eradication offers considerable advantages over disease control when eradication is undertaken against appropriate, carefully chosen targets (Duffy et al., 1990). The benefits of eradication are permanent and accrue after a finite cost, whereas the costs of controlling the same disease must be maintained indefinitely. For example, the United States invested $32 million in SEP over a 10-year period; this amount is equivalent to former U.S. costs and expenditures every 3 months for routine vaccination (discontinued in 1971) and management of its complications (Ginsberg and Woodroffe, 1997). The United States government is investing >$50 million annually to maintain its polio-free status and an estimated $25-50 million to keep domestic measles at low level (Hinman et al., 1985 reviewed by Ginsberg and Woodroffe, 1997). These figures do not reflet the cost of vaccination in the private sector or the annual occurrence of vaccine-associated polio (Duffy et al., 1990).

Mobilization for support

A time-limited goal of eradication allows mobilization of support more readily than a control program. An important corollary requirement for global eradication is that unaffected countries will need to provide material assistance where needed, including geographic areas where small residual foci might not otherwise warrant use of scarce national resources (Duffy et al., 1990).

Effective wild dog management

While a great deal of information about wild dog ecology has become available recently, further research will allow more effective wild dog management. Surveys are needed, especially in central Africa, to give a better picture of wild dog distribution. Simple, effective monitoring techniques are needed to track the status of known populations. Long term studies of larger populations should be continued; such studies will identify new threats as they arise, and will also determine wild dog populationsμ ability to recover from natural perturbations, a crucial component of their viability which has not yet been quantified in the field (Ginsberg and Woodroffe, 1997).

Research to design strategies for disease control

In several cases, Ginsberg and Woodroffe (1997)

reported that more research would enhance the creation and implementation of effective management strategies. A great deal of research has been carried out on wild dogs recently, so that wildlife managers are now much better equipped to conserve wild dogs than they were ten, or even five years ago. Nevertheless, there are still areas where more information would be extremely valuable. In their study (Ginsberg and Woodroffe, 1997), they summarized researches that would facilitate wild dog conservation. Research to help resolve conflicts between wild dogs and farmers is urgently needed, since persecution represents an extremely serious threat. This must involve work on: 1) the true economic losses caused by wild dog predation on livestock, 2) the circumstances under which wild dogs take livestock, and 3) the degree to which public attitudes reflect a real or perceived assessment of the damage caused. Such information will help to determine the combination of husbandry practices, local legislation, compensation and education needed to allow wild dogs and people to coexist (Ginsberg and Woodroffe, 1997). However, a substantial volume of research is also needed into disease control - it was not until the wild dog study populations disappeared from the Serengeti ecosystem that it became clear just how severe a threat disease could pose to wild dogs. We still cannot determine the best strategy for controlling disease - and at present we are not fully equipped to carry out any of them (Ginsberg and Woodroffe, 1997).

Mass vaccination of domestic dogs and destruction of stray dogs approach

Some developed countries have virtually eliminated rabies in humans by mass vaccination of domestic dogs and destruction of stray dogs (MMWR, 1993). This approach is difficult to apply in rural areas of most developing countries, where animals may not be privately owned, destruction may be unacceptable, and such campaigns may be expensive. Some Latin American countries are conducting successful campaigns in cities, however. Attempts are being made to control rabies in wildlife by development of oral vaccines that can be safely distributed in baits (MMWR, 1993).

New and improved vaccines

Eradication of rabies is not feasible, primarily because of the extensive, varied animal reservoirs of the virus and the inability to eliminate those reservoirs through available technology. It is possible to eliminate human rabies in urban areas, although the costs and benefits of doing so should be considered (MMWR, 1993). A review of the technical feasibility of eradicating other diseases preventable by vaccines currently licensed for civilian use in the United States indicates that measles, hepatitis B, mumps, rubella, and possibly disease caused by

Haemophilus influenzae type b are potential candidates (Hinman, 1999). From a practical point of view, measles seems most likely to be the next target. Global capacity to undertake eradication is limited, and care must be taken to ensure that a potential measles eradication effort does not impede achievement of polio eradication (Hinman, 1999). Even in the absence of eradication, major improvements in control are both feasible and necessary with existing vaccines. New and improved vaccines may give further possibilities of eradication in the future (Hinman, 1999).

CURRENT SITUATION AND FUTURE TRENDS

Rabies is probably the oldest recorded infection of mankind. Rabies is an enzootic viral disease widespread throughout the world (Sugiyama and Ito, 2007). Rabies, being a major zoonotic disease, significantly impacts global public health. It is invariably fatal once clinical signs are apparent. The majority of human rabies deaths occur in developing countries (Nagarajan et al., 2008). Although it is a vaccine-preventable disease, the annual number of human deaths caused by rabies is estimated to be 32,000 in Asia (Sugiyama and Ito, 2007). India alone reports more than 50% of the global rabies deaths (Nagarajan et al., 2008). The development of the first rabies vaccine by Pasteur surely had been hoped to eliminate or at least drastically reduce its incidence. To date, the only survivors of the disease have received rabies vaccine before the onset of illness. Many years of research and observations on bait uptake, efficacy, behavioural studies of foxes, reduction and elimination of rabies, population dynamics of foxes following oral vaccination, as well as annual exchanges of researchers' experiences, formed the scientific background for field trials extending over many years. The approach to management of the rabies normally should be palliative (Jackson et al., 2003).

It is a vaccine-preventable disease. Cheap and safe vaccines for animals as well as humans have been developed. No single therapeutic agent is likely to be effective, but a combination of specific therapies could be considered, including rabies vaccine, rabies immunoglobulin (RIG), monoclonal antibodies, ribavirin, interferon-alpha, and ketamine. Corticosteroids were discouraged from use (Jackson et al., 2003). As research advances, new agents may become available in the future for the treatment of human rabies. However, effective rabies prevention in humans with category III bites requires the combined administration of RIG and vaccine (Nagarajan et al., 2008). Vaccination of stray dogs could lead to the eradication of rabies in countries where dog rabies is the sole source of human exposure. The development of safe and effective rabies virus vaccines applied in attractive baits resulted in the first field trials in Switzerland in 1978. Thereafter, technical improvements occurred in vaccine quality and production, including the design of recombinant viruses, as well as in the ease of mass distribution of millions of edible baits over large geographical areas (Rupprecht et al., 2004). Oral vaccination of wildlife with recombinant rabies virus vaccines is beginning to reduce the incidence of rabies among foxes and raccoons. Over the past few decades, extensive oral vaccination programmes focusing upon the red fox, using hand and aerial distribution of vaccine-laden baits, have resulted in the virtual disappearance of rabies in Western Europe. The same dramatic observation held true for southern Ontario (Rupprecht et al., 2004).

The European fox rabies epizootic starting in 1939 at the eastern border of Poland reached Switzerland on March 3, 1967. Rabies spread over large parts of the country until 1977, the year it caused three human deaths (Zanoni et al., 2000). In 1978 the first field trial world-wide for the oral immunization of foxes against rabies was conducted in Switzerland. Initially, the expansion of the vaccination area led to a rapid reduction in rabies cases. However, the 1990s were characterized by a recurdescence of rabies in spite of regular oral immunization of foxes (Zanoni et al., 2000). The last endemic case of rabies was diagnosed in 1996 after an adaptation of the vaccination strategy. During the 1990s in the United States, oral vaccination programmes concentrated upon raccoons, grey foxes, and coyotes, with similar success. For example, raccoon rabies has not spread west of the current focus in the eastern states, grey fox rabies is contained in west central Texas, and no recent cases of rabies have been reported from coyotes away from the Mexican border for several years. Despite the progress observed and the absence of substantive adverse environmental or health effects, oral vaccination is not a panacea, and should be viewed as an important adjunct to traditional prevention and control techniques in human and veterinary medicine (Rupprecht et al., 2004).

However, this goal of eradication has not been achieved because rabies is maintained in many animal reservoirs, including both domestic and wild. There are still many aspects of the pathogenicity of rabies that are unknown. For example, we have no explanation for the long incubation period (up to 6 years). Furthermore, new patterns of rabies infection present a problem for epidemiologists and virologists alike. There are several cases of human rabies in which there was no history of a bite. Despite these continuing problems, there has been tremendous progress in the control of rabies. Local outbreak suppression of rabies among free-ranging wildlife is documented, and regional elimination of particular virus variants among specific, targeted carnivore hosts is demonstrable, but true disease eradication is not achievable at the present time by current techniques (Rupprecht et al., 2004). For example, no practical vaccination methods have been designed for bats. Although lyssaviruses appear in relative compartmenta-

lization between the Chiroptera and Carnivora, major spillover events have been detected from bats to carnivores, and phylogenetic analyses suggest a historical basis for extant viral origins due to interactions between these taxa. Thus, bio-political considerations aside, the possibility for pathogen emergence resulting from transmission by rabid bats with subsequent perpetuation among other animals cannot be discounted easily on any continent, with the possible exception of Antarctica (Rupprecht et al., 2004).

Cell culture rabies vaccines for human use, highly immunogenic and well tolerated, are now used for pre-exposure immunization as well as for post-exposure treatment (Sureau, 1988). Presently available cell culture rabies vaccines induce immunity against the SAD modified live rabies virus used for oral immunization of foxes. They also induce immunity against the newly identified European bat rabies virus (Duvenhage) (Sureau, 1988). The use of the techniques and strategies of oral immunisation of foxes against rabies using SAD B19 can eliminate wildlife rabies among foxes and raccoon dogs, as European experience has shown. The disease then also disappears completely in domestic animals and man. However, the currently available modified-live rabies virus vaccines have either safety problems or do not induce sufficient protective immunity in particular wildlife species. Therefore, there is a need for the development of new live rabies virus vaccines that are very safe and highly effective in particular wildlife species (Dietzschold and Schnell, 2002). Meanwhile, new types of vaccines are being developed by applying gene manipulation techniques to rabies virus in order to overcome the disadvantages of current vaccines (Sugiyama and Ito, 2007).

There have been many changes and accomplishments in rabies control and prevention since the first International conference "Rabies in Europe" was held in Kiev on June 15-18, 2005 (Briggs, 2008). Recommendations from the 2005 meeting addressed epidemiology; rabies diagnosis; animal rabies control; human rabies prevention; vaccinology and immunology and bat rabies. Cell culture rabies vaccines have become widely available in developing countries, virtually replacing the inferior and unsafe nerve tissue vaccines. Limitations inherent to the conventional RIG of either equine or human origin have prompted scientists to look for monoclonal antibody-based human RIG as an alternative. Fully human monoclonal antibodies have been found to be safer and equally efficacious than conventional RIG when tested in mice and hamsters (Nagarajan et al., 2008). Clearly, given their biodiversity, distribution, and abundance, novel methods would be necessary to consider meaningful control of rabies in these unique volant mammals (Rupprecht et al., 2004). As suggested by Kuiken et al. (2005), it is time to form "a joint expert working group to design and implement a global animal surveillance system for zoonotic pathogens that gives early warning of pathogen emergence, is closely integrated to public health surveillance and provides opportunities to control such pathogens before they can affect human health, food supply, economics or biodiversity (Chomel et al., 2007)."

Appropriate use of a highly effective vaccine can help eradicate a major disease when humans are the only natural host for the virus, and there is no natural reservoir or intermediate host. During the 1970s, this goal of eradication was achieved for smallpox largely through the use of the live vaccinia virus vaccine and ranks as one of the major public health achievements in all history. Two other viruses have been similarly targeted for eradication, namely poliovirus and measles, and significant progress toward this goal has been made for both viral pathogens. Live virus vaccines have played and continue to play a central role in these current eradication efforts (Graham and Crowe, 2007). This goal seems more elusive now than ever before because of setbacks from social instability and the threat of bioterrorism, as well as the potential for intentional reintroduction or resurrection of previously eradicated virus pathogens. Therefore, it will be prudent to find ways of maintaining immunity to serious virus pathogens even after eradication of the natural reservoir has been achieved. The new technologies that have improved our ability to identify emerging pathogens and to develop biological for vaccines and therapeutics have also created an intellectual reservoir that is a formidable barrier to true eradication efforts. Elimination of a virus pathogen suggests that epidemic and endemic disease is controlled and that no active cases are present. However, setting the goal of elimination acknowledges the possibility of re-emergence and would include maintenance of active vaccine-induced immunity (Graham and Crowe, 2007).

As a result of the efforts in developing recombinant rhabdoviruses as vaccine vectors and as cytolytic agents, it is likely that clinical trials of genetically engineered rhabdoviruses in humans will take place in the near future. A number of issues need to be considered in the use of such agents, such as their safety for use in humans, as well as the protection of animal populations that may be exposed to such viruses. Nonetheless, the advances in understanding virus replication and pathogenesis should make it feasible to address these issues, so that these viruses that have long been a burden to humanity can instead be a benefit (Lyles and Rupprecht, 2007). Newer approaches in biotechnology may be envisaged some day for eventual extension to bats, as well as more widespread application to global canine rabies remediation in developing countries.

Conclusion

A number of countries throughout the world have been free from rabies for many years; some are reported to have eliminated the disease and in others it has reappeared

after variable periods of time. Apart from these minor variations, however, the global distribution of rabies over the last five to ten years appears unchanged and the disease continues to pose both public health and economic problems of varying severity in all continents except Australia and Antarctica. Most of the data presented here are abstracted from the most recent surveys and study by many researchers/authors. According to Turner (1976), the vagaries of international reporting, accurate in some cases and undoubtedly imprecise in others, demand a cautious interpretation of the available data and suggest that the world picture be viewed as an impression rather than as a precise record.

Some challenges and potential solutions to the ability of national governments to adhere to the global health surveillance requirements detailed in the International Health Regulations (IHR) and some practical challenges such as inadequate surveillance and reporting infrastructure and legal enforcement and maintenance of individual human rights has been reviewed by Sturtevant et al. (2007). Among these causal factors affecting rabies eradication are the burgeoning human population, the increased frequency and speed of local and international travel, the increase in human-assisted movement of animals and animal products, changing agricultural practices that favour the transfer of pathogens between wild and domestic animals, and a range of environmental changes that alter the distribution of wild hosts and vectors and thus facilitate the transmission of infectious agents (Bengis et al., 2004). Animals, particularly wild animals, are thought to be the source of >70% of all emerging infections (Kuiken et al., 2005). Leading factors for emergence of zoonoses are unbalanced and selective forest exploitation and aggressive agricultural development associated with an exponential increase in the bushmeat trade (Wolfe et al., 2005a; Chomel et al., 2007). Similarly, the increase of ecotourism, often in primitive settings with limited hygiene, can be associated with the acquisition of zoonotic agents. Therefore, development of appropriate programs for surveillance and for monitoring emerging diseases in their wildlife reservoirs is essential (Chomel et al., 2007).

According to Chomel et al. (2007), major tasks that should be taken by the international community include better integration and coordination of national surveillance systems in industrialized and developing countries; improved reporting systems and international sharing of information; active surveillance at the interface of rural populations and wildlife habitats, especially where poverty and low income increase risks for pathogen transmission; training of professionals, such as animal scientists, biologists, parasitologists, veterinarians, virologists, zoologists, in wildlife health management; and establishment of collaborative multidisciplinary teams ready to intervene when outbreaks occur. Although debate on the possible negative effect of mass vaccination campaigns on routine health services has gone on for decades, Wiysonge et al. (2006) reports points to an overall positive effect. High-quality mass campaigns usually achieve high vaccination coverage because of high-level political commitment and adequate planning and monitoring of vaccination activities (Wiysonge et al., 2006). Any failure of vaccination and PEP should be investigated thoroughly and independently to trace potential errors in the protocol. A national vaccine adverse-event reporting system should be established to track suspected problems for safety and efficacy (Wu et al., 2009). Surveillance should be heightened to monitor efficacy of vaccines in current use in the country. Seroconversion testing after vaccination is not necessary in either humans or animals (Wu et al., 2009).

REFERENCES

Abazeed ME, Cinti S (2007). Rabies prophylaxis for pregnant women. Emerg. Infect. Dis.; 13:1966–1967.

Adeija A (2007). Vaccine-derived polio spreads in Nigeria. Science and Development Network. SciDev.Net, October 8, 2007. http;//www.scidev.net/News/ News/index.cfm?fuseaction-readNews&itemid=3958& language=1.

Agbeyegbe L (2007). Risk communication: The over-looked factor in the Nigeria polio immunization boycott crisis. Nig. Med. Pract. 51(3):40-44.

Ajayi BB, Baba SS (2006). Rabies in apparently healthy dogs: histological and immunohistochemical studies. The Nig. Postgraduate Med. J. 13(2): 128-134.

Arguin PM, Mandel E, Guzi T, Childs JE (2000). Survey of rabies preexposure and postexposure prophylaxis among missionary personnel stationed outside the United States. J. Travel Med. 7(1): 10-14.

Asselbergs M, (2007). Rabies awareness. Vet. Rec. 161(12):4322.

Awoyomi O, Adeyemi IG, Awoyomi FS, (2007). Socioeconomic Factors Associated With Non-Vaccination of Dogs against Rabies in Ibadan, Nigeria. Nig. Vet. J. 28(3): 59-63.

Bengis RG, Leighton FA, Fischer JR, Artois M, Mörner T, Tate CM (2004). The role of wildlife in emerging and re-emerging zoonoses. Rev. Sci. Tech. 23(2): 497-511.

Böhm M, White PC, Chambers J, Smith L, Hutchings MR (2007). Wild deer as a source of infection for livestock and humans in the UK. Vet. J. 174(2): 260-276.

Briggs DJ (2008). What have we achieved since Kiev? Looking forward. Dev. Bio. (Basel). 131: 517-521.

Cabello CC, Cabello CF (2008). Zoonoses with wildlife reservoirs: a threat to public health and the economy. [Article in Spanish] Rev. Med. Chil. 136(3): 385-393.

Calisher CH, Childs JE, Field HE, Holmes KV, Schountz T (2006). Bats: important reservoir hosts of emerging viruses. Clin. Microbiol. Rev. 19(3): 531-545.

Castrodale L, Hanlon C (2008). Rabies in a puppy imported from India to the USA, March 2007. Zoonoses Public Health. 55(8-10):427-430.

Centers for Disease Control (CDC, 1992a). Update: International Task Force for Disease Eradication, 1990 and 1991. Morbidity and Mortality Weekly Reports 41: 40-42.

Centers for Disease Control (CDC, 1992b). Update: International Task Force for Disease Eradication, 1992. Morb. Mortal. Wkly. Rep. 41: 691, 697-698.

Centers for Disease Control (CDC, 2004). Draft Centers for Disease Control and Prevention's Immunization Safety Office Scientific Agenda: Draft Recommendations. [cited 2009 October 6] available at http://www.cdc.gov/vaccinesafety/00_pdf/draft_agenda_recommendations_080404.pdf.

Chang HG, Noonan-Toly C, Rudd R, Smith PF, Morse DL (2002). Public health impact of reemergence of rabies, New York. Emerg. Infect. Dis. 8(9): 909-913.

Chomel BB, Belotto A, Meslin FX (2007). Wildlife, exotic pets, an

emerging zoonoses. Emerg. Infec. Dis. 13(1):6-11.

Clifton M (2007). How to eradicate canine rabies in 10 years or less. Animal People Newspaper October 26.

Cohen C, Sartorius B, Sabeta C, Zulu G, Paweska J, Mogoswane M et al. (2007). Epidemiology and viral molecular characterization of reemerging rabies, South Africa. Emerging Infectious Diseases [cited 2009 October 6]. Available from http://www.cdc.gov/EID/content/13/12/1879.htm.

Dietzschold B, Schnell MJ (2002). New approaches to the development of live attenuated rabies vaccines. Hybrid. Hybridomics, 21(2): 129-134.

Daniels PW, Halpin K, Hyatt A, Middleton D (2007). Infection and disease in reservoir and spillover hosts: determinants of pathogen emergence. Curr. Top. Microbiol. Immunol. 315:113-131.

Domenech J, Lubroth J, Martin V (2006). Regional and international approaches on prevention and control of animal transboundary and emerging diseases. Ann N Y Acad. Sci. 1081: 90-107.

Duffy J, Long GW, deQuadors CA, Duke BOL, Henderson RH, Meheaus A, Hopkins DR0 (1990). "International Task Force for Disease Eradication". Morbidity and Mortality Weekly Report. 1990/FindArticles.com. 15 August 2009. Cited 2009 October 08. http://findarticles.com/p/articles/mi_m0906/is_n13_v39/ai_8373080/ .

Durosinloun AbdulKareem (2009). Federal ministry of Agriculture and Water Resources Department of Livestock; Kaduna Nigeria.

Eisinger D, Thulke HH (2008). Spatial pattern formation facilitates eradication of infectious diseases. J. Appl. Ecol. 45(2): 415–423.

Faber M, Pulmanausahakul R, Rice AB, Schnell MJ, Dietzschold B (2004). Identification of viral genomic elements responsible for rabies virus neuroinvasiveness. Proc. Natl. Acad. Sci. USA. 101(46): 16328-16332.

Fagbo S (2009). Updates on this situation in countries other than Asia. Dept. of Trop. Vet Dis. University of Pretoria South Africa.

Fenner L, Weber R, (2007). Imported infectious disease and purpose of travel, Switzerland. Emerg Infect Dis. Feb [Cited 2009 October 08]. Available from http://www.cdc.gov/EID/content/13/2/217.htm.

Fu ZF, (2008). The rabies situation in Far East Asia. Dev. Biol. (Basel). 131: 55-61.

Ginsberg JR, Woodroffe R (1997). African Wild Dog Status Survey and Action Plan (1997). The IUCN/SSC Canid Specialist Group's Research and Monitoring: Information for Wild Dog Conservation.

Graham BS, Crowe Jr JE (2007). Immunization Against Viral Diseases. In: Knipe DM and Howley PM (Editors). Fields Virology, 5th Edition. Philadelphia: Lippincott Williams and Wilkins, pp.488-536.

Halpin K, Hyatt AD, Plowright RK, Daszak P, Field HE, Wang L, Daniels PW, Henipavirus Ecology Research Group (2007). Emerging viruses: coming in on a wrinkled wing and a prayer. Clin. Infect. Dis. 44(5): 711-717.

Hayman DTS, Fooks AR, Horton D, Suu-Ire R, Breed AC, Cunningham AA (2008). Antibodies against Lagos bat virus in megachiroptera from West Africa. Emerg. Infect. Dis. [cited 2009 October 8]. Available from http://www.cdc.gov/EID/content/14/6/926.htm.

Heeney JL (2006). Zoonotic viral diseases and the frontier of early diagnosis, control and prevention. J. Intern. Med., 260(5): 399-408.

Hinman A (1999). Eradication of Vaccine-Preventable Diseases. Ann. Rev. Public Health. 20: 211-229

Hinman AR (2006). Vaccine Shortages: History, Impact, and Prospects for the Future. Ann. Rev. Public Health, 27: 235-259.

Hu RL, Fooks AR, Zhang SF, Liu Y, Zhang F (2008). Inferior rabies vaccine quality and low immunization coverage in dogs (Canis familiaris) in China. Epidemiol. Infect., 136: 1556–1563.

Jackson AC (2003). Rabies virus infection: an update. J. Neurovirol. 9(2): 253-258.

Jackson AC, Warrell MJ, Rupprecht CE, Ertl HC, Dietzschold B, O'Reilly M, Fu ZF, Wunner WH, Bleck TP (2003). Management of rabies in humans. Clin. Infect. Dis. 36(1): 60-63.

Jakel V, König M (2008). Factors influencing the antibody response to vaccination against rabies. Dev. Biol. (Basel), 131: 431-437

Jebara KB (2004). Surveillance, detection and response: Managing emerging diseases at national and international levels. Rev. Sci. Technol. 23(2): 709-715.

Jenkins PT, Genovese K, Ruffler H (2007). Broken screens: the regulation of live animal importation in the United States. Washington

DC: Defenders of Wildlife. [Cited 2009 October 08]. Available from http://www.defenders.org/resources/publications/programs_and_polic y/international_conservation/broken_screens/broken_screens_report. pdf.

Knobel DL, Cleaveland S, Coleman PG, Fevre EM, Meltzer MI, Miranda ME et al. (2005). Re-evaluating the burden of rabies in Africa and Asia. Bull. World Health Organ., 83: 360–368.

Kuiken T, Leighton FA, Fouchier RA, LeDuc JW, Peiris JS, Schudel A et al. (2005). Public health: pathogen surveillance in animals. Science 309: 1680–1681.

Leder K, Tong S, Weld L, Kain KC, Wilder-Smith A, von Sonnenburg F et al. For the GeoSentinel Surveillance Network (2006). Illness in travelers visiting friends and relatives: a review of the GeoSentinel Surveillance Network. Clin. Infect. Dis. 43: 1185–1193.

Lubroth J (2006). International cooperation and preparedness in responding to accidental or deliberate biological disasters: lessons and future directions. Rev. Sci. Technol. 25(1): 361-374.

Luyckx VA, Steenkamp V, Rubel JR, Stewart MJ (2004). Adverse effects associated with the use of South African traditional folk remedies. Cent. Afr. J. Med. 50: 46–51.

Lyles DS, Rupprecht CE (2007). Rhabdoviridae. In: Knipe DM and Howley PM (Editors). Fields Virology, 5th Edition. Lippincott Williams & Wilkins, pp.1364- 1408

Madhusudana SN, Tripathi KK (1990). Oral infectivity of street and fixed rabies virus strains in laboratory animals. Indian J. Exp. Biol. 28(5): 497-499.

Mallewa M, Fooks AR, Banda D, Chikungwa P, Mankhambo L, Molyneux E et al. (2007). Rabies encephalitis in malaria-endemic area, Malawi, Africa. Emerg. Infect. Dis. 13: 136–139.

Marano N, Arguin PM, Pappiaoanou M (2007). Impact of globalization and animal trade on infectious disease ecology. Emerg. Infect. Dis. [serial on the Internet]. 2007 Dec [cited 2009 September 28]. Available from http://www.cdc.gov/EID/content/13/12/1807.htm.

Markotter W, Randles J, Rupprecht CE, Sabeta CT, Wandeler AI, Taylor PJ et al. (2006a). Recent Lagos bat virus isolations from bats (suborder Megachiroptera) in South Africa. Emerg. Infect. Dis. 12: 504–506.

Markotter W, Kuzmin I, Rupprecht CE, Randles J, Sabeta CT, Wandeler AI et al. (2006b). Isolation of Lagos bat virus from water mongoose. Emerg. Infect. Dis. 12:1913–1918.

Merianos A (2007). Surveillance and response to disease emergence. Curr. Top. Microbiol. Immunol. 315: 477-509.

Meslin FX, Stöhr K, Heymann D (2000). Public health implications of emerging zoonoses. Rev. Sci. Technol. 19(1): 310-317.

Morbidity and Mortality Weekly Report (MMWR, 1983). Mokola virus: experimental infection and transmission studies with the shrew, a natural host. Morb. Mortal. Wkly Rep. 32(6): 78-80, 85-86.

Morimoto K, McGettigan JP, Foley HD, Hooper DC, Dietzschold B, Schnell MJ (2001). Genetic engineering of live rabies vaccines. Vaccine, 19(25-26): 3543-3551.

Müller T (2005). Fox rabies in Germany – an update. Euro Surveill. 2005; 10 (11): pii=581; cited 2009 September 28. Available online: http://www.eurosurveillance.org/ViewArticle.aspx?ArticleId=581.

Nadin-Davis SA, Turner G, Paul JPV, Madhusudana SN, Wandeler AI (2007). Emergence of Arctic-like rabies lineage in India. Emerg. Infect. Dis. [cited 2009 September 28]. Available from http://www.cdc.gov/ncidod/EID/13/1/111.htm.

Nagarajan T, Rupprecht CE, Rangarajan PN, Thiagarajan D, Srinivasan VA (2008). Human monoclonal antibody and vaccine approaches to prevent human rabies. Curr. Top. Microbiol. Immunol., 317:67-101.

Nel LH, Rupprecht CE (2007). Emergence of lyssaviruses in the Old World: the case of Africa. Curr. Top. Microbiol. Immunol. 315: 161-193.

Njoku Geoffrey (2006). Measles immunization campaign targets 29 million Nigerian children.

UNICEF (2008). Determination of Measles Haemagglutination Inhibiting Antibody Levels among Secondary School Students in Ibadan Nigeria. M.Sc. Project in the Department of Virology, Faculty of Basic Medical Sciences, College of medicine, University of Ibadan, Ibadan Nigeria. p.62.

Ogundipe GAT (1989). The development and efficiency of the animal health information system in Nigeria. Prevent. Vet. Med. 7: 121-135.

Okoh AE (1982). Canine rabies in Nigeria, 1970 - 1980 reported case in vaccinated dogs. Int. J. Zoonoses 9(2): 118-125.

Okolo MI (1989). Vaccine-induced rabies infection in rural dogs in Anambra State, Nigeria. Microbiology 57(231): 105-112.

Okonko IO, Udeze AO, Adedeji AO, Ejembi J, Onoja BA, Ogun AA, Garba KN (2009). Global Eradication of Measles: A Highly Contagious and Vaccine Preventable Disease-What Went Wrong In Africa? J. Cell Anim. Biol. 3(8): 119-140

Okonko IO, Fajobi EA, Ogunnusi TA, Ogunjobi AA, Obiogbolu CH (2008). Antimicrobial Chemotherapy and Sustainable Development: The Past, the Current Trend, and the future. Afr. J. Biomed. Res. 11(3): 235-250.

Olugbode Michael (2007). Nigeria: Measles Outbreak - Borno's Harvest of Death. This Day (Lagos) OPINION 21 June 2007, on the web 22 June 2007 by AllAfrica Global Media (http://www.allAfrica.com).

Opaleye OO, Adesiji YO, Olowe OA, Fagbami AH (2006). Rabies and antirabies immunization in South Western Nigeria: knowledge, attitude and practice. Trop. Doctor, 36(2): 116-117

Oucho JO (2006). Cross-border migration and regional initiatives in managing migration in southern Africa. In: Kok P, Gelderblom D, Oucho JO, van Zyl J, editors. Migration in South and southern Africa. Cape Town (South Africa): HSRC Press. pp. 47–70.

Perry BD, Wandeler AI. (1993). The delivery of oral rabies vaccines to dogs: an African perspective. Onderstepoort J. Vet. Res. 60(4):451-7.

Polley L. (2005). Navigating parasite webs and parasite flow: emerging and re-emerging parasitic zoonoses of wildlife origin. International J. Parasitol. 35(11-12): 1279-1294.

Roseveare CW, Goolsby WD, Foppa IM (2009). Potential and actual terrestrial rabies exposures in people and domestic animals, upstate South Carolina, 1994-2004: a surveillance study. BMC Public Health 9: 65.

Rupprecht CE, Hanlon CA, Slate D (2004). Oral vaccination of wildlife against rabies: opportunities and challenges in prevention and control. Dev. Biol. (Basel). 119: 173-184.

Rweyemamu M, Paskin R, Benkirane A, Martin V, Roeder P, Wojciechowski K (2000). Emerging diseases of Africa and the Middle East. Ann. NY Acad. Sci. 916: 61-70.

Schneider MC, Belotto A, Ade MP, Leanes LF, Correa E, Tamayo H et al. (2005). Epidemiologic situation of human rabies in Latin America in 2004. Epidemiol. Bull. 26:2–4.

Shaw MT, O'Brien B, Leggat PA (2009). Rabies postexposure management of travelers presenting to travel health clinics in Auckland and Hamilton, New Zealand. J Travel Med., 16(1):13-17. Comment in: J. Travel Med., 16(3): 227; author reply 227.

Shope RE (1982). Rabies-related viruses. Yale J. Biol. Med. 55(3-4): 271-275.

Shwiff SA, Sterner RT, Jay-Russell M, Parikh S, Bellomy A, Meltzer MI, et al. (2007). Direct and indirect costs of rabies exposure: a retrospective study in southern California (1998–2003). J. Wild. Dis. 43: 251–257.

Smith NH, Gordon SV, Rua-Domenech R, Clifton-Hadley RS, Hewinson RG (2006). Bottlenecks and broomsticks: the molecular evolution of Mycobacterium bovis. Nat. Rev. Microbiol. 4: 670–681.

Solomon T, Marston D, Mallewa M, Felton T, Shaw S, McElhinney L et al. (2005). Paralytic rabies after a two week holiday in India. BMJ. 331: 501–503.

Spicuzza L, Spicuzza A, La Rosa M, Polosa R, Di Maria G (2007). New and emerging infectious diseases. Allergy Asthma Proc. 28(1): 28-34.

Srinivasan A, Burton EC, Kuehnert MJ, Rupprecht C, Sutker WL, Ksiazek TG, Paddock CD, Guarner J, Shieh WJ, Goldsmith C, Hanlon CA, Zoretic J, Fischbach B, Niezgoda M, El-Feky WH, Orciari L, Sanchez EQ, Likos A, Klintmalm GB, Cardo D, LeDuc J, Chamberland ME, Jernigan DB, Zaki SR; Rabies in Transplant Recipients Investigation Team. (2005). Transmission of rabies virus from an organ donor to four transplant recipients. N. Engl. J. Med. 17;352(11): 1103-1111. N Engl J Med. 2005 352(24): 2551-2; author reply 2552.

Stantic-Pavlinic M (2005). Public health concerns in bat rabies across Europe. Euro. Surveill. 10(11): 217-220.

Sterner RT, Meltzer MI, Shwiff SA, Slate D (2009). Tactics and economics of wildlife oral rabies vaccination, Canada and the United States. Emerg. Infect. Dis. [cited 2009 September 28] Available from http://www.cdc.gov/EID/content/15/8/1176.htm.

Sturtevant JL, Anema A, Brownstein JS (2007). The new International Health Regulations: considerations for global public health surveillance. Disas. Med. Public Health Prep. 1(2): 117-121.

Sugiyama M, Ito N (2007). Control of rabies: epidemiology of rabies in Asia and development of new-generation vaccines for rabies. Comp. Immunol. Microbiol. Infect. Dis., 30(5-6): 273-286.

Sureau P (1988). New vaccines for immunization of man: new approaches towards the prevention of rabies in man. Parassitologia, 30(1): 141-148.

Suzuki K, González ET, Ascarrunz G, Loza A, Pérez M, Ruiz G, Rojas L, Mancilla K, Pereira JA, Guzman JA, Pecoraro MR (2008a). Antibody response to an anti-rabies vaccine in a dog population under field conditions in Bolivia. Zoonoses Public Health, 55(8-10): 414-420.

Suzuki K, Pecoraro MR, Loza A, Pérez M, Ruiz G, Ascarrunz G, Rojas L, Estevez AI, Guzman JA, Pereira JA, González ET (2008b). Antibody seroprevalences against rabies in dogs vaccinated under field conditions in Bolivia. Trop. Anim Health Prod. 40(8): 607-613.

Swanepoel R, Smit SB, Rollin PE, Formenty P, Leman PA, Kemp A et al. (2007). Studies of reservoir hosts for Marburg virus. Emerg. Infect. Dis. 13:1847–1851.

Tamashiro H, Matibag GC, Ditangco RA, Kanda K, Ohbayashi Y (2007). Revisiting rabies in Japan: is there cause for alarm? Travel Med. Infect. Dis. 5(5): 263-675.

Tao XY, Tang Q, Li H, Mo ZJ, Zhang H, Wang DM et al. (2009). Molecular epidemiology of rabies in southern China. Emerg Infect Dis., [cited 2009 September 28]. Available from http://www.cdc.gov/EID/content/15/8/1192.htm.

Tapper ML (2006). Emerging viral diseases and infectious disease risks. Haemophilia, 12 (Suppl. 1): 3-7; Discussion pp.26-28.

Tomori O (1980). Wild life rabies in Nigeria: experimental infection and transmission studies with the shrew (Crocidura sp.). Ann. Trop. Med. Parasitol., 74(2): 151-156.

Tumpey A (2007). The First World Rabies Day Symposium and Expo. Emerg Infect Dis. [cited 2009 October 08]. Available from http://www.cdc.gov/EID/content/13/12/07-1261.htm.

Turner GS (1976). A review of the world epidemiology of rabies. Trans. R. Soc. Trop. Med. Hyg. 70(3): 175-178.

Wang XC (2006). Rabies epidemiology research and analysis in Hunan [in Chinese]. Zhongguo Dongwu Jianyi. 23: 45–46.

Wiktor TJ, Macfarlan RI, Foggin CM, Koprowski H (1984). Antigenic analysis of rabies and Mokola virus from Zimbabwe using monoclonal antibodies. Dev. Biol. Stand. 57: 199-211.

Wolfe ND, Daszak P, Kilpatrick AM, Burke DS (2005a). Bushmeat hunting, deforestation, and prediction of zoonoses emergence. Emerg. Infect. Dis. 11: 1822–1827.

Wolfe ND, Heneine W, Carr JK, Garcia AD, Shanmugam V, Tamoufe U, et al. (2005b). Emergence of unique primate T-lymphotropic viruses among central African bushmeat hunters. Proc. Natl. Acad. Sci. USA. 102: 7994–7999.

Woolf W (2009). Rabies, via dog/cat butchering – Nigeria. A ProMED-mail post, a program of the International Society for Infectious Diseases. Sunday, March 29, 2009.

World Health Organization (1997). International Notes: A Case of Human Rabies Contracted in Nigeria. WHO Weekly Epidemiological Record, 72(22). Last Updated: 2002-11-08

World Tourism Organisation Facts and Figures. [Cited 2009 October 08]. Available from www.world-tourism.org/facts/menu.html

Wu X, Hu R, Zhang Y, Dong G, Rupprecht, CE. (2009). Reemerging rabies and lack of systemic surveillance in People's Republic of China. Emerg. Infect. Dis. [cited 2009 Sept. 28]. Available from http://www.cdc.gov/EID/content/15/8/1159.htm

Yanagisawa N, Takayama N, Suganuma A. (2008). [WHO recommended pre-exposure prophylaxis for rabies using Japanese rabies vaccine] [Article in Japanese]. Kansenshogaku Zasshi, 82(5): 441-444.

Zanoni RG, Kappeler A, Müller UM, Müller C, Wandeler AI, Breitenmoser U, (2000). Rabies-free status of Switzerland following 30 years of rabies in foxes [Article in German]. Schweiz Arch Tierheilkd, 142(8): 423-429.

Zhang YZ, Fu ZF, Wang DM, Zhou JZ, Wang ZX, Lu TF, et al. (2008). Investigation of the role of healthy dogs as potential carriers of rabies virus. Vector Borne Zoonotic Dis. 8: 313–319.

Zinsstag J, Schelling E, Roth F, Bonfoh B, de Savigny D, Tanner M, (2007). Human benefits of animal interventions for zoonosis control. Emerg. Infect. Dis., 13: 527–531.

10

Intragenotypic diversity in the VP4 encoding genes of rotavirus strains circulating in adolescent and adult cases of acute gastroenteritis in Pune, Western India: 1993 to 1996 and 2004 to 2007

Vaishali S. Tatte and D. Shobha D. Chitambar *

Enteric Viruses Group, National Institute of Virology, Pune, India.

Genetic variability of rotaviruses circulating in Pune, India at the two time points was determined by characterizing VP4 genes in 131 rotavirus strains detected in adolescent and adult cases of acute gastroenteritis. The multiplex RT-PCR classified the VP4 genes in 73 P[4] (43.2%), 69P[8] (40.8%) and 27 P[6] (16.0%) genotypes. Sequencing and phylogenetic analysis revealed increase in the prevalence of P[4]-5 and P[8]-2 and decline of P[8]-3 lineages in 2000s as compared to those identified in 1990s (92.8% Vs 100%, 4.2% Vs 33.3% and 93.7% Vs 66.7%, respectively). The P[4]-1 and P[8]-4 lineages circulated at low levels (7.1% / 2.1%) while presence of only P[6]-1 (100%) lineage was detected at both time. The strains with different VP4 genotype specificities displayed 0.2 to 2.3% amino acid divergence. A significant difference (P<0.01) in their association with common and nontypeable G strains was noted between the two time points studied. This is the first report to describe the intragenotypic diversity in the rotavirus VP4 genes from adolescent and adult patients of acute gastroenteritis from India. VP4 being one of the major protective antigens, monitoring the mutations in this protein would be crucial to understand the evolutionary changes in rotaviruses and devise more effective vaccine strategies in developing countries.

Key words: Group A rotavirus, VP4 genotypes, G and P typing, genetic diversity.

INTRODUCTION

Rotaviruses are well-recognized causative agents of severe gastroenteritis among infants and children, worldwide (Parashar et al., 2006). Taxonomically, these viruses belong to the family Reoviridae whereas in morphological analysis they appear as icosahedral particles consisting of 11 segments of double stranded (ds) RNA encased within a triple layered capsid composed of VP4, VP6, VP7 and VP2 proteins. The outermost layer has two proteins: the spike protein VP4 (P-protease sensitive) and the coat protein VP7 (G-glycoprotein) (Estes, 2001). Based on these proteins 27 G and 35 P types have been identified globally (Matthijnssens et al., 2008; Matthijnssens et al.,

2011).

More than sixty different G-P combinations have been found in the rotavirus strains circulating worldwide. Of these 12G and 15P genotypes have been detected in humans (Patel et al., 2011). The most commonly found G and P types are G1, G2, G3, G4, G9, P[4] and P[8], respectively (Santos and Hoshino, 2005; Matthijnssens et al., 2008). Of these most of the VP7 genotypes but only one VP4 genotype have been included in currently licensed rotavirus vaccines. It is expected that the implementation of the vaccination programs reduce significantly the hospitalizations and mortality related to rotavirus infection. However, existence of vast variations in the rotavirus strains and genotypes causing gastrointestinal infections is evident from the rotavirus surveillance studies conducted worldwide including India (Santos and Hoshino, 2005; Kang et al., 2009; Tatte et al., 2010a; Pietsch et al., 2011). Intragenotypic diversity

---
*Corresponding author. E-mail: drshobha.niv@gmail.com.

in VP7, NSP4 and VP6 encoding genes of such strains has been also reported from India (Arora et al., 2009; Tatte et al., 2010b). However, variations within VP4 genotypes circulating in the Indian population have been rarely reported (Samajdar et al., 2008).

A recent molecular epidemiology study carried out in adolescent and adults has reported an increase of mixed infections and increasing trend of genetic reassortment among rotavirus strains (Tatte et al., 2010a). Since adults frequently infected with rotavirus could act as reservoir of infection in early ages, more studies on the diversity within the rotavirus genotypes are necessary. The present study describes the assessment of temporal variations in the VP4 encoding genes of rotavirus strains recovered from adolescents and adults.

MATERIALS AND METHODS

Specimens

In earlier studies conducted at the two time points (1993 to 1996 and 2004 to 2007) in Pune, India for rotavirus surveillance in adolescent (9 to 18 years) and adult (>18 years) cases of acute gastroenteritis, a total of 131 fecal specimens were detected positive for group A rotavirus by ELISA (n = 118) (Tatte et al., 2010a) and RNA PAGE (n = 13). These consisted 26 (18/8) and 105(66/39) specimens respectively from adolescents and adults from the years, 1993 to 1996 (n = 84) and 2004 to 2007 (n = 47). Typing of rotavirus VP7 genes by multiplex PCR showed multiple G genotypes in the rotavirus strains. In the present study VP4 genes of these strains were scrutinized to classify them into lineages and sublineages. These also included VP4 genes from rotavirus strains (n = 31) detected in mixed infections reported earlier (Tatte et al., 2010a).

RNA extraction, RT-PCR and genotyping

Rotavirus ds RNA was extracted from 10% stool suspension prepared in phosphate buffered saline containing calcium chloride (PBS-CaCl$_2$) of pH 7.2, using Trizol® LS reagent (Invitrogen, Carlsbad, CA, USA), as per the manufacturer's protocol. The VP4 genes of rotavirus strains were subjected to RT-PCR followed by multiplex PCR for genotyping according to the method described earlier (Gouvea et al., 1990; Gentsch et al., 1992; Itturiza-Gomara et al., 2004) using a modified thermal cycling program (Chitambar et al., 2008). All of the PCR products, were analyzed by electrophoresis using 1X Tris Acetate EDTA (TAE) buffer, pH 8.3 on 2% agarose gels, containing ethidium bromide (0.5 ug/ml) and visualized under UV transilluminator.

Nucleotide sequencing and phylogenetic analysis

The multiplex PCR products obtained for different VP4 genotypes (P[8] – 345 bp, P[4] -483 bp and P[6] – 267 bp) were purified on minicolumns (QIAquick, Qiagen, Hilden, Germany) and sequencing was carried out using ABI-PRISM Big Dye Terminator Cycle Sequencing kit v3.1 (Applied Biosystems, Foster city, CA) and a ABI-PRISM 310 Genetic Analyzer (Applied Biosystems Foster city, CA). The sequences were aligned with the corresponding sequences of rotavirus strains available in the GenBank by using Clustal W (Thompson et al., 1994). The phylogenetic analysis was carried out in MEGA 4 by using Kimura –2 parameter and

neighbour-joining method (Tamura et al., 2007). The reliability of different phylogenetic groupings was confirmed by using bootstrap test (1000 bootstrap replications) available in MEGA 4.

Accession numbers

Nucleotide sequences of the VP4 genes of this study have been deposited in the GenBank sequence under the accession numbers HQ260459-HQ260577 and FJ6323188-FJ623237.

Statistical analysis

The proportions across two different periods were compared using the Chi-square test with Yates's correction. P-values less than 0.05 were considered statistically significant.

RESULTS

P-typing of rotavirus strains

In a multiplex RT-PCR performed for amplification of VP4 genes of 131 rotavirus strains, a total of 169 amplicons were obtained indicating 74 single and 95 mixed rotavirus infections. Based on the electrophoretic mobilities of the PCR products obtained in multiplex PCR, VP4 genes were classified into 73 P[4] (43.2%), 69 P[8] (40.8%) and 27 P[6] (16.0%) genotypes. All of these VP4 genotypes were found to be in combination with different G genotypes (Table 1).

Nucleotide sequencing

All of the multiplex PCR products were sequenced to confirm the presence of P[4] / P[8] / P[6] genotypes of VP4 in single or mixed infections. The sequences were compared with those available in GenBank.

Phylogenetic analysis of P[4] strains

Phylogenetic analysis of the 73 P[4] strains recovered from 10 adolescents and 63 adults at two time points 1993 to 1996 (n = 41) and 2004 to 2007 (n = 32) showed clustering of the strains in two distinct lineages P[4]-5 and P[4]-1 of the five lineages known to date for this genotype (Figure 1).

All of the 70 strains placed in P[4]-5 lineage showed 95.3 to 99.5% / 94.8 to 100% and 95.9 to 100% / 96.1 to 100% nucleotide / amino acid identities respectively with the reference strains-TB-Chen, RMC/G66, 107EIB, Py04ASR42, Py05ASR60, KO-2, VN594, SC-185 and TW569 and within themselves (Figure 1). These strains were predominantly associated with mixed infections (n = 24, 34.3%) followed by G2 strains (n = 21, 30.0%), G non-typeable strains (n = 18, 25.7%) and unusual G types [G1 (n = 3, 4.3%), G3 (n = 1, 1.4%), G4 (n = 1, 1.4%) and G9 (n = 2, 2.9%)]. All three strains classified

Table 1. Distribution of rotavirus VP4 genotype combinations in different G types identified in 1993 to 1996 and 2004 to 2007.

G-P Types (%)	P[4]		P[8]		P[6]	
	No. positive (%)					
	1993-1996	2004-2007	1993-1996	2004-2007	1993-1996	2004-2007
Common n = 47 (27.8)	n =17(40.5)	n = 4 (12.9.0)	n = 25(52.1)	n = 1(4.8)	n = 0(0.0)	n = 0(0.0)
G1 (38.3)	0(0.0)	0(0.0)	17(35.4)	1(4.8)	0(0.0)	0(0.0)
G2 (44.7)	17(40.5)	4(12.9)	0(0.0)	0(0)	0(0.0)	0(0.0)
G3 (10.6)	0(0.0)	0(0.0)	5(10.4)	0(0)	0(0.0)	0(0.0)
G4 (6.4)	0(0.0)	0(0.0)	3(6.3)	0(0)	0(0.0)	0(0.0)
Unusual n = 12(7.1)	n = 0 (0.0)	n = 7(22.6)	n = 1(2.1)	n = 0(0.0)	n = 4(16.7)	n = 0(0)
G1 (25.0)	0(0.0)	3(9.7)	0(0.0)	0(0.0)	0(0.0)	0(0.0)
G2 (8.3)	0(0.0)	0(0.0)	1(2.1)	0(0.0)	0(0.0)	0(0.0)
G3 (8.3)	0(0.0)	1(3.2)	0(0.0)	0(0.0)	0(0.0)	0(0.0)
G4 (8.3)	0(0.0)	1(3.2)	0(0.0)	0(0.0)	0(0.0)	0(0.0)
G9 (50.0)	0(0.0)	2(6.5)	0(0.0)	0(0.0)	4(16.7)	0(0.0)
Mixed n = 67(39.6)	n =21(50.0)	n = 6(19.4)	n =17(35.4)	n = 5(23.8)	n =17(70.8)	n =1(33.3)
G1 (31.3)	6(14.3)	1(3.2)	7(14.6)	1(4.8)	6(25.0)	0(0.0)
G1, G2 (13.4)	3(7.1)	1(3.2)	2(4.2)	1(4.8)	2(8.3)	0(0.0)
G1 G4 (1.5)	0(0.0)	0(0.0)	0(0.0)	0(0.0)	1(4.2)	0(0.0)
G2 (5.9)	1(2.4)	1(3.2)	0(0.0)	1(4.8)	1(4.2)	0(0.0)
G2, G4 (4.4)	1(2.4)	1(3.2)	0(0.0)	1 4.8)	0(0.0)	0(0.0)
G2, G3 (4.4)	1(2.4)	0(0.0)	1(2.1)	0(0.0)	1(4.2)	0(0.0)
G3 (4.4)	1(2.4)	0(0.0)	1(2.1)	0(0.0)	1(4.2)	0(0.0)
G4 (14.9)	4(9.5)	0(0.0)	4(8.3)	0(0.0)	2(8.3)	0(0.0)
G9 (12.0)	2(4.8)	2(6.4)	0(0.0)	1(4.8)	2(8.3)	1(33.3)
G1, G2, G4 (7.5)	2(4.8)	0(0.0)	2(4.2)	0(0.0)	1(4.2)	0(0.0)
Nontypeable n = 43 (25.4)	n = 4(9.5)	n = 14(45.1)	n = 5(10.4)	n = 15(71.4)	n = 3(12.5)	n =2(66.7)
GNT (100)	4(9.5)	14(45.1)	5(10.4)	15(71.4)	3(12.5)	2(66.7)
Total	42	31	48	21	24	3

in the lineage P[4]-1 were closer to each other and with the prototype strain, DS-1 in nucleotide (96.6 to 100%) and amino acid (97.7 to 100%) sequences.

Differences within all P[4] strains by 5.4 to 6.4% / 3.7 to 6.0% and 2.5 / 2.3% in their nucleotide / amino acid sequences from the prototype strain, DS-1 of lineage P[4]-1 were observed during the 1990s and 2000s. The alignment of the partial deduced amino acid sequences (11-153 aa) showed variations at positions 32(S→N), 89(N→D), 99(N→S), and 120(I→V) as compared to the prototype strain, DS-1. Multiple strain specific changes were also shared by these strains at positions 35(V→I), 49(N→D), 113(T→A), 130(V→I) and 133(N→S).

Phylogenetic analysis of P[8] strains

The nucleotide sequence analysis of the 69 P[8] strains recovered from adolescents (n = 14) and adults (n = 55)

at two time points, 1993 to 1996 (n = 48) and 2004 to 2007 (n = 21) showed clustering of the strains in three distinct evolutionary lineages P[8]-2, P[8]-3 and P[8]-4 of a total of four lineages described to date for this genotype (Figure 2).

All of the 59 strains classified in P[8]-3 lineage showed 97.5 to 100% / 96.1 to 100% and 96.0 to 100% / 94.7 to 100% nucleotide / amino acid identities respectively with the reference strains- Kagawa/90-544, Kagawa/90-513, Hun 9, OP351, 27B3, Dhaka 25-02, ISO22, PA 25/03, 90-551 and CAU and within themselves. All these strains were predominantly associated with commonly detected rotavirus G genotypes [(25/59, 42.4 %); G1 (17/25, 68.0%), G3 (5/25, 20.0%) and G4 (3/25, 12.0%)] followed by mixed infections (20/59, 33.9%) and G-nontypeable strains (13/59, 22.0%). Only one of the strains (939961) showed unusual combination with rotavirus genotype G2 (1/59, 1.7%).

Four of the 9 strains grouped in lineage P[8]-2 were

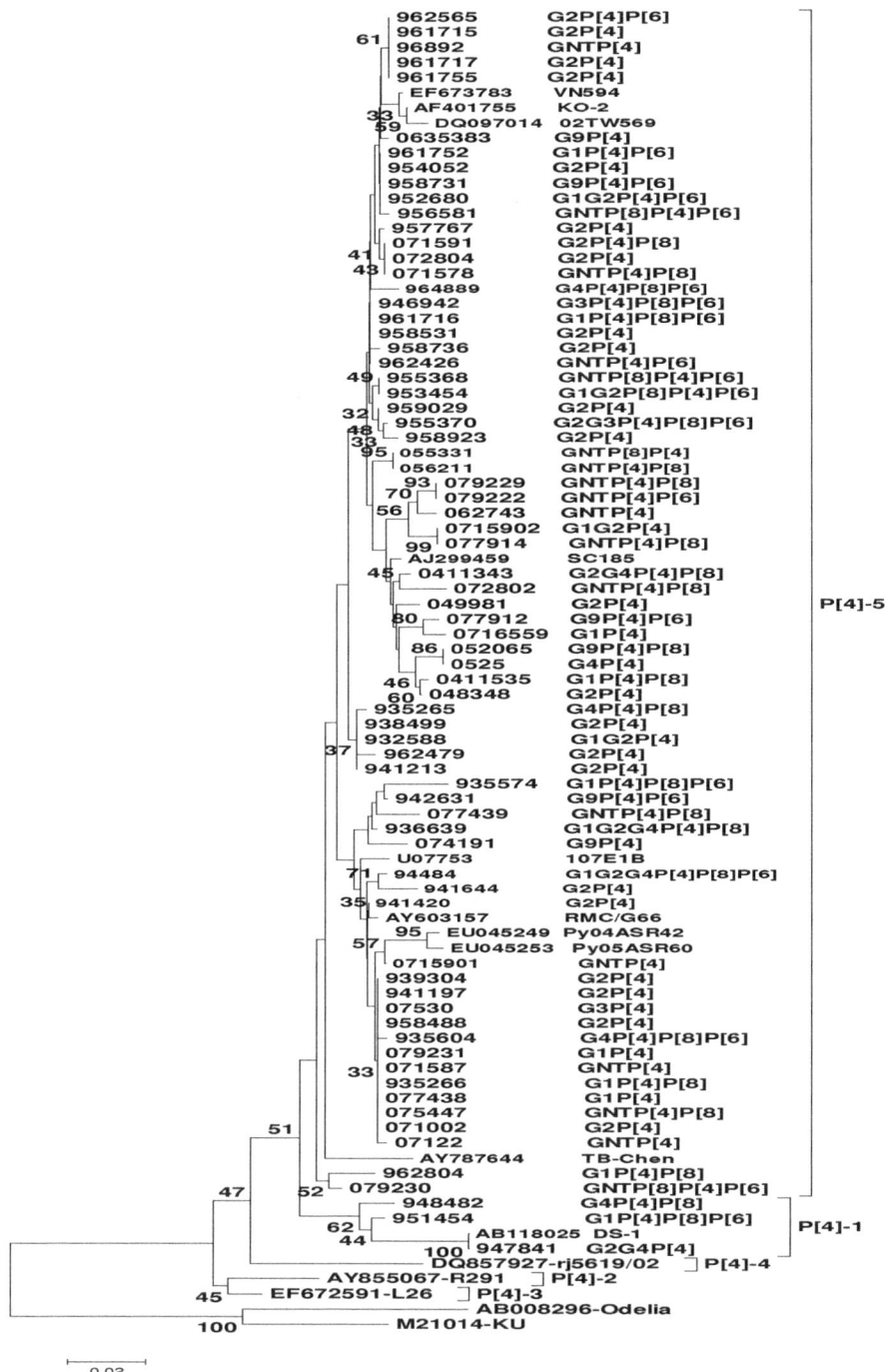

Figure 1. Phylogenetic tree based on partial nucleotide sequences of the VP4 (P[4]) gene (nt 43 to 456) of rotavirus strains recovered from adolescents and adults in 1993 to 1996 and 2004 to 2007. The strains of the present study are identified by the specimen number, followed by the genotypes identified in multiplex PCR. The reference strains are indicated by accession numbers followed by the strain names.

closer to the prototype strain, F45 with 95.7 to 96.8% and 96.7 to 98.9%, nucleotide and amino acid identities respectively. Within these strains nucleotide and amino acid identities were noted to be 96.4 to 100% and 97.8 to 100% respectively. Remaining five strains showed 100% identity with the KU strain at nucleotide and amino acid level. Majority of the strains (7/9) from this lineage were associated with non-typeable G genotypes.

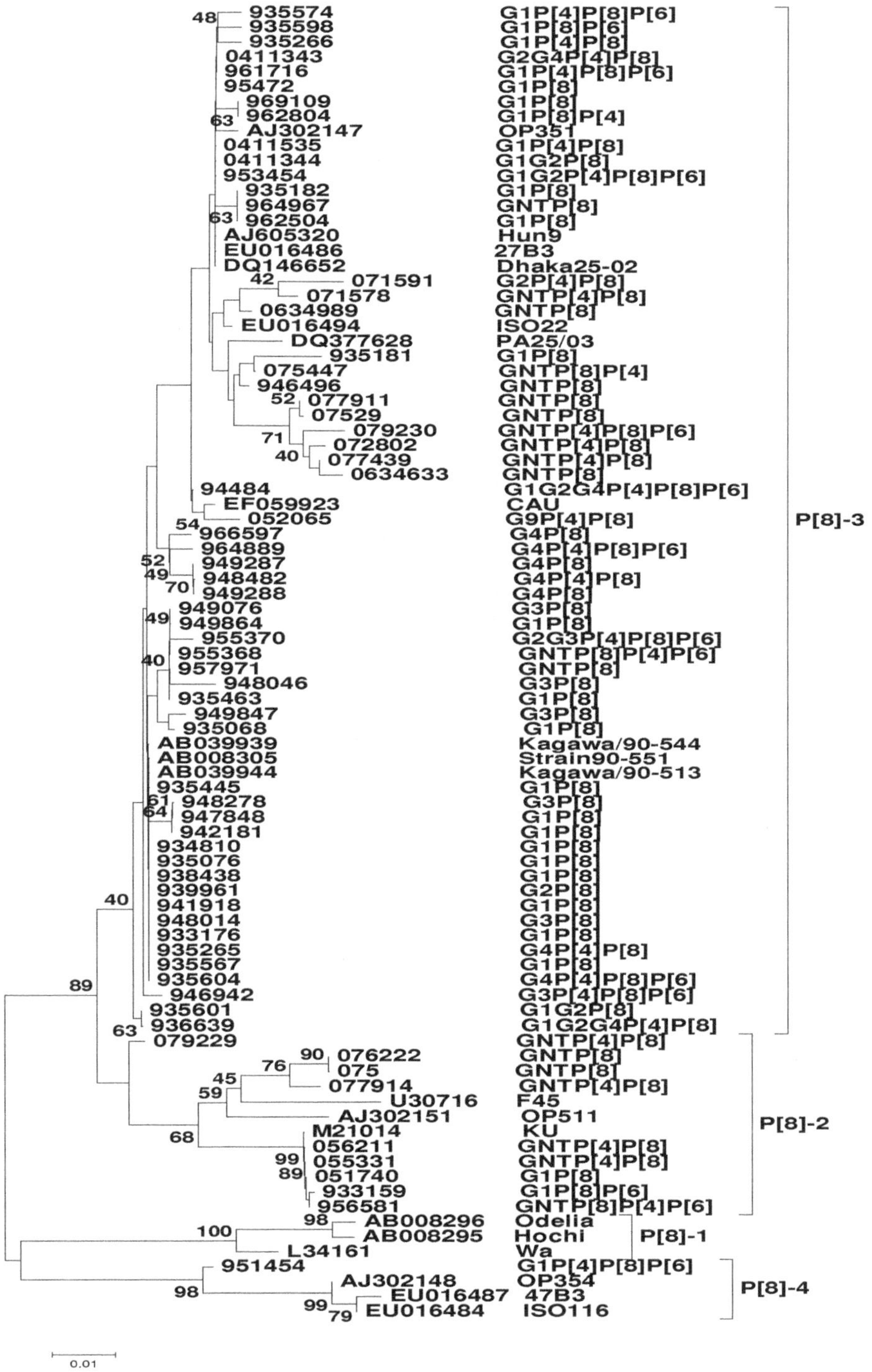

Figure 2. Phylogenetic tree based on partial nucleotide sequences of the VP4 (P[8]) gene (nt 40 to 321) of rotavirus strains recovered from adolescents and adults in 1993 to1996 and 2004 to 2007. The strains of the present study are identified by the specimen number, followed by the genotypes identified in multiplex PCR. The reference strains are indicated by accession numbers followed by the strain names.

The only strain (951454) detected with P[8]-4 lineage specificity showed 97.1 to 97.8 and 97.8% nucleotide and amino acid identities with the reference strains- OP354, 47B3 and ISO116 and was noted to be in the mixed infection to (G1P[4]P8]P[6]) specificity.

The group of strains isolated at two time points, 1993 to

1996 and 2004 to 2007 differed respectively by 6.6 to 8.2% / 4.7 to 6.4% and 3.0% / 2.2% in their nucleotide / amino acid sequences from the prototype strain, Wa of lineage P[8]-1 and from each other. The alignment of the partial deduced amino acid sequences (11-103aa) showed variations at positions 19(H→Y), 35(I →V), 38(S→G), 64(M→I) and 78(N→T) as compared to the prototype strain, Wa. These strains also showed strain specific changes at 20(T→I), 71(P→S), 75(T→K) and 77(P→L) positions.

Phylogenetic analysis of P[6] strains

Phylogenetic analysis of 27 P[6] strains recovered from adolescents (n = 3) and adults (n = 24) at two time points 1993 to 1996 (n = 24) and 2004 to 2007 (n = 3), showed clustering of all of the strains in lineage P[6]-I (Figure 3). All strains showed 94.1 to 100% and 93.9 to 100% nucleotide and amino acid identities respectively with the reference strains, M37, ST-3, SC2, MtA5, RV176, US1205 and Se585. Within these strains 96.1 to 100% nucleotide and 95.5 to 100% amino acid identity was noted. Nucleotide and amino acid divergence respectively of 0.4 to 2.3% and 0.2 to 1.6% was detected between the strains from 1993 to 1996 and 2007. The alignment of the partial deduced amino acid sequences (11-80 aa) showed variations at positions 30(T→S), 73(S→N) and 78(S→N) as compared to the prototype strain, M37.

DISCUSSION

Epidemiological surveys carried out in the late 1970s for the rotavirus strain surveillance used monoclonal antibodies. Though these studies established circulation of G1 to G4 serotypes in Europe, North America and Australia they also reported predominance of nontypeable rotavirus strains in South America, Africa and Asia (Desselberger et al., 2001). A complete assessment highlighting complex picture of the virus emerged with the availability of typing methods for both VP7 (G) and VP4 (P) genes (Gouvea et al., 1990; Gentsch et al., 1992). Rotaviruses identified to have different G and / or P types have indicated that a diverse gene pool contributes to human infections (Desselberger et al., 2001). Further, studies on genetic and antigenic variations in the rotavirus strains of the same genotypes have revealed intragenotypic diversity and rotavirus gene classification on the basis of lineages and sub lineages (Gouvea et al., 1999; Araujo et al., 2007; Espinola et al., 2008). Circulation of different lineages within the P[4], P[8] and P[6] genotypes of rotavirus VP4 genes isolated from UK, Hungary, Brazil and Paraguay (Itturiza-Gomara et al., 2001; Banyai et al., 2004b; Araujo et al., 2007; Espinola et al., 2008) has been reported. However, such

data from India is limited to rotavirus strains recovered from children (Samajdar et al., 2008; Zade et al., 2009). This study analyzes partial sequences of VP4 genes of rotavirus strains recovered from adolescents and adults at two different time points (1993 to 1996 and 2004 to 2007).

Varying frequencies of P[4] genotypes have been reported during rotavirus surveillance conducted in Brazil (21 to 26%), Sao Paulo (1%), India (33%), Korea (15.4%), Tunisia (13%) and Bangladesh (43%) (Araujo et al., 2002; Mascarenhas et al., 2002; Carmona et al., 2006, Volotao et al., 2006; Samajdar et al., 2006; Araujo et al., 2007; Chouikha et al., 2007; Le VP et al., 2008; Paul et al., 2010). Occurrence of five (P[4]-1 to P[4]-5) different lineages of P[4] genotype has been described on the basis of analysis of >100 strains circulating worldwide (Espinola et al., 2008). Three lineages P[4]-1, P[4]-4 and P[4]-5 have been described to combine with either of the G2, G3 and G9 types of VP7 gene. The remaining two lineages, P[4]-2 and P[4]-3 were found in association with G8 and G12 genotypes, respectively (Espinola et al., 2008). As against this, in the present study, P[4]-5 lineage clustered almost in equal proportion with the strains of G2 specificity and mixed / unusual infections / nontypeable G types. These results differed from the recent findings on clustering of G2P[4] strains in the P[4]-2 lineage from Thailand (Khamrin et al., 2010). The P[4] -1 lineage described in the present study clustered only mixed G and / or P infections (3/73). Interestingly, a comparison between the two time points, 1993 to 1996 and 2004 to 2007 showed a significant difference (P<0.01) in the frequencies of association of P[4] with common and nontypeable G strains (Table 1).

Genotype P[8] has been found to have undergone genetic and antigenic drift (Jin et al., 1996) with cocirculation of four distinct P[8] lineages, globally (Gouvea et al., 1999; Cunliffe et al., 2001; Arista et al., 2005, 2006; Ansaldi et al., 2007; Espinola et al., 2008). Several studies have revealed cocirculation of mainly P[8]-1 and P[8]-2 lineages in the late 80's and early 90's and predominant circulation of P[8]-3 lineage in the 2000s (Cunliffe et al., 2001; Itturiza-Gomara et al., 2001; Arista et al., 2005; Araujo et al., 2007; Espinola et al., 2008; Le VP et al., 2010; Tort et al., 2010). In the present study P[8]-3 predominated at both time points (93.7% / 66.7%), thus indicating circulation of this sublineage in Pune, India since 1990s. Cocirculation of P[8]-2 and / or P[8]-4 was also noted at both time points with increase in the circulation of P[8]-2 lineage from 4.2% in 1990s to 33.3% in 2000s. Although P[8]-3 lineage grouped majority of the strains of present study, its association with G genotypes differed with time. In the 1990s it was predominantly associated with common G types (G1, G3 and G4) (25/45, 55.6%) followed by unusual and mixed infections with G and / or P types (16/45, 35.5%) and G nontypeable strains (4/45, 8.9%). On the contrary, strains (n =14) from 2000s associated predominantly with

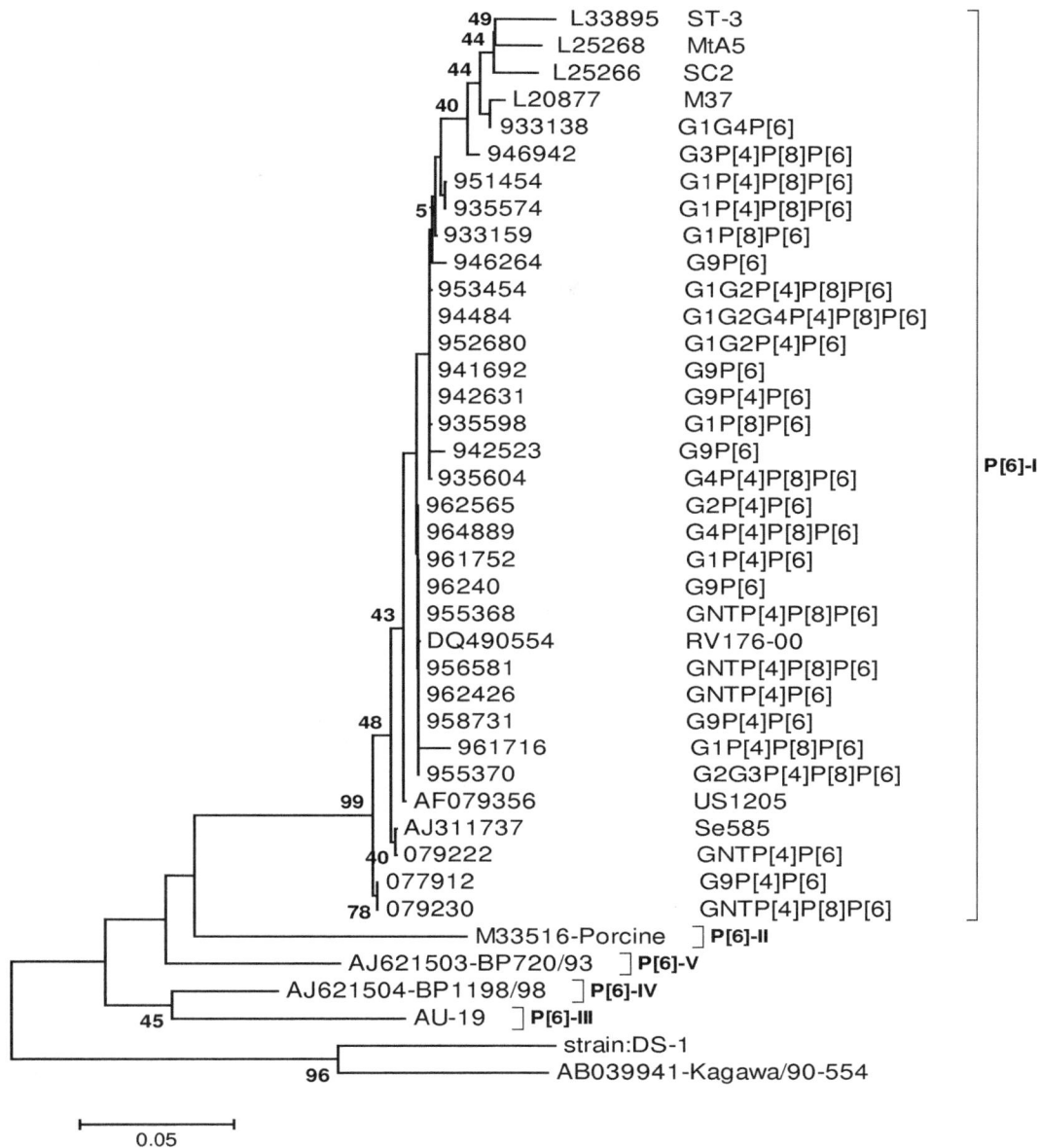

Figure 3. Phylogenetic tree based on partial nucleotide sequences of the VP4 (P[6]) gene (nt 40 to 249) of rotavirus strains recovered from adolescents and adults in 1993 to1996 and 2004 to 2007. The strains of the present study are identified by the specimen number, followed by the genotypes identified in multiplex PCR. The reference strains are indicated by accession numbers followed by the strain names.

nontypeable G types (9/14, 64.3%) followed by mixed G and / or P types (5/14, 35.7%). These results differed from the findings documenting association of P[8]-3 lineage predominantly with common strains (Min et al., 2004; Araujo et al., 2007; Espinola et al., 2008; Samajdar et al., 2008; Le et al., 2010). It was also interesting to note that the common strains (G1, G3 and G4) from the 1990s of this study clustered in the P[8]-3 lineage and that they differed from the contemporary strains of the same types that clustered in P[8]-1 or P[8]-2 lineages (Iturriza-Gomara et al., 2001; Arista et al., 2005; Espinola

et al., 2008). The existence of constraints on the G9 strains circulating worldwide to belong to the P[8]-3 lineage (Iturriza-Gomara et al., 2001; Banyai et al., 2004a) was also noted in this study.

Circulation of P[8]-2 lineage reported at different time points has displayed clustering of rotavirus strains with different G types (G1-G4, G5, G8 and G9) (Itturiza-Gomara et al., 2001; Araujo et al., 2007; Espinola et al., 2008, Samajdar et al., 2008). The rotavirus strains detected in children from Vietnam, Brazil and Bangladesh have shown association of P[8]-2 lineage with common

(G1, G2, and G9) and mixed G strains (Araujo et al., 2007; Nguyen et al., 2008; Dey et al., 2009). In the present study this lineage clustered G non-typeable strains predominantly. However, it is possible that the nontypeable strains could be commonly circulating strains that remained undetected due to less viral load or genomic variations.

OP354 like lineage designated as P[8]-4 has been described to group G1, G3, G4, G9, G1G9, G1G4 and G3G9 strains from different countries (Cunliffe et al., 2001; Banyai et al., 2004a; Nguyen et al., 2008; Samajdar et al., 2008; Nagashima et al., 2010). The association of this lineage with triple P (VP4) type rotavirus infection (G1, P4, P8, P6) was noted for the first time in this study. A significant difference (P<0.01) in the frequencies of association of P[8] with common and nontypeable G strains was evident between the two time points examined in this study (Table 1).

The P[6] genotype previously thought to be restricted to asymptomatic infections has also been identified in combination with G1-G4, G5, G8, G9, G11 and G12 genotypes detected in neonates and children with diarrhea (Adah et al., 1997; Pager et al., 2000; Cunliffe et al., 2002; Mascarenhas et al., 2002; Steele and Ivanoff, 2003; Page and Steele, 2004; Mascarenhas et al., 2006; 2007; Ahmed et al., 2007; Nguyen et al., 2007; Le et al., 2008; Banyai et al., 2009; Mukherjee et al., 2009; Shim et al., 2010). However, no data is available from adolescents and adults. Interestingly, majority of the strains with P[6] genotype in the present study were detected in the mixed infections and placed in lineage P[6]-I (M37 like).

In summary, the findings of this study highlight the intragenotyic diversity and temporal variations in rotavirus VP4 genes. Although it is known that homotypic and heterotypic immune response could be elicited by VP4, antigenic differences in various P[6] lineages identified in human and porcine species have been reported to generate differential neutralizing antibody response (Gorziglia et al., 1990; Li and Gorziglia, 1993; Nakagomi et al., 1999). In this context, it will be important to monitor alterations in the epitopes of P[8] and P[4] rotavirus strains grouped in different lineages and evaluate corresponding antibody responses as VP4 is a major protective antigen . Currently available rotavirus vaccines introduced mainly for the pediatric population constitute VP4 of P[8] genotype. In view of this, vigilance on the intragenotypic diversity in VP4 genes would provide a key to understand the evolution and genetic / antigenic differences in the rotavirus genotypes and assess the need for improvement in rotavirus vaccines.

ACKNOWLEDGEMENTS

The authors thank Dr. A.C. Mishra, Director, National Institute of Virology, Pune for all support. The assistance provided by Mrs L.B. Bhosale and Mrs A.P. Pardesi during sample collection from the hospitals is gratefully acknowledged. Thanks are due to Dr D.V. Barathe from Naidu hospital and Dr A.N. Borhalkar from Shreyas Clinic for extending their co-operation in sample collection. Authors also thank Mr A.M. Walimbe, for statistical analysis of the data.

REFERENCES

Adah MI, Rohwedder A, Olaleya OD, Durojaiye OA, Werchau H (1997). Further characterization of field strains of rotavirus from Nigeria VP4 genotype P6 most frequently identified among symptomatically infected children. J. Trop. Pediatr., 43: 267-274.

Ahmed K, Anh DD, Nakagomi O (2007). Rotavirus G5P[6] in child with diarrhea, Vietnam. Emerg. Infect. Dis., 13: 1232-1235.

Ansaldi F, Pastorino B, Valle L, Durando P, Sticchi L, Tucci P, Biasci P, Lai P, Gasparini R, Icardi G, Paediatric Leghorn Group (2007). Molecular characterization of a new variant of rotavirus P[8]G9 predominant in a sentinel-based survey in Central Italy. J. Clin. Microbiol., 45: 1011-1015.

Arau´jo IT, Assis RM, Fialho AM, Mascarenhas JD, Heinemann MB, Leite JP (2007). Brazilian P[8], G1, P[8], G5, P[8], G9, and P[4], G2 rotavirus strains: nucleotide sequence and Phylogenetic analysis. J. Med. Virol., 79: 995-1001.

Araujo IT, Fialho AM, de Assis RM, Rocha M, Galvao M, Cruz CM, Ferreira MS, Leite JP (2002). Rotavirus strain diversity in Rio de Janeiro Brazil: characterization of VP4 and VP7 genotypes in hospitalized children. J. Trop. Pediatr., 48: 214-218.

Arista S, Giammanco GM, De Grazia S, Colomba C, Martella V (2005). Genetic variability among serotype G4 Italian human rotaviruses. J. Clin. Microbiol., 43: 1420-1425.

Arista S, Giammanco GM, De Grazia S, Ramirez S, Blundo C, Colombo C, Cascia A, Martella, V (2006). Heterogeneity and temporal dynamics of evolution G1 human rotaviruses in a settled population. J. Virol., 80: 10724-10733.

Arora R, Chhabra P, Chitambar SD (2009). Genetic diversity of genotype G1 rotaviruses co-circulating in western India. Virus. Res., 146: 36-40.

Bányai K, Esona MD, Kerin TK, Hull JJ, Mijatovic S, Vásconez N, Torres C, de Filippis AM, Foytich KR, Gentsch JR (2009). Molecular characterization of a rare, human-porcine reassortant rotavirus strain, G11P[6], from Ecuador. Arch. Virol., 154: 1823-1829.

Banyai K, Gentsch JR, Schipp R, Jakab F, Bene J, Melegh B, Glass RI. Szu¨cs G (2004a). Molecular epidemiology of human P[8], G9 rotaviruses in Hungary between 1998 and 2001. J. Med. Microbiol., 53: 791-801.

Banyai K, Martella V, Jakab F, Melegh B, Szucs G (2004b). Sequencing and phylogenetic analysis of human genotype P[6] rotavirus strains detected in Hungary provided evidence for genetic heterogeneity within the P[6] VP4 gene. J. Clin. Microbiol., 42: 4338-4343.

Carmona RCC, Timenetsky MCST, Morillo SG, Richtzenhain L (2006). Human rotavirus serotypes G9, Sao Paulo, Brazil, 1996-2003. Emerging. Infect. Dis., 12: 963-968.

Chitambar SD, Tatte V, Dhongde R, Kalrao V (2008). High frequency of rotavirus viremia in children with acute gastroenteritis :discordance of strains detected in stool and sera. J. Med. Virol., 80: 2169-2176.

Chouikha A, Fodha I, Noomen S, Bouzid L, Mastouri M, Peenze I, de Beer M, Dewar J, Geyer A, Sfar T, Gueddiche N, Messaadi F, Trabelsi A, Boujaafar N, Steele AD (2007). Group A rotavirus strains circulating in the eastern center of Tunisia during a ten-year period (1995-2004). J. Med. Virol., 79: 1002-1008.

Cunliffe NA, Gondwe JS, Graham SM, Thindwa BD, Dove W, Broadhead RL, Molyneux ME, Hart CA (2001). Rotavirus strain diversity in Blantyre, Malawi, from 1997 to 1999. J. Clin. Microbiol., 39: 836-843.

Cunliffe NA, Rogerson S, Dove W, Thindwa BDM, Greensill J, Kirkwood CD, Broadhead RI, Hart CA (2002). Detection and characterization of rotaviruses in hospitalized neonates in Blantyre, Malawi. J. Clin. Microbiol., 40: 1534-1537.

Desselberger U, Iturriza-Go`mara M, Gray J (2001). Rotavirus epidemiology and surveillance. Novartis. Found. Symp., 238: 125-147.

Dey SK, Hayakawa Y, Rahman M, Islam R, Mizuguchi M, Okitsu S, Ushijima H (2009). G2 Strain of Rotavirus among Infants and Children, Bangladesh. Emerg. Infect. Dis., 15: 91-94.

Espinola EE, Amarilla A, Arbiza J, Parra GI (2008). Sequence and phylogenetic analysis of the VP4 gene of human rotaviruses in Paraguay. Arch. Virol., 153: 1067-1073.

Estes MK (2001). Rotavirus and their replication. In Fields Virology, 4th eds. Knipe DM, Griffin DE, Lamb RA, Martin MA, Roizman B, Status SE, Philadelphia, Lippincott Williams and Wilkins, pp. 1747-1785.

Gentsch JR, Glass RI, Woods P, Gouvea V, Gorziglia M, Flores J, Das BK, Bhan MK (1992). Identification of group A rotavirus gene 4 types by polymerase chain reaction. J. Clin. Microbiol., 30: 1365-1373.

Gorziglia M, Larralde G, Kapikian AZ, Chanock RM (1990). Antigenic relationships among human rotaviruses as determined by outer capsid protein VP4. Proc. Natl. Acad. Sci., 87: 7155-7159.

Gouvea V, Glass RI, Woods P, Taniguchi K, Clark HF, Forrester B, Fang ZY (1990). Polymerase chain reaction amplification and typing of rotavirus nucleic acid from stool specimens. J. Clin. Microbiol., 28: 276-282.

Gouvea V, Lima RC, Linhares RE, Clark HF, Nosawa CM, Santos N (1999). Identification of two lineages (WA-like and F45-like) within the major rotavirus genotype P[8]. Virus Res., 59:141-147.

Iturriza-Gomara M, Isherwood B, Desselberger U, Gray J (2001). Reassortment invivo: Driving force for diversity of human rotavirus strains isolated in the United Kingdom between 1995 and 1999. J. Virol., 75: 3696-3705.

Iturriza-Gomara M, Kang G, Gray J (2004). Rotavirus genotyping: keeping up with an evolving population of human rotaviruses. J. Clin. Virol., 31: 259-265.

Jin Q, Ward RL, Knowlton DR, Gabbay YB, Linhares AC, Rappaport R, Woods PA, Glass RI, Gentsch JR (1996). Divergence of VP7 genes of G1 rotaviruses isolated from infants vaccinated with reassortant rhesus rotaviruses. Arch. Virol., 141: 2057-2076.

Kang G, Arora R, Chitambar S, Deshpande J, Gupta MD, Kulkarni M, Naik TN, Mukherji D, Venkatasubramanium S, Gentsch JR, Glass RI, Parashar UD, Indian Rotavirus Strain Surveillance Network (2009). Multicenter, hospital-based surveillance of rotavirus disease and strains among Indian children J. Infect. Dis., 200: S147-153.

Khamrin P, Maneekarn N, Malasao R, Nguyen TA, Ishida S, Okitsu S, Ushijima H (2010). Genotypic linkages of VP4, VP6, NSP4, NSP5 genes of rotaviruses circulating among children with acute gastroenteritis in Thailand. Infect. Genet. Evol., 10: 467-472.

Le VP, Chung YC, Kim K, Chung SI, Lim I, Kim W (2010). Genetic variation of prevalent G1P[8] human rotaviruses in South Korea. J. Med. Virol., 82: 886-896.

Le VP, Kim JY, Cho SL, Nam SW, Lim I, Lee HJ, Kim K, Chung SI, Song W, Lee KM, Rhee MS, Lee JS, Kim W (2008). Detection of unusual rotavirus genotypes G8P[8] and G12P[6] in South Korea. J. Med. Virol., 80: 175-182.

Li B, Gorziglia M (1993). VP4 serotype of the Gottfried strain of porcine rotavirus. J. Clin. Microbiol., 31: 3075-3077.

Mascarenhas JDP, Linhares AC, Gabbay YB, Lima CS, Guerra SFS, Soares LS, Oliveira DS, Lima JC, Mâcedo O, Leite JPG (2007). Molecular characterization of VP4 and NSP4 genes from rotavirus strains infecting neonates and young children in Belém, Brazil. Virus. Res., 126: 149-158.

Mascarenhas JDP, Linarhes AC, Bayma APG, Lima JC, Sousa MS, Araujo IT, Heinemann MB, Gusma-o RHP, Gabbay YB, Leite JPG (2006). Molecular analysis of VP4, VP7 and NSP4 genes of P[6]G2 rotavirus genotype strains recovered from neonates admitted to hospital in Belem, Br. J. Med. Virol., 78: 281-289.

Mascarenhas JDP, Linhares AC, Gabbay YB, Leite JRP (2002). Detection and characterization of rotavirus G and P types from children participating in a rotavirus vaccine trial in Belem, Brazil. Mem. Inst. Oswaldo. Cruz. Rio de Janerio, 97: 113-117.

Matthijnssens J, Ciarlet M, Heiman E, Arijis I, Delbeke T, McDonald SM, Palombo EA, Iturriza-Gomara M, Maes P, Patton JT, Rahman M, Ranst MV (2008). Full genome-based classification of rotaviruses reveals a common origin between human Wa-like and porcine

rotavirus strains and human DS-1 like and bovine rotavirus strains. J. Virol., 82: 204-3219.

Matthijnssens J, Ciarlet M, McDonald SM, Attoui H, Bányai K, Brister JR, Buesa J, Esona MD, Estes MK, Gentsch JR, Iturriza-Gómara M, Johne R, Kirkwood CD, Martella V, Mertens PP, Nakagomi O, Parreño V, Rahman M, Ruggeri FM, Saif LJ, Santos N, Steyer A, Taniguchi K, Patton JT, Desselberge, U, Van Ranst M (2011). Uniformity of rotavirus strain nomenclature proposed by the Rotavirus Classification Working Group (RCWG). Arch. Virol., 156: 1397-1413.

Min BS, Noh YJ, Shin JH, Baek SY, Kim JO, Kyung M, Ryu SR, Kim BG, Kim DK, Lee SH, Min HK, Ahn BY, Park SN (2004). Surveillance Study (2000 to 2001) of G and P Type Human Rotaviruses Circulating in South Korea. J. Clin. Microbiol., 42: 4297-4299.

Mukherjee A, Dutta D, Ghosh S, Bagchi P, Chattopadhyay S, Nagashima S, Kobayashi N, Dutta P, Krishnan T, Naik TN, Chawla-Sarka M (2009). Full genomic analysis of a human group A rotavirus G9P[6] strain from Eastern India provides evidence for porcine-to-human interspecies transmission. Arch. Virol., 154: 733-746.

Nagashima S, Kobayashi N, Paul SK, Ghosh S, Sarkar MC, Hossian MA, Krishnan T (2010). Identification of P[8]b subtype in OP354-like human rotavirus strains by a modified RT-PCR method. Jpn. J. Infect. Dis., 63: 208-211.

Nakagomi T, Horie Y, Koshimura Y, Greenberg HB, Nakagomi O (1999). Isolation of a human rotavirus strain with a super -short RNA pattern and a new P2 subtype analysis. J. Clin. Microbiol., 37: 1213-1216.

Nguyen TA, Hoang LP, Pham LD, Hoang KT, Okitsu S, Mizuguchi M, Ushijima H (2008). Use of sequence analysis of the VP4 gene to classify recent Vietnamese rotavirus isolates. Clin. Microbiol. Infect., 14: 235-241.

Nguyen TA, Khamrin P, Trinh QD, Pham LD, Hoang LP, Hoang KT, Yagyu F, Okitsu S, Ushijima H (2007). Sequence analysis of Vietnamese P[6] rotavirus strains suggests evidence of interspecies transmission. J. Med. Virol., 79: 1959-1965.

Page NA, Steele AD (2004). Antigenic and genetic characterization of serotype G2 human rotavirus strains from the African continent. J. Clin. Microbiol., 42: 595-600.

Pager CT, Alexander JJ, Steele AD (2000). South African G4P[6] asymptomatic and symptomatic neonatal rotavirus strains differ in their NSP4, VP8*, and VP7 genes. J. Med. Virol., 62: 208-216.

Parashar UD, Gibson CJ, Bresse JS, Glass RI (2006). Rotavirus and severe childhood diarrhea. Emerg. Infect. Dis., 12: 304-306.

Patel MM, Steele D, Gentsch JR, Wecker J, Glass RI, Parashar UD (2011). Real-world Impact of Rotavirus Vaccination. Pediatric. Infect. Dis., 30: S1-S5.

Paul SK, Ahmed MU, Hossain MA, Mahmud MC, Bhuiyan MR, Saha SK, Tabassum S (2010). Molecular characterization of group A human rotavirus among hospitalized children and adults in Bangladesh: finding of emerging G12 strains. Mymensingh. Med. J., 19: 16-26.

Pietsch C, Schuster V, Liebert UG (2011). A hospital based study on inter-and intragenotypic diversity of human rotavirus A VP4 and VP7 gene segments, Germany. J. Clin. Virol., 50: 136-141.

Samajdar S, Varghese V, Barman P, Ghosh S, Mitra U, Dutta P, Bhattacharya SK, Narasimham MV, Panda P, Krishnan T, Kobayashi N, Naik TN (2006). Changing pattern of human group A rotaviruses: emergence of G12 as an important pathogen among children in eastern India. J. Clin. Virol., 36: 183-188.

Samajdar S, Ghosh S, Dutta D, Chawla-Sarkar M, Kobayashi N, Naik TN (2008). Human group A rotavirus P[8] Hun9-like and rare OP354-like strains are circulating among diarrhoeic children in Eastern India.. Arch. Virol., 153: 1933-1936.

Santos N, Hoshino Y (2005). Global distribution of rotavirus serotypes/genotypes and its implication for the development and implementation of an effective rotavirus vaccine. Rev. Med. Virol., 15: 29-56.

Shim SY, Jung YC, Le VP, Son DW, Ryoo E, Shim JO, Lim I, Kim W (2010). Genetic Variation of G4P[6] Rotaviruses: Evidence for novel strains circulating between the hospital and community. J. Med. Virol., 82: 700-706.

Steele AD, Ivanoff B (2003). Rotavirus strains circulating in Africa during 1996-1999; emergence of G9 strains and P[6] strains. Vaccine, 17:

361-367.

Tamura K, Dudley J, Nei M, Kumar S (2007). MEGA 4: Molecular Evolutionary Genetics Analysis (MEGA) software version 4.0. Mol. Biol. Evol., 24: 1596-1599.

Tatte VS, Gentsch JR, Chitambar SD (2010a). Characterization of group A rotavirus infections in adolescents and adults from Pune, India: 1993-1996 and 2004-2007. J. Med. Virol., 82: 519-527.

Tatte VS, Rawal KN, Chitambar SD (2010b). Sequence and phylogenetic analysis of the VP6 and NSP4 genes of human rotavirus strains: Evidence of discordance in their genetic linkage. Infect. Gen. Evol., 10: 940-949.

Thompson JD, Higgins DG, Gibson TJ (1994). CLUSTAL W: improving the sensitivity of progressive multiple sequence alignment through sequence weighting, position-specific gap penalties and weight matrix choice. Nucleic. Acid. Res., 22: 4673-4680.

Tort LF, Volotao EM, de Mendonca MC, da Silva ME, Siqueira AA, Assis RM, Moratoria G, Cristina J, Leite JP (2010). Phylogenetic analysis of human P[8]G9 rotavirus strains circulating in Brazil reveals the presence of a novel genetic variant. J. Clin. Virol., 47: 345-355.

Volotao EM, Soares CC, Maranhao AG, Rocha LN, Hoshino Y, Santos N (2006). Rotavirus surveillance in the city of Rio de Janeiro-Brazil during 2000-2004: detection of unusual strains with G8P[4] or G10P[9] specificities. J. Med. Virol., 78: 263-272.

Zade JK, Chhabra P, Chitambar SD (2009). Characterization of VP7 and VP4 genes of rotavirus strains: 1990-1994 and 2000-2002. Epidemiol. Infect., 137: 936-942.

11

Norovirus based viral gastroenteritis in Chennai city of southern India - An epidemiological study

S. Anbazhagi*, S. Kamatchiammal and D. Jayakar Santhosh

National Environmental Engineering Research Institute Chennai Zonal Laboratory, CSIR Madras Complex, Taramani, Chennai – 600 113, India.

The study evaluated the incidence of norovirus associated gastroenteritis in India. During the year 2005 to 2006, stool samples from communicable disease hospital were collected from subjects suspected for non – bacterial gastroenteritis and were examined for norovirus. Samples were analyzed by reverse transcriptase polymerase chain reaction and further confirmed by hybridisation and sequencing. In addition the epidemiology of norovirus was studied by statistical analysis. 44.4% of the samples were positive for norovirus, the isolated strain represented the Genogroup I of norovirus. The study observed that norovirus infection is not only a winter vomiting disease it can also occur in the summer.

Key words: Gastroenteritis, norovirus, reverse transcriptase polymerase chain reaction, stool.

INTRODUCTION

Diarrheal disease is a major cause of childhood morbidity and mortality, especially in developing countries (Bern et al., 1992). Several different groups of viruses have been shown to be responsible for the high incidence of acute viral diarrhea among children during their first few years of life. Rotavirus is the single most important etiological agent in severe dehydrating diarrhea (Parashar et al., 2003). Recent studies investigating calicivirus in sporadic cases of gastroenteritis in children have concluded that caliciviruses comprise the second cause of viral diarrhea after rotavirus (Simpson et al., 2003; Subekti et al., 2002). Direct electron microscopy is used to screen fecal specimens for enteric viruses in the public health laboratories of many countries and it is clear that talented electron microscopists who receive fecal samples collected early in the course of an infection can often detect viruses (Wright et al., 1998). Detection of enteric viruses in stool specimens using direct electron micro-scope requires virus concentrations of at least 10^6 per ml of stool (Doane et al., 1994). The small numbers of viral particles present in fecal samples make direct electron microscopy studies even after concentration, relatively

insensitive. However, this method requires highly highly skilled microscopists and expensive equipment, making it not feasible for large epidemiological or clinical studies. Norovirus is an important cause of acute nonbacterial gastroenteritis in children and adults worldwide. Noro-viruses belong to the family caliciviridae. Calicivi-ruses are small non-enveloped viruses approximately 27 to 35 nm in diameter with a positive-sense, single stranded RNA genome (Green et al., 2001). The first discovered norovirus was associated with a human outbreak of gastroenteritis in Norwalk, Ohio, which gave the name Norwalk Virus to the prototype strain of norovirus, in 1968 (Adler and Zickl et al., 1969). Norovirus is also associating with sporadic and outbreak cases of gastroenteritis in individuals of all ages, with a distinct seasonality linked to the winter months (O'Neill et al., 2002). Infection is characterized by acute onset of nausea, vomiting, abdominal cramps, and diarrhoea, which generally last for about 48 h. Transmission occurs predominantly through ingestion of contaminated water, food (particularly oysters); and person by the fecal-oral route, airborne transmission and contact with conta-minated surfaces (Anbazhagi and Kamatchiammal, 2010; Girish et al., 2002; Parashar and Monroe et al., 2001). Caliciviruses are characterised by stability in the environ-ment (Rzezutka and Cook, 2004) and relative resistance to inactivation (Duizer et al., 2004).

*Corresponding author. E-mail: sanbazhagi@gmail.com.

The ease with which norovirus is transmitted as well as the low infectious dose required to establish an infection results in extensive outbreaks in numerous environments such as hospitals, hotels, schools, nursing homes, and cruise ships (Marks et al., 2000; 2003; Cheesbrough et al., 2000). Several comprehensive reviews on norovirus infections, including recommenda-tions on outbreak control, have been reported (Parashar et al., 2001; Chadwick et al., 2000).

The frequent occurrence of such outbreaks highlights the difficulties and challenges in effective control of norovirus infection in paediatric setting (Isakbaeva et al., 2005). In India the role of noroviruses in causing gastroenteritis is not well defined. Very few studies evaluated the role of viral agents in childhood diarrhoea in India (Preeti and Shobha, 2008). This study was conducted to find out the parameters responsible for norovirus infection by analysing the stool samples collected from the diarrhoeal patients.

MATERIALS AND METHODS

Standard norovirus strain

Norovirus positive stool isolates were obtained from Christian Medical College, Vellore, India. The isolates were confirmed by polymerase chain reaction using respective primers (Green et al., 1995) and were used as positive template. PCR reactions were carried out on a Perkin-Elmer Cetus DNA Thermo cycler 2400 (Perkin- Elmer Cetus Corp., USA).

Stool samples

Fecal specimens were collected in sterile plastic containers from cases of acute nonbacterial gastroenteritis at the Communicable Disease Hospital, Chennai between July 2005 and November 2006 and stored at −20°C. 81 ill subjects who were suspected for non bacterial gastroenteritis were chosen for the present study. A close-ended questionnaire was given to the subjects and was used for analysis purpose. Stratified sampling design was adopted for the study purpose.

RNA extraction

Viral RNA was extracted from 200 µl of a 10 to 20% stool PBS/MEM suspension by binding to size-fractionated silica particles in the presence of guanidinium isothiocyanate as described by Boom et al. (1990). The RNA was eluted in 49 µl of RNAse free distilled water and 1 µl of RNAsin (Biocorporals, India). RNA was either used directly in RT-PCR or stored at -70°C.

Reverse transcriptase PCR evaluation

cDNA synthesis

cDNA synthesis was carried out using cDNA synthesis kit/RevertAid kit (Biogene UK/Fermentas, USA). 5 µl of purified template was added to 1 µl of 10 mM of random hexamer and 0.3 µM each of dATP, dCTP, dGTP and dTTP in a final reaction volume of 12 µl and incubated at 65°C for 5 min and then snap-cooled. 5X RT

buffer (components containing 1× First Strand Buffer® 0.01 M dithiothreitol) 40 units of Rnase inhibitor, were added to the chilled primer/template mixture making a final volume of 19 µl. This mixture was incubated at 42°C for 2 min before the addition of 1 µl of reverse transcriptase (Fermentas, USA). The reactions were then incubated for 50 min at 42°C followed by 70°C for 15 min.

Conventional RT-PCR

5 µl cDNA was added to 35 µl of PCR master mix (10 mM Tris (pH 8.3), 50 mM KCl, 1.5 mM MgCl₂, 0.2 mM of each dNTP, 20 pmol of each primer (NI forward primer and E3 reverse primer) (Green et al., 1995), 1 U of Taq polymerase). After an initial dena-turation at 94°C for 2 min, 30 amplification cycles of 95°C for 1 min, 40°C for 1 min and 72°C for 1 min were performed followed by a final extension of 72°C for 10 min. RT-PCR amplicons were analyzed by electrophoresis of 20 µl of reaction mix in agarose at 10 V/cm for 1.5 to 2 h. PCR amplifies a 113 bp region of the RNA polymerase gene. Molecular weights were determined by comparison with a 100 bp DNA ladder (Genei Bangalore, India).

Gel electrophoresis and slot blot analysis of PCR products

The PCR products were analyzed by electrophoresis on a 2% agarose gel containing ethidium bromide and visualized with UV light. 15 µl of the PCR product was denatured with equal volume of 0.5 N NaOH and spotted onto the hybridization membrane using a vacuum filtration manifold. The DNA transferred to the membrane was cross-linked in a UV Cross-linker (Foto/Prep from Fotodyne). Hybridization was performed with a cDNA (norovirus positive stool samples) labelled with dioxigenin – 11 dTTP was replaced by a mixture of digoxigenin-11 dUTP and dTTP in the molar ratio of 0.35:0.65 and were used as probes. Hybridization was performed as described by Boehringer- Mannheim (Genius TM Kit, Boehringer Mannheim, Indianapolis, IN). The labelled probe (100 µl) was denatured by heating at 95°C for 5 min and snap cooled on ice. The probe was added to the bag containing the membrane and kept at 42°C overnight. Hybridization and chromogenic detection of the hybrids with nitroblue tetrazolium (NBT) and BCIP (5-bromo-4-chloro-3-indolylphosphate) were carried out according to the protocols recommended by the manufacturer.

Norovirus cloning, sequencing and phylogenetic analysis

Each norovirus-positive RT-PCR product was cloned and selected for sequence analysis. PCR products were purified with Minelute PCR purification kit (QIAGEN, Germany). The purified PCR products were cloned using the TOPO TA Cloning® system (Invitrogen, India) according to the manufacturer's instructions. After transformation, at least five positive colonies were selected. Selected colonies were grown in LB medium containing 50 to 100 µg/ml ampicillin for overnight. Plasmid DNAs were isolated and were sequenced to confirm the presence of the insert using automatic sequencer ABI PRISM 310 Genetic Analyzer (Applied Biosystems) (MWG, Bangalore, India). Sequences were compared with other sequences in genbank using basic local alignment search tool (BLAST) family of programs on the World Wide Web service of National Centre for Biotechnology Information (NCBI), USA (http://www.ncbi.nlm.nih.gov) and were read using Tree view.

Statistical analysis

Results obtained from the virological determinations were subjected to statistical analysis using the SPSS package (SPSS Version 11,

GmbH software, Munich, Germany).

RESULTS AND DISCUSSION

Detection of norovirus by conventional RT-PCR

Increasing attention has been paid to viral gastroenteritis outbreaks during the last few years. The development of molecular methods has shown that norovirus is one of the most common causes of gastroenteritis outbreaks in adults and children (Höhne and Schreier, 2004). The increasing number of outbreaks leads to increasing numbers of samples that have to be handled in diagnostic laboratories. Therefore, RT-PCR seems to be the method of choice because it combines a potentially high throughput with reproducibility, sensitivity, and specificity. PCR products of 81 stool samples were analysed by agarose gel electrophoresis. 113 bp product specific for NI/E3 primers confirms the presence of norovirus among 36 number of stool samples. The debate over, whether RT-PCR is a legitimate tool for measuring infectious virus is well rehearsed; fundamentally the detection of nucleic acid does not directly indicate the presence of infectious virus. However, the norovirus capsid is environmentally robust, it is likely that any naked RNA would be quickly inactivated and replication is efficient. Hence detection of norovirus cDNA in a sample strongly suggests its origin from an infectious virion (Battacharya et al., 2004).

An interesting study by Battacharya et al. (2004) using hepatitis A virus, demonstrated that signals generated after RT-PCR amplification of viral genome correlated well with infectivity substantiates the present hypothesis.

Hybridization of PCR products

PCR products of 81 stool samples were analysed by Slot blot analysis. The RNA extracted from the stool samples were amplified using the norovirus specific primers NI/E3, blotted on a nylon membrane and probed using the digoxigenin labelled norovirus positive stool. Analysis of the PCR products by hybridization using the specific probes gave the same result as agarose gel electrophoresis.

PCR amplicon cloning

Representative RT-PCR amplicons for NoV generated from stool samples were cloned using a TA cloning system (TOPO®, Invitrogen, UK) as previously described by Leoni et al. (2003). Duplicate plates of the transformant were made and the colonies were picked and processed directly for PCR. The PCR was performed using the insert specific primers NI and E3. The expected PCR products of 113 bp were obtained.

Nucleotide sequencing

Plasmid DNAs were isolated and were sequenced to confirm the presence of the insert. NoV amplicon (101 bp segment) was characterized for genogroup, genotypes and genetic relationship with the reference strains based on their capsid regions classification scheme (Stephen et al., 1997). Their partial nucleotide sequences were compared to each other as well as to reference NoV strains available in the DDBJ DNA/GenBank database by BLAST. The nucleotide sequence of the 5' ends of the NoV capsid gene was determined by direct sequencing with the amplified fragments. All of the NoV sequences showed highest identity of 86 to 94% and were classified into genogroup I. The Phylogenetic tree illustrating the genetic relationship of norovirus isolates from stool samples of subjects, India is given in Figure 1. Sequenced NoVs were very similar at the amino acid level and were most closely related to the Hu/NLV/GI/684/US, NV/Saitama T83GI/02/JP, NV/Saitama T67GI/02/JP, NV/Saitama T62GI/02/JP, NV/Saitama T61GI/02/JP, NV/Saitama T59GI/02/JP and NV/Saitama KU4aGI/99/JP strain.

Clinical features of norovirus -positive and norovirus-negative patients

Clinical data of the 81 subjects infected with diarrhoea were listed in Table 1. Out of 81 patients admitted for diarrhoeal infection it was found that 44.4% were found positive for norovirus and 55.6% were negative. The most common clinical symptoms were diarrhoea (92%), vomiting (83%) and fever (67%). The percentage of diarrhoea in norovirus-positive patients was higher than norovirus-negative patients. It was found that more patients had symptoms of severe diarrhoea (92%) than vomiting (83%), which is in agreement with Wyatt et al. (1974). It was also found that NoV-infected persons developed symptoms of watery diarrhoea and vomiting typically remain symptomatic for 2 to 3 days.

Seasonal distribution of norovirus

The seasonal pattern of NoV infection is shown in Figure 2. During the study period between July 2005 and November 2006, prominent numbers of NoV infection was observed on April 2006 (6 norovirus-positive) followed by January 2006 (5 norovirus-positive). Whereas in September 2005 and May 2006 showed 4 cases of positive stool samples. Among the 36 positive samples collected throughout the study, fifty two percent (19 of 36) of positive samples were notified during this peak period of infectious time. Norovirus is generally referred to as "winter-vomiting disease", but this study on contrary identified "norovirus" in the summer season especially during April.

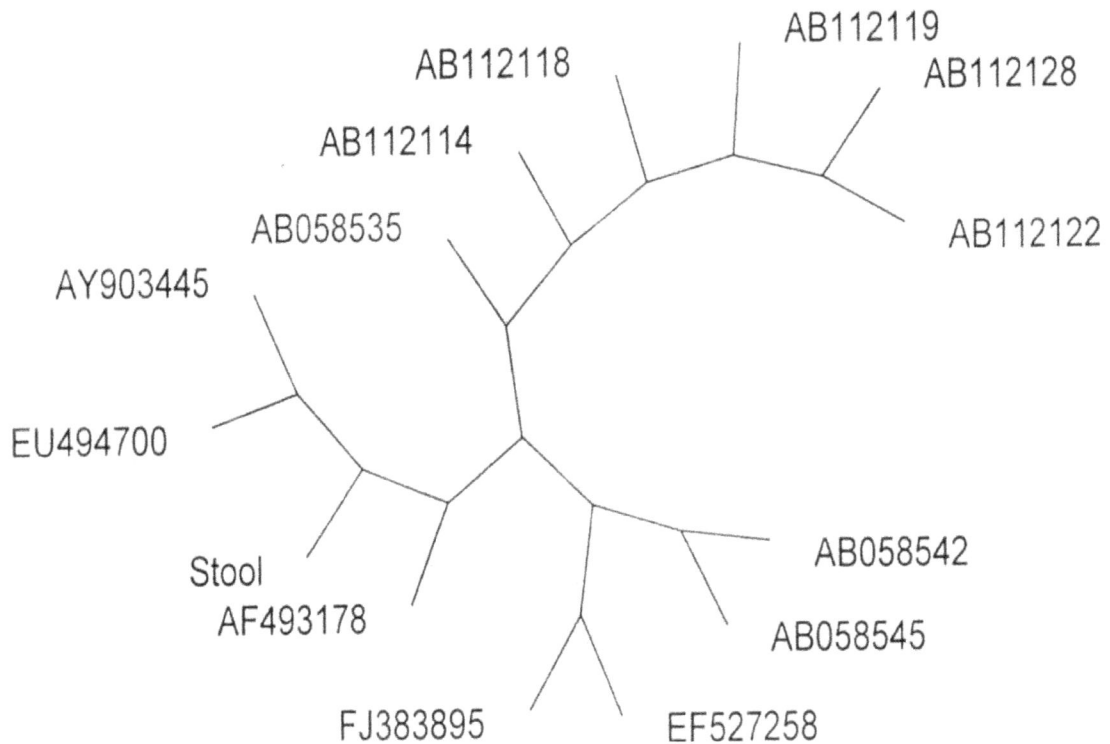

Figure 1. The Phylogenetic tree illustrating the genetic relationship of norovirus isolates from stool samples of gastroenteritis subjects, India.

Statistical analysis of the clinical samples

Theoretically a large number of variables are at force in determining the "norovirus" among subjects admitted due to diarrhea. The list includes nature of vomiting, source of water, number of days hospitalized, vegetable, nature of diarrhoea, fresh or preserved food, dysentery, sex and milk products. It may not be necessary that all the ariables shall have its influence in norovirus uniformly. A certain set of variable may be dominant while certain other is more influencing. This can be a hypothetical situation as to which type of variable exerts its influence significantly as a determinant factor is a matter to be taken for hypotheses testing. The nine variables have been identified to run the correlation with "norovirus" which include dummy (coding) variable such as sex, vegetable, milk produce, fish, source of water, nature of diarrhoea, dysentery and nature of vomiting and the other independent variable namely, number of day hospitalized. Correlation matrices were applied to understand and shortlist the number of variables, which correlated with the norovirus. Of the nine variables, nature of vomiting, number of days hospitalized, fresh or preserved food, dysentery, sex and milk were chosen as variables those which exhibit a positive correlation and rest of the factors like source of water, vegetable, nature of diarrhoea turned negative with norovirus. The short listed factors were considered for running the regression

against norovirus.

Tables 2a and b shows the regression and coefficients for the prediction of norovirus from the collected stool samples. The regression analysis brings forth the factors like number of days hospitalized due to diarrhoea and sources of drinking water turned out to be significant (P<0.05). The R square value turns at to be 0.732. This shows that 73% of the variations in dependent factor are explained by these two independent variables.

Conclusion

From the study, it was observed that out of 81 stool samples collected from diarrhoeal patients, 44.4% were positive for norovirus and 55.6% were negative for Norovirus. Presence of norovirus among the stool samples were detected by reverse transcriptase PCR and were confirmed by both agarose gel electrophoresis and by hybridization. This study has monitored norovirus distribution in peak summer month (April) also which is in contradiction with norovirus outbreaks which are generally more prevalent in winter.

ACKNOWLEDGEMENT

This work was supported by grants from Council of

Table 1. Details of the stools samples collected from the patients infected with diarrhoea.

Patient No.	Sex	Age (years)
2941/c	M	52
1282/m	M	0.67
1284/m	M	24
2950/c	M	35
2940	F	25
1285/m	M	1
3257/c	F	32
3251/c	M	38
3252/c	M	20
3244/c	M	21
3246/c	M	24
2984/c	M	48
1299/m	F	22
1298/m	F	28
2942	M	1.5
3472/c	M	7
1569/m	F	1.25
3421/c	M	1
3254/c	M	22
3439	F	70
3507/c	F	1.5
1621/m	F	40
3516/c	M	1.17
3528/c	F	36
3480/c	M	1.25
3501/c	M	75
3499/c	M	1.5
23/m	M	30
17/m	F	37
25/m	M	1
58/m	M	0.92
71/m	M	4
114/c	F	52
127/m	F	25
157/m	F	1.5
31/m	F	49
33/c	F	40
266/m	F	35
270/c	M	1.5
422/c	F	63
614/c	F	24
1038/c	F	47
1036/c	M	20
876/c	F	60
883/c	M	33
1507/c	M	0.92
1459/c	M	21
642/m	F	0.67
643/m	F	40
706/m	F	50
1407/c	M	30

Table 1. Cont.

1408/c	F	38
1406/c	F	6
732/m	M	25
734/m	M	4
746/m	M	31
1503/c	F	0.83
752/m	M	55
1505/c	M	39
751/m	M	35
769/m	F	1.5
1697/c	M	52
842/m	F	35
789/m	F	50
794/m	F	38
793/m	M	56
1602/c	F	25
2144/c	M	35
2487/c	M	45
2516/c	F	55
4089/c	M	15
4064/c	M	62
1766/m	M	20
1793/m	F	50
1844/m	F	46
1848/m	F	35
3784/c	F	51
4267/c	M	23
4317/c	M	38
4523/c	M	0.75
796/m	M	17

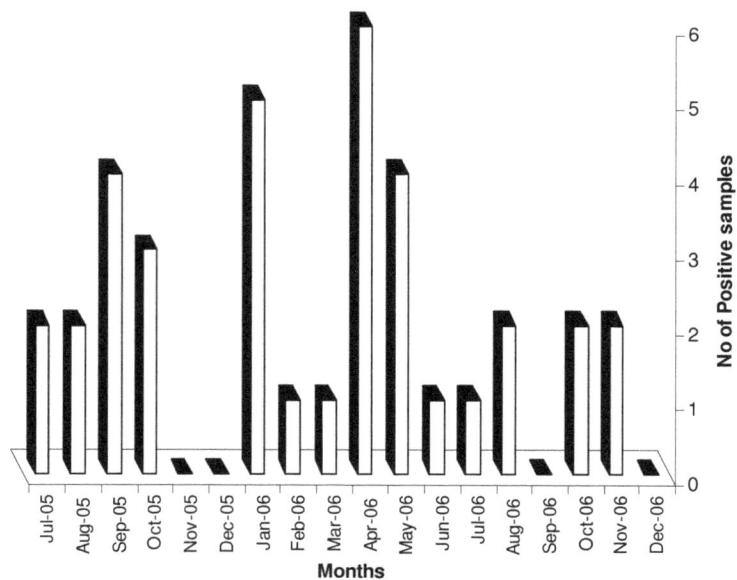

Figure 2. Seasonal distribution of norovirus positive stool samples collected from July '05 to November '06.

Table 2a. Prediction of norovirus – Regression results model summary.

Model	R	R square	Adjusted R square	Standard error of the estimate
1	0.855	0.732	0.698	0.275

Predictors: (constant), nature of vomiting, source of water, no. of days hospitalized, vegetable, nature of diarrhoea, fresh or preserved fish, dysentery, sex and milk.

Table 2b. Coefficients.

	Unstandardized coefficients		Standardized coefficients	t	Sig.
	B	Std. error	Beta		
(Constant)	-0.708	0.169		-4.181	0.000
Sex	3.088E-02	0.067	0.031	0.462	0.646
No. of days hospitalized	0.315	0.025	0.823	12.512	0.000
Vegetable	-4.728E-02	0.042	-0.074	-1.113	0.269
Milk	2.529E-02	0.040	0.044	0.628	0.532
Fresh/Preserved fish	2.279E-02	0.043	0.036	0.527	0.600
Source of water	-0.104	0.053	-0.129	-1.962	0.054
Nature of Diarrhoea	-1.040E-02	0.021	-0.033	-0.500	0.618
Dysentery	2.976E-02	0.027	0.072	1.095	0.277
Nature of vomiting	6.948E-02	0.060	0.078	1.153	0.253

Dependent variable: norovirus.

Scientific and Industrial Research (CSIR), New Delhi-India in the form of Senior Research Fellowship.

REFERENCES

Adler JL, Zickl R (1969). Winter vomiting disease. J. Infect. Dis., 119: 668–673.

Anbazhagi S, Kamatchiammal S (2010). A Comparative Study for the Efficient Detection of Norovirus from Drinking Water by RT-PCR and Real-Time PCR. Water Air Soil Pollution, InPress.

Battacharya SS, Kulka M, Lampel KA, Cebula TA, Goswami BB (2004). Use of reverse transcription and PCR to discriminate between infectious and non-infectious hepatitis A virus. J. Virol. Mtd., 116: 181– 187.

Bern C, Martinez J, de Zoysa I, Glass RI (1992).The magnitude of the global problem of diarrheal disease: a ten year update. Bull WHO, 70: 705-14.

Boom R, Sol CJ, Salimans MM, Jansen CL, Wertheim-van DPM, Van der NJ (1990). Rapid and simple method for purification of nucleic acids. J. Clin. Microbiol., 28: 495–503.

Chadwick PR, Beards G, Brown D, Caujl EO, Cheesbrough J, Clarke I (2000). Management of hospital outbreaks of gastroenteritis due to small round structured viruses. J. Hospital Infect., 45: 1-10.

Cheesbrough JS, Green J, Gallimore CI, Wright PA, Brown DW (2000). Widespread environmental contamination with Norwalk-like viruses (NLV) detected in a prolonged hotel outbreak of gastroenteritis. Epidemiol. Infect., 12: 93- 98.

Doane FW (1994). Electron microscopy for the detection of gastroenteritis viruses. In A. Z. Kapikian (ed.), Viral infections of the gastrointestinal tract. p. 101–130. Marcel Dekker, Inc, New York, N.Y.

Duizer E, Bijkerk P, Rockx B, de GA, Twisk F, Oopmans M (2004). Inactivation of caliciviruses. Appl. Environ. Microbiol., 7: 4538-4543.

Girish R, Broor S, Dar L, Ghosh D (2002). Foodbrone outbreak caused by a Norwalk-like viruses in India. J. Med. Virol., 67: 603-607.

Green J, Gallimore CI, Norcott JP, Lewis D, Brown DWG (1995). Broadly reactive reverse transcriptase polymerase chain reaction in the diagnosis of SRSV-associated gastroenteritis. J. Med. Virol., 47: 392– 398.

Green KY, Chanock RM, Kapikian AZ (2001). Human caliciviruses. In Griffin, Field virology, 4th Edition. Eds. Knipe PM, Howeley DE pp. 841-874. Philadelphia: Lippincott Williams and Wilkins.

Höhne M, Schreier E (2004). Detection and characterization of Norovirus outbreaks in Germany: application of a one-tube RT-PCR using a fluorogenic real-time detection system. J. Med. Virol., 72: 312-319.

Isakbaeva ET, Bulens SN, Beard RS, Adams S, Monroe SS, Chaves SS (2005). Norovirus and child care: challenges in outbreak control. Pediatrics Infect. Dis., 24: 561-3.

Leoni F, Gallimore CI, Green J, McLauchlin J (2003). A rapid method for identifying diversity within PCR amplicons using the heteroduplex mobility assay and synthetic nucleotides: application to characterisation of dsRNA elements associated with Cryptosporidium. J. Microbiol. Mtd., 54: 95–103.

Marks PJ, Vipond IB, Carlisle D, Deakin D, Fey RE, Caul EO (2000). Evidence for airborne transmission of Norwalk –like virus (NLV) in a hotel restaurant. Epidemiol. Infect., pp. 481-487.

Marks PJ, Vipond IB, Regan FM, Wedgwood K, Fey RE, Caul EO (2003). A school outbreak of Norwalk-like virus: evidence for airborne transmission. Epidemiol. Infect., 131: 727- 736.

O'Neill HJ, McCaughey C, Coyle PV, Wyatt DE, Mitchell F (2002). Clinical utility of nested multiplex RT-PCR for group F adenovirus, rotavirus and Norwalk-like viruses in acute viral gastroenteritis in children and adults. J. Clin. Virol., 25: 335–343.

Parashar UD, Monroe SS (2001) 'Norwalk-like viruses' as a cause of Foodborne disease outbreaks. Rev. Med. Virol., 11: 243-252.

Parashar U, Quiroz ES, Mounts AW, Monroe SS, Fankhauser R, Ando T (2001). Norwalk-like viruses, Public health consequences and outbreak management. MMWR Recomm. Rep., 50: 1-17.

Parashar UD, Bresee JS, Glass RI (2003). The global burden of diarrheal disease in children. Bull. World Health Organ. 81, 236- 240.

Preeti C, Shobha DC (2008). Norovirus genotype IIb associated acute gastroenteritis in India. J. Clin. Virol., 42: 429-432.

Rzezutka A, Cook N (2004). Survival of human enteric viruses in the environment and food. FEMS Microbio. Rev., 28: 441–453.

Simpson R, Aliyu S, Iturriza-GM, Desselberger U, Gray J (2003). Infantile viral gastroenteritis: on the way to closing the diagnostic gap. J. Med. Virol., 70: 258–262.

Subekti D, Lesmana M, Tjaniadi P, Safari N, Frazier E, Simanjuntak C, Komalarini S, Taslim J, Campbell JR, Oyofo BA. (2002). Incidence of Norwalk-like viruses, rotavirus and adenovirus infection in patients with acute gastroenteritis in Jakarta, Indonesia. FEMS Immuno. Med. Microbio., 33: 27-33.

Stephen F, Altschul TL, Madden AA, Schäffer JZ, Zheng Z, Webb M, David LJ (1997). "Gapped BLAST and PSI-BLAST: a new generation of protein database search programs", Nucleic Acids Res., 25: 3389-3402.

Wright PJ, Gunesekere IC, Doultree JC, Marshall JA (1998). Small round-structured (Norwalk-like) viruses and classical human caliciviruses in Southeastern Australia, 1980–1996. J. Med. Virol., 55: 312–320.

Wyatt RG, Dolin R, Blacklow R, Dupont HL, Buscho RF, Thornhill TS (1974). Comparison of three agents of acute infectious nonbacterial gastroenteritis by cross-challenge in volunteers. J. Infect. Dis., 129: 709–14.

12

Degree of liver injury in Dengue virus infection

Ali K. Ageep

Department of Pathology, Faculty of Medicine, Red Sea University, Port Sudan, Sudan.
E- mail: aleykh@yahoo.com.

This study evaluated the degree of liver damage during an extensive Dengue Virus epidemic in Port Sudan, Sudan, extending from July to December 2009. In 633 confirmed Dengue cases, the degree of hepatic injury was assessed as follows: Grade 0 - normal levels of liver enzymes; Grade 1 - mild elevation in the liver enzymes, not exceeding the double of the reference value; Grade 2 - elevated liver enzymes, with the levels of the enzymes increased to more than three times the reference values; Grade 3 - acute hepatitis, with liver enzymes levels increased to at least 10 times their normal values; Grade 4 - evidence of hepatic failure (high prothrombin time) or renal involvement (high creatinin). It was observed that 63.8% of Grade 1, 17.9% of Grade 2, 3.9% of Grade 3 and 1.1% of Grade 4 had liver damage. In this study, the severe degree of liver injury existed with the presence of the complications. So an aspartate aminotransferase (AST), at least, should be done regularly in the follow up of Dengue patients.

Key words: Dengue virus, hepatitis, liver failure, Port Sudan.

INTRODUCTION

Dengue virus (DENV) is a member of the Flaviviridae family, which include West Nile virus, yellow fever virus, Japanese encephalitis virus, and tick-borne encephalitis virus, among others (Lindenbach and Rice, 2003). The most serious manifestations of the infection are Dengue hemorrhagic fever (DHF) and Dengue shock syndrome (DSS). Meanwhile, no effective vaccine or antiviral drug therapy is currently available against Dengue virus (Gubler, 2002).

Dengue viral infection has been recognized as one of the world's biggest emerging epidemic. Throughout the tropics, this infection has an annual incidence of 100 million cases of DF with another 250,000 cases of DHF and mortality rate of 24,000 to 25,000 per year (Gubler, 2002; Halstead, 1999). In the last years, Port Sudan has faced many outbreaks; one was reported in 2005. After this outbreak, Port Sudan has become an endemic area with Dengue virus (Ageep et al., 2006). Dengue virus type 3 is the common circulating serotype in the region (Amal et al., 2010).

Typically, people infected with Dengue virus are asymptomatic (80%) or have mild symptoms such as an uncomplicated fever (Whitehorn, 2010; Reiter, 2011). Others have more severe illness (5%), and in a small proportion it is life-threatening (Whitehorn, 2010; Reiter, 2011). The incubation period ranges from 3 to 14 days,

but most often it is 4 to 7 days (Chen et al., 2010). The characteristic symptoms of Dengue are: a sudden-onset fever, headache (typically behind the eyes), muscle and joint pains, and a rash. The alternative name for Dengue, "break-bone fever", comes from the associated muscle and joints pains (Varatharaj, 2010). Severe disease is marked by two problems: dysfunction of endothelium and disordered blood clotting (Martina et al., 2009). Endothelial dysfunction leads to the leakage of fluid from the blood vessels into the chest and abdominal cavities, while coagulation disorder is responsible for the bleeding complications. Higher levels of virus in the blood and involvement of other organs (such as the liver) are also associated with more severe disease (Seneviratne et al., 2006).

Dengue may occasionally affect several other body systems (George, 1997). This may be either in isolation or along with the classic Dengue symptoms (Seneviratne et al., 2006). Hepatic dysfunction is common in Dengue infection, and is attributed to a direct viral effect on liver cells or as a consequence of dysregulated host immune responses against the virus. Other contributing factors include race, diabetes, hemoglobinopathies, pre-existing liver damage and the use of hepatotoxic drugs (George, 1997; Chen et al., 2004). Although there are isolated case reports of fulminant hepatic failure, the derangements in the transaminases are usually mild and self-limiting (George, 1997).

Although, the number of patients affected by the virus is increasing each year, little work has been done in the studied area (regarding the pathogenesis, the liver changes and the complications of Dengue infection). Hence, this study was aimed at evaluating the degree of liver injury by measuring the level of the liver enzymes, prothrombin time (PT) and creatinine in Dengue-infected patients. These parameters were compared with the clinical presentations of the patients evaluate how the degree of liver damage is related to the complications of the disease.

This study has many significant values. First, it will increase the awareness of the local health staff about the importance of evaluating the degree of liver damage in Dengue infected patients. It also highlighted the importance of measuring the liver enzymes (at least the aspartate aminotransferase (AST)) in the follow-up of Dengue virus infection. Finally, this research form a base for future studies in the region regarding the outcome, the mortality, the hospital stay and the prognosis of Dengue infection according to the level of liver damage.

METHODOLOGY

Study area/setting

This study was conducted in the outpatient and inpatient departments of Port Sudan Teaching Hospital, Port Sudan, Sudan, during the outbreak of Dengue infection in the period from July to December, 2009. Port Sudan city is the capital of the Red sea state and it is the major sea port of the Sudan. The total number of the whole population was (739,300) according to the national census of 2002, with adjusted growth rate. There are four localities in this area (Port Sudan, Sinkat, Tokar and Halayib). Port Sudan Teaching Hospital is a governmental hospital, which is regarded as a tertiary care hospital. The total number of beds is about 380, and the medical services are opened for all population with the aid of the best professional staff in the region.

Study design

This is a descriptive, hospital-based study.

Study subjects

This included all confirmed cases during the outbreak of Dengue infection in the period from July to December 2009, who were seen in Port Sudan Teaching Hospital. This study constituted all age groups from infants to old patients.

Exclusion criteria

All of the patients were tested first for malaria, typhoid, leptospira and brucellosis. Any patient infected with these diseases was excluded from the research. Patient with positive tests of hepatitis A, B, C, E or even had recent history of infection with these hepato-tropic viruses was also excluded from this research.

Data collection

The clinical information were taken from the patients and registered in predesigned questionnaires. The following information was included:

1. Clinical presentation: classic or not.
2. Hemorrhage: present or not.
3. Dengue shock: present or not.
4. Features of encephalopathy: present or not.
5. Features of renal impairment: present or not.
6. Gastrointestinal bleeding: present or not.
7. Features of cholecystitis: present or not.

Ethical clearance

Informed consent was taken from all patients participating in this research. Ethical clearance was approved from the local ethical review committee (ERC).

Enzyme-linked immunosorbent assay (ELISA) technique

Antibodies against Dengue virus antigens were test by ELISA technique (ELISA test NovaTec Germany). This test had 98% sensitivity and 95% specificity. In the procedure of test the following steps was taken:

1. Microtiter strip wells were pre-coated with Dengue virus antigen type 2 to bind to the corresponding antibodies of the specimens.
2. Afterward, the wells were washed to remove all the unbound sample material.
3. Then, horseradish peroxidase (HRP) labeled anti-human IgM conjugate was added. The conjugate bound to the captured Dengue virus-specific antibodies.
4. A second step of washing was formed.
5. The immune complex formed by the bound conjugate was visualized by adding tetramethylbenzidine (TMB) substrate, which gave blue reaction product. The intensity of the product was proportional to the amount of Dengue virus-specific antibodies in the specimen.
6. Sulfuric acid was added to stop the reaction. This produced a yellow end point color.
7. The test was red using ELISA micro-well plate reader at 450 nm.

Biochemical and coagulation tests

To study the degree of liver damage, samples were collected in to two blood containers. The first was a plain container into which serum was extracted for the assessment of the liver enzymes - aspartate aminotransferase (AST) and alanine transaminase (ALT), and creatinine levels. The second was tri-sodium citrate container from which plasma was used to detect the prothrombin time (PT) level. A semi-automated spectrophotometer (Bio-system) was used for the measurement of these biochemical tests.

Liver injury scoring system

The degree of liver damage was assessed according to the levels of the liver enzymes, PT and creatinine as follows: Grade 0 - normal levels of liver enzymes; Grade 1 – mild elevation in the liver enzymes, not exceeding the double of reference value; Grade 2 - elevated liver enzymes, with the levels of the enzymes increased to more than three times the reference values; Grade 3 - acute

Table 1. Relation between clinical presentation and the degree of liver injury in patients infected with Dengue virus.

Clinical presentation	N	F (%)	Males (%)	Females (%)	Grade 0 (%)	Grade 1 (%)	Grade 2 (%)	Grade 3 (%)	Grade 4 (%)
Dengue fever	567	89.6	248 (39.2)	319 (50.4)	82 (12.9)	394 (62.2)	91 (14.4)	0	0
Dengue hemorrhage	28	4.5	15 (2.3)	13 (2.2)	2 (0.3)	6 (0.9)	12 (1.9)	8 (1.3)	0
Dengue shock syndrome	10	1.6	4 (0.6)	6 (0.9)	0	2 (0.3)	3 (0.5)	5 (0.8)	0
Gastro-intestinal bleeding	16	2.5	10 (1.6)	6 (0.9)	0	1 (0.2)	6 (0.9)	9 (1.4)	0
Encephalopathy	4	0.6	2 (0.3)	2 (0.3)	0	0	0	0	4 (0.6)
Cholecystitis	2	0.3	1 (0.2)	1 (0.2)	0	1 (0.2)	1 (0.2)	0	0
Renal impairment	6	0.9	0	6 (0.9)	0	0	0	3 (0.5)	3 (0.5)
Total (%)	633	100	280 (44.2)	353 (55.8)	84 (13.2)	404 (63.8)	113 (17.9)	25 (4.0)	7 (1.1)

N, Number of the patients; F, frequency (%).

hepatitis, with liver enzymes levels increased to at least 10 times their normal values; Grade 4: evidence of hepatic failure (high PT) or hepato-renal involvement (high creatinine).

The reference values for the tests

The normal ranges for the blood samples according to control normal people in the area, were as follows:

1. ALT: 7 to 56 Unit/L
2. AST: 5 to 40 Unit/L
3. Creatinine: 0 to 1 mg/dl
4. PT: 11 to 15 s

Statistical analysis

Statistical analysis was done using SPSS program. Chi square test was used to compare categorical variables and Fischer exact test were applicable. For none normally distributed quantitative variables, median and Inter quartile ranges (IQR) were used. Chi square test was used to compare categorical variables and Fischer exact test were used when numbers were too small to perform the Chi-square testing. Results were presented as frequency and percentage.

RESULTS

Six hundred and thirty three confirmed Dengue patients were included in this study. 248 were male and 319 were female (male to female ratio was approximately 3:4).

Degree of liver injury

Table 1 shows that, 13.2% of the patients had no increase in the transaminases level (Grade 0), 63.8% presented mild alterations in the liver enzymes levels (Grade 1), 17.9% presented (Grade 2) liver involvement, 3.9% of the patients had progressed to acute hepatitis (Grade 3) and 1.1% had severe liver damage with fulminant hepatic failure (Grade 4).

Changes in the liver enzymes

In 86% of the patients there was elevation of the liver enzymes. All of them (549 patients) had increase in the AST level. The change in the ALT was seen in 82% of the patients.

Uncomplicated Dengue infection

Most of the patients (567 patients) had features of Dengue fever without complications. In this group, the common degree of liver damage (69%) was Grade 1 (that is. mild elevation in the liver enzymes).

Complicated Dengue infections

Dengue hemorrhage

Twenty eight patients presented with Dengue hemorrhage and 12 of them (43%) had Grade 2 liver injury.

Dengue shock syndrome

Ten patients presented with Dengue shock syndrome. In this group, 5 patients (50%) had (Grade 3) liver injury.

Gastro-intestinal bleeding

From the 16 patients presented with gastro-intestinal bleeding, 56% had (Grade 3) liver injury.

Encephalopathy

4 patients presented with encephalopathy and allof them had (Grade 4) liver damage.

Renal impairment

Of the 6 patients who were complicated by renal impairment, 50% had Grade 3 and 50% had (Grade 4) liver damage.

DISCUSSION

Dengue virus is among the commonest causes of febrile illnesses in Port Sudan, Sudan (Ageep et al., 2006). The importance of this study lies in the fact that it is the first documented research in this region of Africa, which studied the severity of liver damage in Dengue infection. In this research, we included Dengue patients with mild symptoms seen in the outpatient department and severe cases who were admitted in the impatient units; so this study covers the mild as well as the severe cases of Dengue virus infections.

To date, there are two hypotheses that explain the damage of the liver in Dengue patients. The first is the immune enhancement hypothesis. Chen et al. (2004) reported that strong correlation was found between T cell activation and hepatic cellular infiltration in immuno-competent mice infected with Dengue virus. They noted that the kinetics of liver enzyme elevation also correlated with that of T cell activation and suggested a relationship between T cell infiltration and elevation of liver enzymes. Chaturvedi et al. (1999) in their study detected the appearance of different helper cells cytokines in human white blood cells cultures infected in vitro with Dengue virus type 2. In their study, they have reported that during Dengue infection, monocytes, B cells, T cells and mast cells produce large amounts of cytokines. Despite all this, the role of host immunity in Dengue infection is still very unclear. Unregulated host immune response may play a role in severity of Dengue infection by modifying the immune response; severe infection can be prevented. The second hypothesis relates the damage in the liver to direct virulence of the virus (Seneviratne et al., 2006). According to these studies, we can hypothesize the same mechanism responsible for the liver damage that occurred in our patients.

Liver damage with elevation of aminotransferases and reactive hepatitis is a common complication of dengue virus infection. Hence, measurement of AST and ALT is mandatory to ascertain the liver involvement (Souza et al., 2004). In this study, 86% of our patients had high AST level and 82% had high ALT level. However, Wong and Shen (2008) reported that AST abnormality was predominantly higher compared to ALT; 91 and 72%, respectively. Another study done by Kuo et al. (1992) has shown approximately 90% of the AST abnormality in Dengue patients. A different retrospective study from Wichman et al. (2004) among Thai patients in 2001 outbreak, reported liver dysfunction in 20 from 347 patients (5.8%) with Dengue infection. Our study is consistent with the results from previous studies. The difference from Wichman et al. (2004) results may be explained by the difference in the Dengue virus type involved in their outbreak, or its hepatotoxicity. Other differences may be in the immune status of their patients or the days of collection of their serum samples (Wichman et al., 2004).

In this study, we noticed relation between the degree of liver damage and the presence of the complications. In 71% of the patients having Dengue hemorrhage, severe degree of liver damage (Grade 2 and 3) occurred. The deranged liver functions may participate in the causation of bleeding in these patients. Severe degree of liver injury (Grade 2 and 3) also was found in 80% of Dengue shock syndrome. All of the patients having encephalopathy had (Grade 4) liver damage. Encephalopathy in our patients may be due to fulminant hepatic failure or a high level of the virus that directly damage the brain. Involvement of the kidneys was also related to the severity of liver damage; 50% with Grade 3 and 50% with Grade 4. Again, this may be a part from hepato-renal syndrome or direct virus virulence. Similar results to our work, in complicated Dengue infection, were also seen in other countries. In Saudi Arabia, Khan et al. (2008) had made an association between high AST level and complications of Dengue virus. In Taiwan, Kuo et al. (1992) also reported higher bleeding episodes in those who had high levels of AST and ALT. In Vietnam, Nguyen et al. (1997) reported that DHF may cause mild to moderate liver dysfunction in most cases; only few patients may suffer from acute liver failure leading to encephalopathy and death. Additionally, a report from India done by Shah (2008) pointed to a high mortality in Dengue patients with hepatitis and encephalopathy.

Since Port Sudan is one of the endemic areas with Dengue infection, Dengue virus should be added to the differential diagnosis of hepatitis in the local hospitals protocol. AST can be a useful surrogate marker to predict disease severity and bleeding outcome in Dengue infection, therefore it should be measured in all Dengue patients. The level of the other liver enzymes, PT and creatinine should also be assessed in severe cases of Dengue infection. We suggest the grading system presented in this study to be applied in Dengue virus management protocol.

REFERENCES

Ageep AK, Aml AM, Mubarak SE (2006). Clinical presentations and laboratory findings in suspected cases of Dengue virus. Saudi Med. J. 27(11):1711-1713.

Amal M, Kenneth E, Emad M, Magdi S, Mubarak S, Ali KA (2010). Dengue hemorrhagic fever outbreak in children in Port Sudan. J. Infect. Pub. Health 4(1):1-6.

Chaturvedi UC, Elbishbishi EA, Agarwal R, Raghupathy R, Nagar R, Tandon R, Pacsa AS, Younis OI, Azizieh F (1999). Sequential production of cytokines by Dengue virus-infected human peripheral blood leukocyte cultures. J. Med. Virol. 59(3):335-340.

Chen HC, Lai SY, Sung JM, Lee SH, Lin YC, Wang WK (2004). Lymphocyte activation and hepatic cellular infiltration in immuno-competent mice infected by Dengue virus. J. Med. Virol. 73(3):419-431.

Chen LH, Wilson ME (2010). Dengue and chikungunya infections in travelers. Curr. Opin. Infect. Dis. 23(5):438-444.

George R (1997). LLCS. Clinical spectrum of Dengue infection. Washington: Cab Int. pp. 23-25.

Gubler DJ (2002). Epidemic Dengue/ Dengue hemorrhagic fever as a public health, social and economic problem in the 21st century. Trends microbial. 10(2):100-103.

Halstead SB (1999). Is there an inapparent Dengue explosion? Lancet 353(9158):1100-1101.

Khan NA, Azhar EI, El-Fiky S, Madani HH, Abuljadial MA, Ashshi AM, Turkistani AM, Hamouh EA (2008). Clinical profile and outcome of hospitalized patients during first outbreak of Dengue in Makkah, Saudi Arabia. Acta tropica 105(1):39-44.

Kuo CH, Tai DI, Chang-Chien CS, Lan CK, Chiou SS, Liaw YF (1992). Liver biochemical tests and Dengue fever. Am. J. Trop. Med. Hyg. 47(3):265-270.

Lindenbach BD, Rice CM (2003). Molecular biology of flaviviruses. Adv. Virus Res. 59:23-61.

Martina BE, Koraka P, Osterhaus AD (2009). Dengue virus pathogenesis: An integrated view. Clin. Microbiol. Rev. 22(4):564-581.

Nguyen TL, Nguyen TH, Tieu NT (1997). The impact of Dengue haemorrhagic fever on liver function. Res Virol. 148(4):273-277.

Reiter P (2011). Yellow fever and Dengue: a threat to Europe? Eur. Surveill. 15(10):19509.

Seneviratne SL, Malavige GN, de Silva HJ (2006). Pathogenesis of liver involvement during Dengue viral infections. Trans. R Soc. Trop. Med. Hyg. 100:608-614.

Shah I (2008). Dengue and liver disease. Scand. J. Infect. Dis. 40(11-12):993-994.

Souza LJ, Alves JG, Nogueira RM, Gicovate NC, Bastos DA, Siqueira EW, Souto FJT, Cezário TA, Soares CE, Carneiro RC (2004). Aminotransferase changes and acute hepatitis in patients with Dengue fever: analysis of 1,585 cases. Braz. J. Infect. Dis. 8(2):156-163.

Varatharaj A (2010). Encephalitis in the clinical spectrum of Dengue infection. Neurol. India 58(4):585-591.

Whitehorn J, Farrar J (2010). "Dengue". Br. Med. Bull. 95:161-173.

Wichman O, Hongsiriwon S, Bowonwatanuwong C, Chotivanich K, Sukthana Y, Pukrittayakamee S (2004). Risk factors and clinical features associated with severe Dengue infection in adults and children during the 2001 epidemic in Chonburi, Thailand. Trop. Med. Int. Health 9(9):1022-1029.

Wong M, Shen E (2008). The utility of liver function tests in Dengue. Ann. Acad. Med. Singap. 37(1):82-83.

13

Serodiagnosis evaluation of rabies and animal bites in North of Iran, 2010

Behzad Esfandiari[1], Mohammad Reza Youssefi[2]* and Ahmad Fayaz[3]

[1]Pasteur Institute of Iran-Amol Research Center, Iran.
[2]Department of Veterinary Parasitology, Islamic Azad University, Babol – Branch, Iran.
[3]Rabies Section in Pasteur Institute of Iran.

Rabies disease is one of the most important public health problems in some countries of the world such as those in the Eastern Mediterranean region. According to the world health organization reports, more than 10 million people who are bitten by animals are annually treated by prophylactic treatment regimen of rabies in the world. The present study was undertaken to evaluate the prevalence and other information's about rabies as well as the variables related to the bitten persons during 2010 in North of Iran. After sending samples to north unit of Pasteur Institute located in Amol City, all samples were analyzed by indirect immunofluorescent technique and in the case of observing negri-bodies, samples were announced as positive. Also, negative samples were injected to mice. During the study period, 22 exposed persons treated for animal bites were included in our study that 14 (63.63%) male and 8 (36.37%) female. Injuries take place in hand, leg, face and abdomen that most injury observed in hand. Because rabies is endemic in wild life of Iran, infection of domestic animals is occurring repeatedly, also increasing population of stray dogs and developing statistics of animal bite cases and dissipation of rabies in most provinces of Iran, notifies a need to pay more attention on controlling this disease.

Key words: Rabies, animal bitten, Pasteur Institute, Iran.

INTRODUCTION

Rabies is an acute fatal viral encephalitis that usually transmitted from animals to man followed by domestic and wild animal bites (Rad et al., 1999). Rabies disease is one of the most important public health problems in some countries of the world such as those in the Eastern Mediterranean region (WHO, 2010). According to the world health organization reports, more than 10 million people who are bitten by animals are annually treated by prophylactic treatment regimen of rabies in the world. About 50,000 human deaths are annually reported due to rabies (Simani, 2004). In Asia, most of the mortality cases of human rabies were reported from the under-developed countries such as India, Pakistan and Bangladesh which have high populations and have no specific strategies for controlling rabies. The real numbers of human deaths due to rabies in these

countries are more than these numbers, because there is no advanced surveillance system of disease control to find out the real numbers of infected and fatal human cases (WHO, 2010). The present study was undertaken to evaluate the prevalence of rabies as well as the variables related to the bitten persons during 2010 in North of Iran.

MATERIALS AND METHODS

In this Cross-Sectional study, data related to morbidity of human rabies cases and all of recorded data related to persons who were bitten by animals during 2010 in Ardabil, Talesh, Rodsar, Lahijan, Behshahr, Bandargaz and Aliabad katol cities were analyzed. Suspicious cases of rabies were transferred to Pasteur Institute of North in two ways. In initial years, the head of the animal was transferred to this institute for security issues. During 2010, samples taken from occipital area of the brain was transferred to this unit in kits manufactured by Pasteur Institute Company (Pas-78) which contained preservative solution. The samples all were analyzed by Indirect Immunoflorescent technique and in case of observing negri-bodies the samples were announced as positive. But if the

*Corresponding author. E- mail: youssefi@baboliau.ac.ir.

Table 1. Prevalence of rabies and animal bites in north of Iran, 2010 according to morbidity, sex, kind and location of injury.

	Number of samples	Sex		Location of injury					Kind of injury		
		Male	Female	Hand	Leg	Face	Abdomen	Hand and Leg	Surface	Moderate	Deep
Numbers	22	14	8	10	4	1	2	5	14	5	3
Percent (%)	100	63.63	36.37	45.45	18.18	4.54	9.09	22.74	63.63	22.72	13.65

Table 2. Determination of animals bites in north of Iran, 2010 according to type of the animal.

	Stray dog	Wolf	Jackal	Cow	Squirrel
Numbers	14	2	1	4	1
Percent (%)	63.63	9.09	4.54	18.18	4.54

sample was negative a small amount of the sample was pounded in porcelain pounder and poured in 20 cc vials and then serum saline containing 500 Penicillin and 1560 µg/ml Streptomycin was added to it and a suspension of 10 to 20% was prepared. After 30 min, 30 µl of this suspension was injected to 12 Balb/c mice intracerebrally using gage 26 syringe and these mice were kept for 28 days. Fatalities up to fifth day were not due to rabies and after that were attributed to rabies. Then, wet mounts from the brain of dead mice were prepared and analyzed by Indirect Immunofluorescent technique. Conducted data was statically analyzed using SPSS software and T and Chi square tests.

RESULTS

During the study period, 22 exposed persons treated for animal bites were included in study that 14 (63.63%) were male and 8 (36.37%) were female. Injuries took place in hand, leg, face and abdomen and most of injuries were observed in hand (Table 1).

In this study animals suspicious to rabies that had attacked human were also analyzed. In positive cases, dogs with 63.63% and jackal and squirrel with 4.54% respectively had the highest and lowest attack rates (Table 2).

The highest number of infected cases was in range of 21 to 40 years old individuals (45.45%) and the lowest was in those in range of fewer than 20 years old (22.72%) (Figure 1).

DISCUSSION

Rabies has a special place in the history of medical research and is one of the diseases that can be important in human health (Janani et al., 2008). Europe and North America have successfully controlled rabies in domestic animals and only wild animals are the source of rabies in these countries and dog bite is still the main way of transmission of disease to human (Alavi and Alavi, 2008; Krebs et al., 2004). As also shown in this study stray dog bite allocates the highest rate of incidence of rabies disease to itself (57.39% of the cases) which in comparison to other animals is significantly higher, that in development country among Iran basal problem is lack of control of stray dogs.

According to Zeynali's report, more than 50,000 individuals each year receive anti rabies treatments in Iran because of getting bitten by animals suspicious to rabies. Also, according to their analysis young individuals are at greater risk and more than 90% of cases are male (Zeynali et al., 1999) which is consistent with data obtained in present study.

In domestic animals the highest numbers of positive cases were seen in cows (21.73%) and according to Pasteur Institute's, reports positive cases of rabies in cows in year 2003 and 2002, respectively were 56.3 and 52.4%. Higher infected cases of cows in comparison to other domestic animals can be due to higher sensitivity to infection and also distinct symptoms of the disease in this animal. Whereas, sensitivity of sheep and goat is less than cow and symptoms of the disease in these animals is not distinct like cow and many cases of rabies in sheep and goats can be misdiagnosed with other diseases and so will not be reported (WHO, 2000; Simani, 2003).

In a study conducted by Rezaeinasab (1904 to 2003) in Kerman province of Iran, 10 individuals were infected with rabies disease which 2 of them were female and 8 were male and half of them

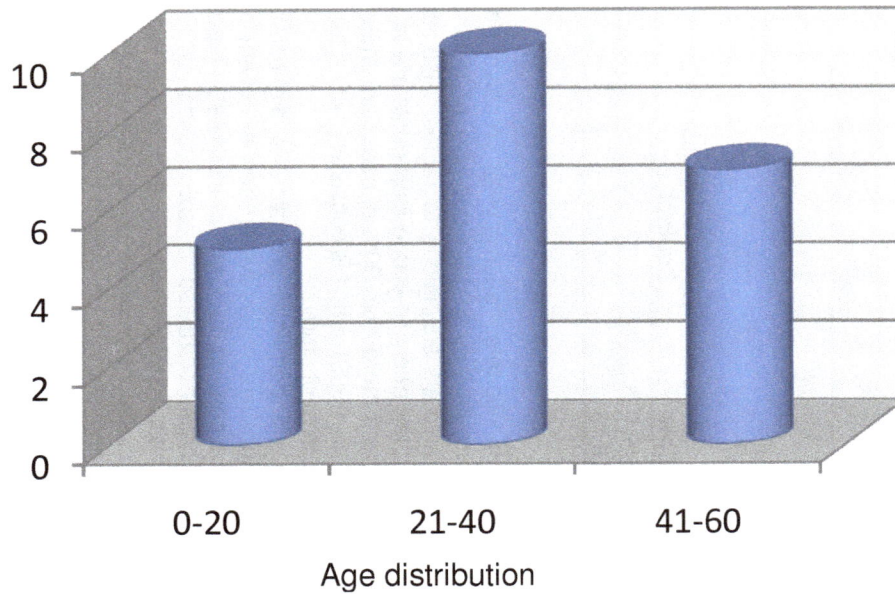

Figure 1. Age distribution of persons who were bitten by animals in North of Iran, 2010.

were attacked by dogs (Rezaeinasab et al., 2007).

Alavi demonstrated that of 894 patients 62.0% were male, and average age of males and females were 24.4 years and 26.2 years, respectively. Dogs, scorpions, mice and snakes were the most commonly involved animal species, causing injuries with a frequency of 69, 12.5, 8.8 and 4.4%, respectively. Feet (58.1%) and hands (30.6%) were the most commonly affected body parts, followed by the face and other parts. Infectious complications were seen in 127 patients, among them 94 soft tissue infections (74.1%), 28 cases of sepsis (22.0%) and five of endocarditis (3.9%). Thirty-five cases (3.9%) died following animal bites and stings, among them 28 (80%) due to scorpion stings, 4 (11.4%) related to dogs and 3 (8.6%) from snake bites. No cases of rabies were observed in these patients (Sheikholeslami et al., 2009).

Because rabies is endemic in wild life of Iran, infection of domestic animals is occurring repeatedly, also increasing population of stray dogs and developing number of animal bite cases and dissipation of rabies in most provinces of Iran, notifies a need to pay more attention on controlling this disease.

ACKNOWLEDGEMENTS

The authors are grateful to Mr. Behzadi for their time and help, and also for financial support of Pasteur Institute of Iran-Amol Research Center.

REFERENCES

Alavi SM, Alavi L (2008). Epidemiology of animal bites and stings in Khuzestan, Iran, 1997- 2006. J. Inf. Pub. Health, pp. 51-55.

Janani AR, Fayaz A, Simani S, Farahtaj F, Eslami N, Howaizi N, Biglari P, Sabetghadam M (2008). Epidemiology and control of rabies in Iran. Dev. Biol., 131: 207-211.

Krebs JW, Wheeling JT, Childs JE. (2004). Rabies surveillance in the United States during 2002. J. Am. Vet. Med. Assoc., 224(5): 705-708.

Rad MA, Firoozbakhsh F, Hemmat K (1999). *Zoonoses updates*. 1st. ed. (compiled in Persian language from AVMA articles edited by William Clark), Published by University Press Center of Iran, Tehran, Iran. Pp. 224-231.

Rezaeinasab M, Rad I, Bahonar AR, Rashidi H, Fayaz A, Simani S (2007). The prevalence of rabies and animal bites during 1994 to 2003 in Kerman province, southeast of Iran. Iranian J. Vet. Res., pp. 343-350.

Sheikholeslami NZ, Rezaeian M, Salem Z (2009). Epidemiology of animal bites in Rafsanjan, southeast of Islamic Republic of Iran, 2003-2005. East. Med. Health. J., 15(2): 455-457.

Simani S (2003). Rabies situation in Iran. J. Vet. Med., 2: 275-278.

Simani S (2004). Rabies disease book. 1st. ed. compiled in Persian language, Published by Pasteur Institute of Iran. pp. 150-151.

World Health Organization in the Eastern Mediterranean Region (2010). Annual reports of regional director (1990-2010), Alexandria, Regional Office for Eastern Mediterranean Region. pp. 29-35.

Zeynali M, Fayaz A, Nadim A (1999). Animal Bites and Rabies: Situation in Iran. Archi. Iran. Med., 2(3): 120-124.

14

Detection of human T-cell lymphotropic virus Type-1 among patients with malignant hematological diseases in Capital of Iran, Tehran

Seyed Hamidreza Monavari[1]*, Hossein Keyvani[1], Hamidreza mollaie[2], Mehdi fazlalipour[2], Farzin sadeghi[2], Mostafa Salehi-Vaziri[1], Roghaeh mollaie[3] and Farah Bokharaei-Salim[1]

[1]Department of Virology and Antimicrobial Resistance Research Center, Tehran University of Medical Sciences, Tehran, Iran.
[2]Department of Medical Virology, Tehran University of Medical Sciences, Tehran, Iran.
[3]Department of Medical Technology, Hazrat rasool Hospital, Tehran University of Medical Sciences, Tehran, Iran.

Human T-cell lymphotropic virus type-1 (HTLV-1) is a deltaretrovirus linked causally to adult T-cell leukemia or lymphoma (ATL), and HTLV-1-associated myelopathy/tropical spastic paraparesis (HAM/TSP). The aim of this study was to detect HTLV-1 infection in patients with malignant hematological diseases and also determining the prevalence of HTLV-1 in these patient groups. Sixty patients with malignant hematological diseases were included in the study and tested by enzyme-linked immunosorbent assay (ELISA) for anti-HTLV-1, and Real time-PCR for the sequences from HTLV-1 tax gene. The mean age of patients was 33.9 ± 18.3 years. 18 of the subjects were found HTLV-1 seropositive using ELISA and the viral prevalence by Real time-PCR was 12%. HTLV-1 was found in 25% of patients with acute myelogenous leukemia (AML), 58.3% of patients with chronic myelogenous leukemia (CML), 16.7% of patients with acute lymphoblastic leukemia (ALL), and no detected in patients with lymphoma. The present study revealed that HTLV-1 is prevalent in patients with malignant hematological diseases and in our study. The major HTLV-1 associated syndromes were chronic myelogenous leukemia and acute lymphoblastic leukemia.

Key words: Human T-cell lymphotropic virus type-1, malignant hematological diseases, prevalence, Iran.

INTRODUCTION

A type C retrovirus human T-cell lymphotropic virus type 1 (HTLV-1) is the causative agent of two distinct human diseases, adult T-cell leukemia or lymphoma (ATL), and a chronic progressive demyelinating disorder known as HTLV-1-associated myelopathy/tropical spastic paraparesis (HAM/TSP) (Matsuoka and Jeang, 2007). HTLV-1 infection has also been associated with a variety of chronic inflammatory diseases such as uveitis (Mochizuki et al., 1996), Sjo¨gren's syndrome (Eguchi et al., 1995), chronic arthropathy (Hasunuma, 1997), infective dermatitis (Lee and Schwartz, 2011), polymyositis synovitis (Sowa, 1992), thyroiditis

(Desailloud and Hober, 2009), and bronchioalveolar pneumonitis (Sugimoto et al., 1993). The role of HTLV-1 infection in these disorders is still under investigation. It is estimated that 10 to 20 million people world-wide are infected with HTLV-1 (Johnson et al., 2001). This infection is endemic in southern Japan, the caribbean basin, central Africa, central and south America, the melanesian islands in the Pacific basin, and in the aboriginal population in Australia (Proietti et al., 2005). In Iran, this virus has been found in isolated pockets that HTLV-1 infection is endemic (Khorasan, the northeastern province of Iran). The prevalence of HTLV-1 infection in Mashhad was 0.77% among blood bank donors (Tarhini et al., 2009). However, little is known on the prevalence of HTLV-1 in patients with malignant hematological diseases in Iran, including a possible HTLV-1 association with other malignancies (Table 1).

*Corresponding authors. E-mail: hrmonavari@yahoo.com.

Table 1. Prevalence of HTLV-1 among patients with malignant hematological diseases in Iran.

Hematological malignancies	Patient No	Mean age	Male / female	HTLV-I antibody positive (%)	HTLV-I PCR positive (%)
Acute myelogenous leukemia (AML)	26	29.8±12.2	19/7	8 (30.8)	3 (11.5)
Chronic myelogenous leukemia (CML)	21	51±14.6	16/5	7 (33.3)	7 (33.3)
Lymphoma	3	27.3±2.5	2/1	1 (33.3)	0 (0)
Acute lymphoblastic leukemia (ALL)	10	12.3±6.2	7/3	2 (20)	2 (20)
Total	60	33.9±18.3	44/16	18 (30)	12 (20)

The association between retroviruses and hematologic malignancies is also described. There have been few studies on the association between human T cell lymphotropic virus type 1 (HTLV-1) infection and malignancy risk (Inoue et al., 2008). It is still controversial whether or not HTLV-1 infection affects the incidence of several malignancies. Adedayo et al. (2004) found an association between HTLV-1 and lymphoid malignancies in Dominican population. There are case reports of ATL (Starkebaum et al., 1987). Little is known on the prevalence of HTLV-1 in patients with various hematologic. The association between retroviruses and hematologic malignancies is also described. There have been few studies on the association between human T cell lymphotropic virus type 1 (HTLV-1) infectionand malignancy risk (Inoue et al., 2008). It is still controversial whether or not HTLV-1 infection affects the incidence of several malignancies.

Adedayo et al. (2004) found an association between HTLV-1 and lymphoid malignancies in Dominican population. There are case reports of ATL HTLV-1 in lymphoid malignancies except ATL (Starkebaum et al., 1987). Little is known on the prevalence of HTLV1 in patients with various hematologic malignancies, therefore we studied the prevalence of HTLV1 carriers among patients myelogenous leukemia (CML), acute lymphoblastic leukemia with acute myelogenous leukemia (AML), chronic (ALL) and lymphoma.

MATERIALS AND METHODS

Study design

In this cross-sectional study 60 patients with established malignant hematological diseases who were admitted to oncology Unit of Hazrate Rasul Hospital, Tehran, Iran, from 2009 to 2010 were enrolled. Diagnosis of malignancy was confirmed based on pathology (histology) findings.

The malignancies were as follows: acute myelogenous leukemia (AML) (26 cases), chronic myelogenous leukemia (CML) (21 cases), acute lymphoblastic leukemia (ALL) (10 cases) and lymphoma (3 cases).

Collection and preparation of samples

About 5 ml of peripheral blood were collected from each patient into EDTA-containing vacutainer tubes. Plasma was stored at -70°C until anti HTLV-1 antibody analysis. Blood buffy coat were isolated from EDTA-treated blood by centrifugation and stored at -70°C for later detection. All patients gave written consent to participate in this study, which conforms to the guidelines of the 1975 Declaration of Helsinki.

Immunoassay for anti-HTLV-1

Serum samples were examined for anti HTLV-1 antibody by enzyme-linked immunosorbent assay (ELISA) method using anti HTLV-1 antibody kit (ELISA; Dia. Pro Diagnostic Bioprobes, Milan, Italy). Assay procedures and the interpretation of the results were performed in accordance with the instructions provided by the manufacturer.

Real time polymerase chain reaction

To detect HTLV-1 provirus in peripheral blood cells DNA was extracted from 200 µl blood Buffy coat using the High pure extraction kit (Roche Diagnostics GmbH, Mannheim, Germany). Quantitative determination of the amplified products was done with the Rotor Gene 6000 (Corbett Research, Australia) Real-time detection system in accordance with the instructions provided by the manufacturer and subjected to PCR with a Maxima probe qPCR Master Mix (2X) kit (Fermentas, Germany). The primer set for the HTLV- 1 tax gene was PXF (5'-CAAACCGTCAAGCACAGCTT-3') positioned at 7163 to 7182 and PXR (5'-TCTCCAAACACGTAGACTGGGT- 3') positioned at 7385 to 7364 and the probe for HTLV-1 tax gene was PXT (5'-TTCCCAGGGTTGGACAGAGTCTTCT- 3') positioned at 7331 to 7355 (Gabet et al., 2003). The thermal cycler profile is optimized and validated with heat activation (15 min at 95°C) of hot-start Taq polymerase was followed by 40 cycles of denaturation (30 s at 95°C), annealing (30 s at 50°C), and extension (30 s at 72°C). For positive control, DNA was extracted from a HTLV-1 producing human T-cell line (MT-2).

Statistical analysis

Data analyses were done by SPSS software version 11 (SPSS, Chicago, IL). Descriptive results were presented as frequencies, and 95% confidence intervals.

Table 2. Demographic characteristics of patients positive to HTLV-1.

Case	ELISA	PCR	Syndrome	Duration of blood transfusion in month	Age/gender
1	+	-	Lymphoma	Unknown	25/M
2	+	+	CML	2	69/M
3	+	+	AML	2	50/F
4	+	-	AML	3	31/M
5	+	+	AML	1	31/M
6	+	+	CML	2	42/F
7	+	+	CML	3	42/M
8	+	+	CML	3	36/M
9	+	+	AML	5	21/M
10	+	+	CML	2	54/M
11	+	+	CML	5	43/F
12	+	+	CML	6	65/M
13	+	-	AML	4	31/M*
14	+	+	ALL	5	21/M
15	+	+	ALL	1	4/F
16	+	-	AML	2	41/M
17	+	-	AML	4	25/M
18	+	-	AML	3	32/M

The correlation between different factors was evaluated by chi-square test (x^2), or Fisher's exact test when an expected value was less than 5.

RESULTS

Sixty patients with established malignant hematological diseases were recruited in this study. The mean age of patients was 33.9 ± 18.3 years. Out of 60 patients, 44 (73.3%) were male. According to the type of hematological malignancy, 26 (43.3%) with acute myelogenous leukemia (AML), 21 (35%) with chronic myelogenous leukemia (CML), 3 (5%) with lymphoma, and 10 (16.6%) with acute lymphoblastic leukemia (ALL) consist our study population (Table 2).

Eighteen of sixty cases of malignant hematological diseases were positive with ELISA for HTLV-1 antibody, obtaining an HTLV-1 seroprevalence of 30% (18/60). The Molecular method of Real time-PCR that amplifies sequences from the tax region provided a viral prevalence of 20% (12/60). Statistical comparisons showed that ELISA detected higher positive results (P < 0.05) than Real time-PCR. HTLV-1 antibody was found in 30.8% of patients with acute myelogenous leukemia, 33.3% of patients with chronic myelogenous leukemia, 20% of patients with acute lymphoblastic leukemia and 33.3% in lymphoma patients. In our study the major HTLV-1 associated syndromes were chronic myelogenous leukemia and acute lymphoblastic leukemia.

In this study, a significant difference was seen; the history of blood transfusion (p=0.04) between patients with positive and negative results for HTLV-1 infection.

DISCUSSION

HTLV-1 causes adult T-cell leukemia (ATL) and HTLV I-associated myelopathy (tropical spastic paraparesis), a nononcogenic neurologic disease, arthropathy, and Sjogren's syndrome, infective dermatitis of childhood, hyperinfective strongyloidiasis (Gotuzzo et al., 1999), and Norwegian scabies (Blas et al., 2005). HTLV-1 is cell associated and is spread in cells after blood transfusion, sexual intercourse, or breastfeeding. The information and understanding of HTLV-1 prevalence in different population and patients groups is crucial because it may be useful in establishing prophylactic measures to decrease rates of viral transmission from infected individuals.

In the present study, we demonstrate that the prevalence of HTLV-1 infection in patients with malignant hematological diseases in Iran is 20% and HTLV-1 seroprevalence is 30%. Whereas the gold standard method for the diagnosis of HTLV-1 infection is the detection of HTLV-1 genome in the specimen of patients, it seems that the prevalence of HTLV-1 infection in our study population is about 20%.

There are several reports which demonstrated a comparable HTLV-1 prevalence to our study. Farias de Carvalho et al. (1997) found a seroprevalence of 28.9% among patients with T-cell lymphoid malignancies in Brazil. Adedayo and Shehu (2004) found a 38.6% of HTLV-1 seropositives in all hematological malignancies in India. Miyagi et al. (2002) found a HTLV-1 prevalence of 26.1% in 88 cases of non-Hodgkin's lymphoma in Japan. Barrientos et al. (2005) in southern Chile found an HTLV-1/2 viral prevalence in patients with malignant hematological diseases 18%, and in chronic lymphoproliferative disorders 27% (Barrientos, 2005). The overall HTLV-1 prevalence rate found in our study group is greater than that seen in some studies but is closed to the others.

On the other hand, HTLV-1 is among the infectious agents that can be transmitted via blood transfusion (Matsuoka and Jeang, 2007). In the present study, significant difference was seen between patients with and without HTLV-1 infection. These patients with HTLV-1 infection may acquire this infection from blood transfusion, despite all of the requirements for screening the blood supply. Therefore, the present study suggests that serious consideration must be given to prevent HTLV-I infection via transfusion in hematological malignanant patients. Routine serological screening for HTLV-I antibody and detection of HTLV-1 genome in blood donors is indicated to permit deferral of blood product donations by asymptomatic HTLV-1 carriers.

In conclusion, the results of this study show an association between HTLV-1 and malignant hematological diseases. Therefore, the possibility of HTLV-1 infection should be considered in patients who suffer malignant hematological diseases.

ACKNOWLEDGEMENTS

This investigation has been funded by Tehran University of Medical Sciences, and had not other financial support. The code of this project was MT 392.

REFERENCES

Adedayo OA, Shehu SM (2004). Human T cell lymphotropic virus type 1 (HTLV 1) and lymphoid malignancies in Dominica: A seroprevalence study. Am. J. hematol., 77(4): 336-339.

Barrientos A, Lopez M, Sotomayor C, Pilleux L, Calderń S, Navarrete M (2005). Prevalence of human T Cell lymphotropic virus type 1 and 2 among patients with malignant hematological diseases in South Chile. J. Med. Virol., 83(4): 745-748.

Blas M, Bravo F, Castillo W, Castillo WJ, Ballona R, Navarro P (2005). Norwegian scabies in Peru: The impact of human T cell lymphotropic virus type I infection. Am. J. Trop. Med. Hyg., 72(6): 855.

de Carvalho SMF, de Oliveira MSP, Thuler LCS, Rios M, Coelho RCA, Rubim LC (1997). HTLV-I and HTLV-II infections in hematologic disorder patients, cancer patients, and healthy individuals from Rio de Janeiro, Brazil. JAIDS J. Acquired Immune Defic. Syndromes, 15(3):238.

Desailloud R, Hober D (2009). Viruses and thyroiditis: An update. Virol J., 6(5).

Eguchi K, Mizokami A, Katamine S (1995). HTLV-I infection in primary Sjgren's syndrome--epidemiological, clinical and virological studies]. Nippon rinsho Japanese J. Clin. Med., 53(10): 2467.

Gabet AS, Kazanji M, Couppie P, Clity E, Pouliquen JF, Sainte Marie D (2003). Adult T cell leukaemia/lymphoma like human T cell leukaemia virus 1 replication in infective dermatitis. Br. J. Haematol., 123(3): 406-412.

Gotuzzo E, Terashima A, Alvarez H, Tello R, Infante R, Watts DM (1999). Strongyloides stercoralis hyperinfection associated with human T cell lymphotropic virus type-1 infection in Peru. Am. J. Trop. Med. Hyg., 60(1): 146.

Hasunuma T (1997). Pathomechanism of HTLV-I associated arthropathy and the role of tax gene]. Nippon rinsho Jpn. J. Clin. Med., 55(6): 1482.

Inoue H, Matsushita K, Arima N, Hamada H, Uozumi K, Ozaki A (2008). High prevalence of human T-lymphotropic virus type I carriers among patients with myelodysplastic syndrome refractory anemia with excess of blasts (RAEB), RAEB in transformation and acute promyelocytic leukemia. Leukemia and lymphoma, 49(2): 315-321.

Johnson JM, Harrod R, Franchini G (2001). Molecular biology and pathogenesis of the human T cell leukaemia/lymphotropic virus Type 1 (HTLV 1). Intl. J. Exper. Pathol., 82(3): 135-147.

Lee R, Schwartz RA (2011). Human T-lymphotrophic virus type 1-associated infective dermatitis: A comprehensive review. J. Am. Acad. Dermatol.,

Matsuoka M, Jeang KT (2007). Human T-cell leukaemia virus type 1 (HTLV-1) infectivity and cellular transformation. Nat. Rev. Cancer, 7(4): 270-280.

Miyagi J, Toda T, Uezato H, Ohshima K, Miyakuni T, Takasu N (2002). Detection of Epstein-Barr virus and human T-cell lymphotropic virus type 1 in malignant nodal lymphoma, studied in Okinawa, a subtropical area in Japan. Intl. J. hematol., 75(1): 78-84.

Mochizuki M, Ono A, Ikeda E, Hikita N, Watanabe T, Yamaguchi K (1996). HTLV-I Uveitis. JAIDS J. Acquired Immune Defic. Syndromes, 13: S50.

Proietti FA, Carneiro-Proietti ABF, Catalan-Soares BC, Murphy EL (2005). Global epidemiology of HTLV-I infection and associated diseases. Oncogene, 24(39): 6058-6068.

Sowa JM (1992). Human T lymphotropic virus I, myelopathy, polymyositis and synovitis: an expanding rheumatic spectrum. J. Rheumatol., 19(2): 316.

Starkebaum G, Kalyanaraman VS, Kidd PG, Loughran TP (1987). Serum reactivity to human T-cell leukaemia/lymphoma virus type I proteins in patients with large granular lymphocytic leukaemia. The Lancet, 329(8533): 596-599.

Sugimoto M, Imamura F, Matsumoto M, Sonoda E, Cho I, Ando M (1993). Pulmonary involvement in patients with human T lymphotropic virus type 1-associated myelopathy: The presence of specific IgA antibody in bronchoalveolar lavage fluid. Am. J. Trop. Med. Hyg., 48(6): 803.

Tarhini M, Kchour G, Zanjani DS (2009). Declining tendency of human T-cell leukaemia virus type I carrier rates among blood donors in Mashhad, Iran. Pathology, 41(5): 498-499.

15

Incidence of okra mosaic virus at different growth stages of okra plants (*Abelmoschus esculentus* (L.) *Moench*) under tropical condition

A. A. Fajinmi[1]* and O. B. Fajinmi[2]

[1]Department of Crop Protection, COLPLANT, University of Agriculture Abeokuta, P.M.B. 2240 Alabata, Ogun State, Nigeria.
[2]National Institute of Horticultural Research and Training, Idi-Ishin, Ibadan, Oyo state, Nigeria.

The degree of Okra mosaic virus (OKMV) at different growth stages of okra plants was studied using a netted barrier method. In a two factor RCB design with three replications, a 2 m high netted barrier were laid out in unit plots of 2 x 2 m using eight treatments: T1- Netting up to 7 days after seedling emergence (DAE); T2- Netting up to 14 DAE; T3- Netting up to 21 DAE; T4- Netting up to 28 DAE; T5- Netting up to 35 DAE; T6- Netting up to 42 DAE; T7- Netting up to last harvest; T8- No netting (untreated control). The number of *Podagrica unifoma* (Jac.) and *Podagrica sjostedti* (Jac.) were recorded weekly and the number of virus infected plants from all the plants of each replication. It was observed that by preventing the vectors (*P. unifoma* (Jac.) and *P. sjostedti* (Jac.)) of okra mosaic virus (OKMV) by the use of 2 m high net barrier around the okra plots, until the plants became more than 21 days old after emergence, decreased the populations of *P. unifoma* (Jac.) and *P. sjostedti* (Jac.) and virus infected plants of both the resistant and susceptible okra varieties. Low virus infection in plots netted for 21 days after seedling emergence or more resulted in 25 - 50% increased yields in both tolerant and susceptible varieties. These observations by this study showed that virus infection in okra plants at growth stages earlier than four weeks has more severe effect on the physiological performance of okra plant and subsequent reduction in growth performance and yield of okra. Therefore some effective control measure is very necessary at early growth stages of okra plant.

Key words: *Abelmoschus esculentus*, *Hibiscus esculentus*, Okra Mosaic virus, infection, netted barrier, control, Nigeria.

INTRODUCTION

Okra [*Abelmoschus esculentus* (L.) *Moench* or *Hibiscus esculentus* (Linné)] is an important vegetable crop in much of the tropics including Nigeria (Schippers, 2002). Young fruit are consumed fresh or cooked. Okra is a good source of vitamin A, B, C and protein, carbohydrates, fats, minerals, iron and iodine (Diaz and Ortegon, 1997). Fresh fruit are harvested when 3 - 7 days old. Consumption of 100 g of fresh okra fruit provides 20, 15 and 50% of the daily requirement of calcium, iron and ascorbic acid, respectively (Hamon, 1988; Schippers, 2002). Old fruit are used in processed products (Schippers, 2002).

Okra is a warm, rainy season crop, requiring high soil and high day and night air temperatures, but growers start cultivation in January when average temperatures are below 37 °C as an early crop for better returns (Simmone et al., 2004). Pods grow rapidly are ready for harvest in about 60 days when grown from seed. Pods must be picked about 4 - 5 days after flowering, when ~ 10 cm in length and before they mature and toughen. Okra comes in varying shades of green (there is also a new red variety), and can be smooth or have a ribbed surface (Jha and Dubey, 1998). Okra can be picked every other day during fruiting, and several times more if the crop is mowed and allowed to re-grow (Schippers, 2002).

A number of viruses infect okra (Kucharek, 2004) including: that causing Okra leaf curl disease (OLCD), is suspected of being associated with a whitefly-transmitted

*Corresponding author. E-mail: ayofaji@yahoo.com.

Table 1. Effect of netting on incidence of vectors (*Podagrica unifoma* (Jac.) and *P. sjostedti*) and Okra mosaic virus in tolerant (okra, cv. 47-4) and susceptible (okra cv. Jokoso) okra varieties.

Treatments Days after emergence (DAE)	*Podagrica unifoma* (Jac.)/10 leaves (no.)		*P. sjostedti* (Jac.)/10 leaves (no.)		Virus infected plants (%)	
	okra, cv. 47-4	okra cv. Jokoso (s)b	okra, cv. 47-4	okra cv. Jokoso (s)b	okra, cv. 47-4	okra cv. Jokoso (s)b
Netting up to 7 DAE	4.2 ab	4.6 ab	2.9 b	3.2 b	21.2 b	24.4 b
Netting up to 14 DAE	4.3 ab	4.6 ab	2.8 b	3.8 b	20.4 b	24.7 b
Netting up to 21 DAE	4.2 ab	3.4 b	2.6 b	3.6b	17.1 b	24.2 b
Netting up to 28 DAE	2.8 b	3.0 b	2.4 b	2.8 bc	17.1 b	22.1 b
Netting up to 35 DAE	2.4 b	2.6 b	1.4 c	2.7 bc	17.4 b	20.6 c
Netting up to 42 DAE	1.9 b	2.2 b	1.2 c	1.7 c	16.0 b	19.2 c
Netting up to harvests	1.4 b	1.9 b	1.0 c	1.2 c	14.4 b	18.5 c
No netting (control).	6.4 a	7.2 a	4.5 a	8.4 a	30.8 a	38.2 a

a. Data are averages of 3 replications from 8 observations; values followed by the same letters do not differ significantly at 5% by DMRT.
b. s = susceptible.

geminivirus (Genus *Begomovirus*), and Yellow vein mosaic virus (OYVMV), transmitted by the whitefly (*Bemisia tabaci*) (Ali et al., 2000).

Okra mosaic virus (OKMV) has always been a serious problem in okra (Kucharek, 2004). Yield reductions of 20 - 50% have occurred (Kucharek, 2004). This loss may increase to 90% (Pullaiah et al., 1998; Kucharek, 2004). *Okra mosaic virus* symptoms are characterized by a homogenous interwoven network of yellow mosaic pattern enclosing islands of green tissue in leaf blades. In extreme cases, infected leaves become yellowish or creamy color (Kucharek, 2004). The virus is not seed transmitted (Koenig and Givord, 1974), but it is mainly transmitted by the beetles of *Podagrica* spp. (Lana and Taylor, 1975; Atiri, 1984, 1990; Alegbejo, 2001a, b).

The integrated pest management constraints are that vectors usually attack the young okra plants at the vegetative stage for virus transmission. Frequent use of pesticides by the farmers, without recognizing the vector(s), its incidence patterns and the virus infection time, create poisonous residues in the food chain. Understanding the growth stage critical for virus transmission can help greatly to undertake appropriate control measures to prevent virus transmission.

The objective of this study therefore is to identify the degree of *Okra mosaic virus* (OKMV) at different growth stages of okra plants, so that appropriate control measures can be undertaken at the critical stages of vector infestation and virus transmission.

MATERIALS AND METHODS

The experiment was carried out in 2005 and 2006 rainy season at an on farm adaptive experimental field in University of Agriculture Abeokuta Ogun state Nigeria. In a two factor RCB design with three replications, eight treatments were laid out in unit plots of 2m x 2m and using a 2 m high netted barrier: T1- Netting up to 7 days after

seedling emergence (DAE); T2- Netting up to 14 DAE; T3- Netting up to 21 DAE; T4- Netting up to 28 DAE; T5- Netting up to 35 DAE; T6- Netting up to 42 DAE; T7- Netting up to last harvest; T8- No netting (untreated control). Standard cultural practices and recommended rates of fertilizers were applied; no control measures were taken for pest infestation. Weekly observations were made to record the number of *Podagrica unifoma* (Jac.) and *P. sjostedti* (Jac.) from the upper ten leaves of 10 randomly selected plants and the number of virus infected plants from all the plants of each replication. Data were also taken on the height of the plants, number of fruits per plant and yield. Averages for the two year data was given.

Data regarding Okra mosaic virus and *P. unifoma* (Jac.) and *P. sjostedti* (Jac.) population were recoded on weekly basis and subjected to statistical analysis. All possible interactions were determined through ANOVA and treatments mean were compared by LSD or DMR test at 5% level of probability (Steel et al., 1997).

The two cultivars of okra (okra, cv. 47-4 and the susceptible variety okra cv. Jokoso), used for this experiment were collected from National Institute for Horticultural Research, Ibadan, Nigeria. At the on set of pods starting to grow, fertilizer NPK 15:15:15 fertilizer was applied at 150 NPK kg ha^{-1} on the ground around the plants. The fertilizer was watered into the soil, keeping both the fertilizer and the water off the plant directly.

Weeding commenced at two weeks after sowing of okra seed and subsequent weeding was carried out as at when due. Thinning was done two weeks after sowing of okra seed.

RESULTS AND DISCUSSION

Okra plots under netting for more than 28 DAE reduced the number of *P. unifoma* (Jac.) and *P. sjostedti* (Jac.) as well as virus infection considerably when com-pared with that of the un-netted plots or plots netted up to 21 DAE (Table 1). Although no consistent trend was evident, plant height and fruit bearing capacity of the plants increased in the plots having a net barrier up to 28 DAE and above. As a result there was a significant increase in okra fruit yields (Table 2). The results strongly indicated that okra plants under 28 DAE were more prone to vector infesta-

Table 2. Effect of netting on plant height, number of fruits per plant and yields of okra in tolerant (okra, cv. 47-4) and susceptible (okra cv. Jokoso) okra varieties.

Netting Days after emergence (DAE)	Pt. ht. at harvest (cm)		Fruits/ plant (no.)		Yield (t/ha)	
	okra, cv. 47-4	okra cv. Jokoso (s)b	okra, cv. 47-4	okra cv. Jokoso (s)b	okra, cv. 47-4	okra cv. Jokoso (s)b
Netting up to 7 DAE	45.5	40.8	23.8	24.0	6.1 b	4.2 b
Netting up to 14 DAE	48.6	45.8	25.4	22.0	6.9 a	4.6 b
Netting up to 21 DAE	50.6	47.2	25.5	23.6	7.4 a	5.9 a
Netting up to 28 DAE	52.8	50.5	26.6	24.8	7.5 a	6.4 a
Netting up to 35 DAE	53.8	54.5	24.2	20.0	7.3 a	6.3 a
Netting up to 42 DAE	64.5	58.3	20.4	21.5	6.0 b	5.8 a
Netting up to harvests	68.8	62.8	22.5	18.2	6.8 a	4.8 b
No netting (control).	70.4	64.8	15.2	14.2	3.6 c	3.1 c

a. Data are averages of 3 replications from 8 observations; values having the same letters do not differ significantly at 5% by DMRT.
b. s = susceptible.

tion and virus infection, and therefore, it is necessary to protect okra crops from virus infection at least up to 28 DAE of plant growth. There was no signify-cant difference in the intensity of virus infection between the tolerant variety (okra, cv. 47-4) and the susceptible variety (okra cv. Jokoso), indicating that okra, cv. 47-4 is not all that tolerant to OKMV.

The use of pesticide in the control of insect pest could be avoided and improved okra plants growth and yields, and low vector and virus infestation achieved by preventing insect vectors of the virus from reaching the plant as observed by this study. In prevention of insect vectors of viruses on okra, similar work was carried out in the management of Okra yellow vein mosaic virus (OYVMV) by Kulat et al. (1997) where aqueous plant leaf extracts of tobacco (2%) *Ipomoea cornea* (5%) and a seed extract of *Azadirachta indica* and *Pongamia bragla* (5%) were used for the control of whitefly (*Bemisia taabaci*) and *Aphis gossypii,* vectors of the virus on okra. The vectors of the virus were targeted and controlled from transmitting the virus resulting in very low incidence of the virus and subsequent improved crop yield.

Also Adiroubane and Letchoumanane (1998) used plant extracts, sacred basil (*Ocimum sanctum*), malbar nut (*Adhutoda vesica*), Chinese chaste tree (*Vitex negudo*) and synthetic insecticides (Endosulfan and Carbaryl) and their combinations products in controlling okra jassid, whitefly and fruit borers, vectors of viruses of Okra during rainy season. All the treatments suppressed insect's population. Sprays with leaf extracts of *Prosposchilensis* and *Bougainvillea spectabilis* has been found highly effective in reducing yellow vein mosaic virus in okra by suppressing insect population (Pun et al., 1999).

Okra fields need to be protected up to at least 21 DAE from the attack of virus vectors and viruses for satisfactory yields as observed by this study. Yield loss of okra can be minimized by 25 - 50% if the plants are protected up to 28 DAE.

Conclusion

The easiest and method of reducing *Okra mosaic* disease of okra is planting of resistant varieties against this disease. However okra plants protected up to 28 days after germination also reduced the spread of OKMV by checking its vector *P. unifoma* (Jac.) and *P. sjostedti* (Jac.). Therefore if virus vectors on okra plants are checked and controlled, viral diseases incidence on cultivated okra plant will be greatly minimized and subsequent healthy crop and increased fruit yield.

ACKNOWLEDGEMENT

The authors are grateful for support received from the virology laboratory of the National Institute of Horticultural research and Training, Idi-Ishin, Ibadan, Oyo state, Nigeria.

REFERENCES

Adiroubane D, Letachoumanane S (1998). Field efficacy of botanical extracts for controlling major pests of okra. Indian J. Agric. Sci., 68: 168–70

Alegbejo MD (2001a). Effect of sowing date on the incidence and severity of Okra mosaic Tymovirus. J. Veg. Crop Prod. 8:9-14.

Alegbejo MD, (2001b). Reaction of okra cultivars screened for resistance to okra mosaic virus in Samaru, Northern Guinea Savanna, Nigeria. J. Sustain. Agric. Environ. 3:315-320.

Ali M, Hossain MZ, Sarkern NC (2000). Inheritance of Yellow Vein Mosaic Virus (YVMV) tolerance in a cultivar of okra (*Abelmoschus esculentus* (L.) Moench). Euphytica 111(3):205-209.

Atiri GI, (1984). The occurrence of *okra mosaic virus* in Nigerian weeds. Ann. Appl. Biol. 104: 261-265.

Atiri GI, (1990). Relationship between growth stages, leaf curl development and fruit yield in okra. Sci. Hortic. 54:49-53.

Diaz F A, Ortegon MAS (1997). Fruit characteristics and yield of new okra hybrids. Subtrop. Plant Sci. 49:8-11.

Hamon S (1988). Evolutionary organization of its kind *Abelmoschus* (okra). Co-adaptation and evolution of two species grown in West Africa, *A. esculentus* and *A. caillei*. Paris, ORSTOM, DTP Works and Documents. 46:191.

Jha AK, Dubey SC (1998). Effect of plant age and weather parameters on collar rot of okra caused by *Macrophomina phaseolina*. J. Mycol. Plant Pathol. 28(3):351-353.

Koenig R, Givord L (1974). Serological interrelationships in the turnip yellow mosaic virus group. Virology 58:119.

Kucharek T (2004). 2004 Florida plant disease management guide: Okra. Plant Pathology Department document PDMG-V3-41. Florida Cooperative Extension Service, Institute of Food and Agricultural Sciences, University of Florida, Gainesville, FL.

Kulat SS, Nimbalkar SA, Hiwase BJ (1997). Relative efficacy of some plant extracts against *Bemisia tabaci* and *Aphis gossypii* Glover and *Amrasca devastans* of okra. PKV Res. J., 21:146–8.

Lana AO, Taylor TA (1975). The insect transmission of an isolate of okra mosaic virus occurring in Nigeria. Ann. Appl. Biol. 82:361-364.

Pullaiah N, Reddy TB, Moses GJ, Reddy BM, Reddy DR (1998). Inheritance of resistance to yellow vein mosaic virus in okra (*Abelmoschus esculentus* (L.) Moench). Ind. J. Gene. Plant Breeding 58(3):349-352.

Pun KB, Sabitha D, Jeyaran R, Doraiswammy S (1999). Screening of plant species for presence of antiviral principles against okra yellow vein mosaic virus. Indian Phytopathol., 52:221–3.

Schippers R (2002). African indigenous vegetables. An overview of the cultivated species. Natural Resources Institute/ACP-EU, Technical Centre for Agricultural and Rural Cooperation, CD ROM 214, Chatham, UK.

Simmone EH, Maynard GJ, Hochmuth DN, Vavrina CS, Stall WM, Kucharek TA, Webb SE (2004). Okra production in Florida. Horticultural Sciences Department document HS729. Florida Cooperative Extension Service, Institute of Food and Agricultural Sciences, University of Florida, Gainesville, FL. Available on-line at: http://edis.ifas.ufl.edu/CV128.

Steel RGD, Torrie JH, Dicky D (1997). Principles and Procedures of Statistics. A Biometrical Approach. 3rd ed. McGraw Hill Book Co. Inc. New York.

16

Pathological, serological and virological findings in goats experimentally infected with Sudanese Peste des Petits Ruminants (PPR) virus isolates

Nussieba A. Osman[1]*, A. S. Ali[2], M. E. A/Rahman[3] and M. A. Fadol

[1]Department of Pathology, Parasitology and Microbiology, College of Veterinary Medicine and Animal Production, Sudan /Sudan University of Science and Technology, P.O.Box 204, Khartoum North, Sudan.
[2]Department of Preventive Medicine and Veterinary Public Health, Faculty of Veterinary Medicine, University of Khartoum, Post code 13314, Khartoum North, Sudan.
[3]Department of Virology, Central Veterinary Research Laboratories, Soba, P.O.Box 8067, Khartoum, Sudan.
[4]Viral Vaccine Production Unit, Central Veterinary Research Laboratories, Soba, P.O.Box 8067, Khartoum, Sudan.

Four Peste des Petits Ruminants virus (PPRV) Isolates were collected from clinical cases of three goats and one sheep from Khartoum State; Soba/Khartoum State and Bashaier/River Nile State. These PPR viruses were isolated in Lamb kidney cells (LKC) and Lamb testis cells (LTC) and identified by Agar Gel Precipitation Test (AGPT) and Hemagglutinition (HA) tests. Four PPRV isolates were used for experimental infection in four groups (n = 4) of Sudanese goats. Goats of group A and B were inoculated with the 4th passage of two Sudanese PPRV cultured in lamb testis cells, 6 × 10 TCID /ml of Bashaier and 6 × 10 TCID /ml of Soba isolates, isolated from sheep and goat respectively. Whereas, group C and D were received 6ml of the fifth passaged 20% infected tissue suspensions of Khartoum and Soba PPR isolates propagated in goats. The inoculated goats showed typical PPR clinical signs, gross lesions and histopathological changes while control animals (group E) appeared healthy. Two goats from group A died on the 16th and 19th days post inoculation. PPR viruses were detected by HA test from lacrymal fluid and nasal swabs on the 6th and 7th dpi. Serum samples were collected and tested for PPRV antibodies by C-ELISA from the sixteen experimentally infected goats and from control animals. Traces of PPRV antibodies were shown on day 7 and they were continued to rise till the 28th day and dropped on the 30th day which is the 9th day post challenge. The observed clinical signs, post mortem lesions and the detectable antibodies indicated that the tissue culture propagated PPR viruses and the infected tissue homogenate were effective for initiation of infection.

Key words: PPRV isolates, tissue culture virus, goat adapted virus, experimental infection, immune response.

INTRODUCTION

Peste des petits ruminants virus (PPRV) is a member of genus *Morbillivirus*, closely related to rinderpest virus (RPV), causes an acute febrile disease of small ruminants with morbidity and mortality rates as high as 100 and 90% respectively (Abu-Elzein et al., 1990).

Clinically, the disease is characterized by severe pyrexia, ocular and nasal discharges, necrotizing and stomatitis, conjunctivitis, gastroenteritis, diarrhoea and pneumonia (Ismail et al., 1995 and Jones et al., 1993). The prin-

ciple host of PPR are sheep and goats, with goat being more susceptible to infection and erosive subsequent disease (Ezeokoli et al., 1986). In goats and sheep PPR and RP viruses produce clinical disease and pathology that are indistinguishable. However, for the diagnosis of rinderpest in small ruminants it is essential to differentiate it clearly from PPR. Infection rates in sheep and goats rise with age and the disease, which varies in severity, is rapidly fatal in young animals (Lefevre and Diallo, 1990 and Wosu, 1994). The disease appears with a higher incidence in the rainy season. The infection is transmitted by close contact between infected and susceptible animals (Lefevre and Diallo, 1990).

*Corresponding author. E-mail: nussieba@yahoo.com.

Experimentally, the virus has been transmitted parenterally through different routes: nasal, oral, subcutaneous, intraocular, intratracheal and intravenous or by contact (Durtnell, 1972 and Durojaiye, 1980).

The aim of this study is to determine the pathological, virological and serological findings in goats experimentally infected with tissue culture virus and infected tissue suspensions of PPRV. On the other hand, is to investigate the possibilities of identification of PPR virus from the samples of experimentally infected goats.

MATERIALS AND METHODS

Animals

Twenty one healthy goats of local breed aged between 4-5 months were purchased from the local market. These animals were grouped in 5, A, B, C, D, and E. First 4 groups contain 4 animals in each and group E contains 5 animals as control. Goats were quarantined for 7 days and kept under close observations for any signs of the disease. All animals were free from detectable PPR antibodies as judged by C-ELISA.

Collection of samples

Four Lymph nodes and spleen samples were collected from sheep and goats suspected to be infected by PPRV. The first sample originated from goats was collected from Khartoum State; the second and the third originated from goats were collected from Soba, Khartoum State and the fourth originated from sheep was collected from River Nile State.

10 - 20% (w/v) suspensions of lymph nodes and spleen samples were prepared by grinding with sterile sand using mortar and pestle in PBS pH 7.4 supplemented with antibiotics. The supernatant was used for PPRV isolation after identification by AGPT (White, 1958) and HA test (Nussieba et al., 2008).

Virus isolation and titration

10 - 20% suspension of spleen and lymph nodes from suspected animals were used for initial PPRV isolation which was carried in primary lamb kidney (LK) and lamb testis (LT) cells. The 2nd passage of LK and LT cells were used for propagation of PPR virus. For experimental inoculation of goats each PPRV isolate was adapted to LT cells with a minimum of 4 passages following the technique described previously by Plowright and Ferris (1959). The tissue culture dose end point $TCID_{50}$ (50% Tissue Culture Infective Dose) of a virus suspension was determined as described by Plowright and Ferris (1962). The titre was calculated by the Spearman-Karber method (Spearman, 1908 and karber, 1931) and expressed as $\log_{10} TCID_{50}$ / ml.

PPRV isolates

The 4th passage of PPRV Bashaier (LTC P4) and Soba (LKC P3, LTC P4) isolates were used for inoculation of goats as tissue culture adapted virus. These isolates originated from sheep and goat, respectively. The 3rd passage of another two PPRV isolates, Khartoum and Soba (LKC P3) originated from sheep was subjected to further two passages in goats (Goat P5). Viruses from the 5th goat passage were used as 20% infected tissue suspensions.

Virus propagation in goats

The procedure of the propagation of PPRV in goats was followed as described earlier by Durtnell (1972). Two goats were inoculated with 6 ml subcutaneously (s/c) accompanied by 1 ml intranasally (i/n) of the 3rd virus passage. On the 12th dpi goats were slaughtered and organs were collected aseptically. Spleen and lymph nodes were prepared as 20% suspension in PBS for further passage in goats. The 5th passage of PPRV (LTC. P3 Goat P5) was used for inoculation of goats.

Experimental inoculation of goats with PPRV isolates

Group A and B were treated with tissue culture isolates (TCV) while group C and D were treated with infected tissue suspensions (ITS). Goats of group A and B were inoculated with $6 \times 10^{4.5}$ $TCID_{50}$ /ml of PPRV Bashaier (LTC P4) and $6 \times 10^{4.3}$ $TCID_{50}$ /ml PPRV Soba (LKC P3 LTC P4) isolates, respectively. Goats of group C and D were inoculated with 20% infected spleen and lymph nodes suspensions of PPRV Khartoum (LTC P3 Goat P5) and PPRV Soba (LTC P3 Goat P5) isolates, respectively. Each goat received 5 ml subcutaneously and 1 ml intranasally following a combinations of the procedures described by Mann et al. (1974) and Bundza et al. (1988).

Challenge experiment of immunity was carried out on the 21st dpi with $10^{5.4}$ $TCID_{50}$ /ml of virulent PPRV Sinnar strain (72/1) at a dose of 5 ml/animal subcutaneously. Rectal temperatures were recorded at 8.30 am daily. Blood was collected from all the infected goats at peak of temperature for virus isolation. Recovered goats were slaughtered after a week of the challenged experiments and postmortem findings were recorded. Slices of mesenteric lymph nodes, spleen, lung, small and large intestine were collected for histopathological examinations following the procedure of Carleton (1967).

For determining the humoral immune response of PPRV isolates, serum samples were collected at 7, 10, 14, 18, 21, 28 and 30 days post inoculation. PPR C-ELISA (BDSL, 2000) was carried out to determine the antibody titres induced by PPRV isolates.

Detection of PPRV in samples of inoculated goats

PPR viruses were detected in nasal and lacrymal swabs using haemagglutination (HA) test as described previously by Nussieba et al. (2008). Lacrymal and nasal swabs were collected on 6 and 7 dpi, respectively in 150 μl of PBS pH 7.2 for antigen detection. The swab fluid was centrifuged 1 h after collection, at 3000 rpm for 20 - 30 min and then stored at -40°C.

RESULTS

Clinical response to PPR viral isolates in infected goats

Inoculated goats developed clinical signs similar to those of the naturally infected animals while uninoculated (control) goats remained apparently healthy. Inoculated goats of all groups remained healthy for 3 - 4 days following inoculation passaged tissue culture and infected tissue suspensions PPR viruses. Clinical signs began with an elevated temperature up to 40 - 40.6°C in infected goats

Table 1. Clinical observation of the goats after infection with PPRV.

Type of virus	Fever	Stomatitis	Pneumonia	Diarrhoea	Ocular involvement	Nasal involvement	Labial scabs	Death
TCV[1]	8 G	3 G	4 G	4 G	6 G	8 G	3 G	2 G
%	100%	37.5%	50%	50%	75%	100%	37.5%	25%
ITS[2]	8 G	2 G	3 G	4 G	6 G	8 G	2 G	0
%	100%	25%	37.5%	50%	75%	100%	25%	0%

Table 2. Days of the onset of clinical signs in goats infected with PPRV.

Experiment No.	Type of virus	Incubation period	Days of onset of					Days killed (k) or dead (D)
			Viraemia	Lacrymal excretion	Nasal excretion	Mouth lesions	Diarrhoea	
Exp. 1	TCV[1]	5 - 7th	5 - 6th	5th	5th	7th	6 - 13th	16 and 19th
Exp. 2	ITS[2]	5 - 7th	5 - 6th	5th	5th	7th	9 - 13th	—

Notes:
TCV[1] : Tissue culture virus.
ITS[2] : Infected tissue suspension.

while animals in the control group showed no thermal response. Pyrexia was detected in all experimentally infected goats on the 5 - 7th dpi. Goats developed a very slight superficial necrosis of the lips on the 7th day. On the 6th day, the fever was sustained. On the 5th day, serous nasal and lacrymal dischargesappeared. Serous nasal and lacrymal discharges involved all the 16 goats (100%) on the 7th day. During this phase animals in general were dull, depressed and anorexic with congested mucous membranes. A cough was usually noticed early in the disease on the 8th and the 9th day. The respiration was usually fast and shallow. On the 9th day, conjunctivitis appeared and involved goats on the 13th day. Mucoid nasal discharges were observed in goats on the 13th day and on the 15th day.

Diarrhoea with abdominal pain was a common feature of the disease and it occurred on the 9th day following the onset of fever. Goats showed evidence of diarrhoea on the 13th day. The signs persisted for 6 days and the animals become progressively weaker.

Between the 11 and 13th days, the fever regressed and the oral and the encrusted lip lesions began to resolve. The nasal discharges and crust formation occurred while the oral lesions extended to cross the muco-cutaneous portion of the lip with scab formation at mouth commissure at the 12 and 15th day respectively. Death was usually preceded by severe emaciation, dehydration, subnormal temperature and collapse. Two goats (25%) of group (A) which were inoculated with PPRV Bashaier isolate died on day 16 and 19th post inoculation respectively. The incidence of the different clinical signs is shown in Table 1 and the onset of the disease is summarized in Table 2.

Immune response to PPR viral isolates in infected goats

Serum samples were collected from goats experimentally inoculated with PPRV at 7, 10, 14, 18, 21, 28 and 30th dpi (Figure 1) and examined using C-ELISA to detect antibodies against PPRV. 4 sera (25%) out of the 16 goats experimentally inoculated with PPRV showed detectable antibodies at 7 dpi while all the goats (100%) showed detectable antibodies at 10 dpi. The titre of antibodies induced by PPRV started from weak positive (PI 51 - 70%) at 7 dpi to moderate (PI 73 - 82%) and strong positive (PI 85-90%) at the following days till 21 dpi. Following challenge of experimental animals at 21 dpi, a rise in antibodies titre was observed at 28 dpi. In 30 dpi the antibody titres were less than in 28 dpi.

Detection of PPRV antigen in samples from inoculated goats

The HA titres of PPRV antigen detected in nasal and lacrymal swabs of experimentally infected goats ranged between 4 and 16.

Post-mortem examination of inoculated goats

Gross pathology: The most characteristic post mortem lesions were found in the gastrointestinal and respiratory tract. There was evidence of emaciation in all the inoculated goats (100%). In some goats of the two groups the rumen, reticulum, omasum and abomasum were filled with foetid watery fluid. The small intestine showed evidence of severe inflammation. The small intestine was

PI values

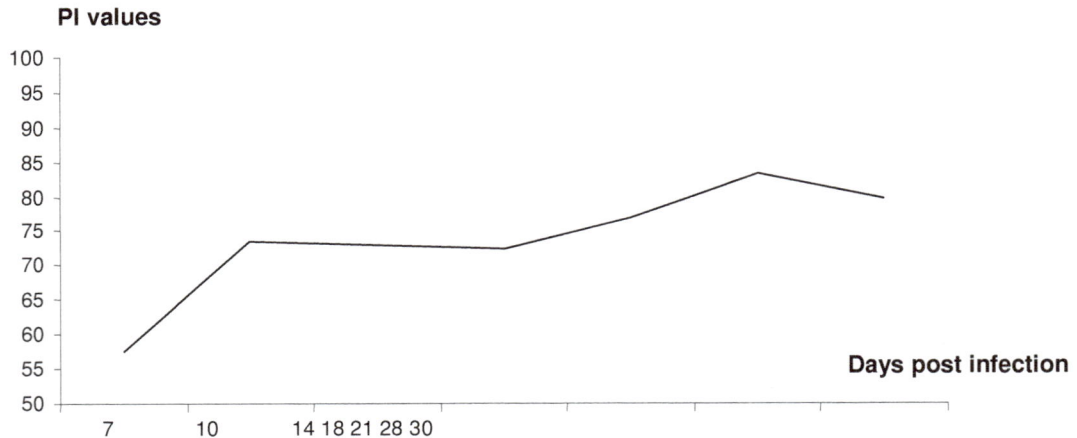

Figure 1. PI values for PPRV antibodies.

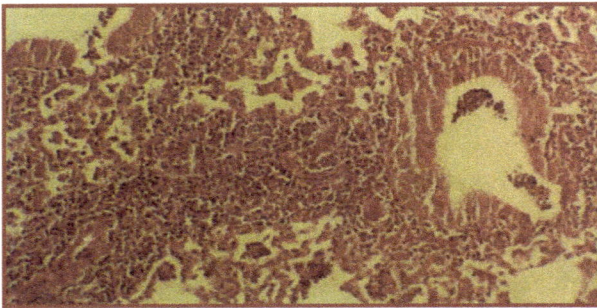

Figure 2a. Lung: note proliferation of bronchiolar epithelium, intense diffusion of mononuclear cells. H and E X10.

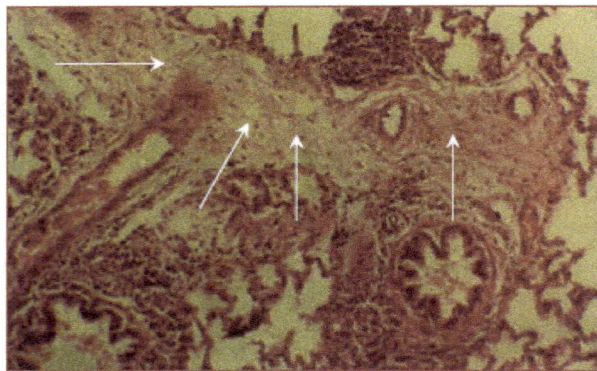

Figure 2b. Lung: note interlobular proliferation of fibrous tissue (white arrows). H&E X10.

filled with watery fluid as found in the abomasum. The colon and caecum showed evidence of linear haemorrhage and areas of ulcerations. Lymph nodes were engorged and oedematous. Spleen was enlarged and oedematous. In the respiratory system, the trachea showed evidence of inflammation and filled with frothy exudates. Pneumonia was usually observed in a few lung lobes. The most involved was the right apical lobe and

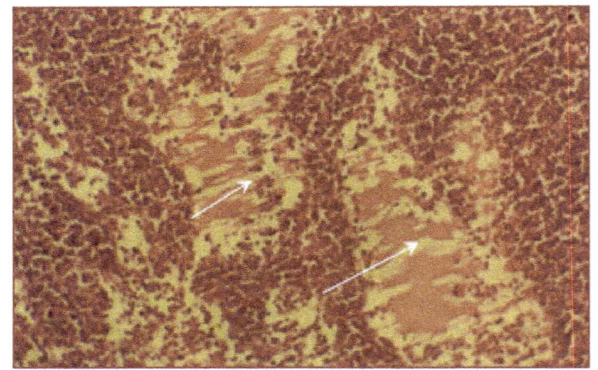

Figure 3a. Lymph node: note oedema (white arrows). H and E X10.

the less involved were the intermediate and cardiac lobes. Post mortem examination showed evidence of pneumonia, congestion of the lung lobes and hydrothorax. Fatty change in kidneys was observed.

Histopathology

The lung showed bronchiointerstitial pneumonia characterized by proliferation of bronchiolar lining epithelium, intense diffusion of mononuclear cells mainly lymphoid, macrophages and plasma cells in the periductal, the interstitial tissue and alveoli lumina (Figure 2a). Areas of scarring were seen in interlobular connective tissue (Figure 2b). In lymph nodes, there was oedema in the cortical and medulla (Figure 3a) and infiltration of mononuclear cells and some giant cells in subcapsular areas and medullary sinuses (Figure 3b). The spleen showed haemorrhage and haemosidren pigment deposition (Figure 4a and 4b). The intestine showed atrophic villi with partial denudation of epithelial lining and intense diffusion of mononuclear cells in the lamina propria and submucosa (Figure 5).

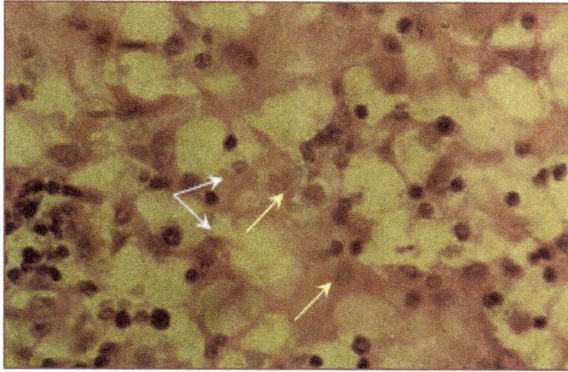

Figure 3b. Lymph node: note proliferation of mononuclear and giant cells (white and yellow arrows). H and E X40.

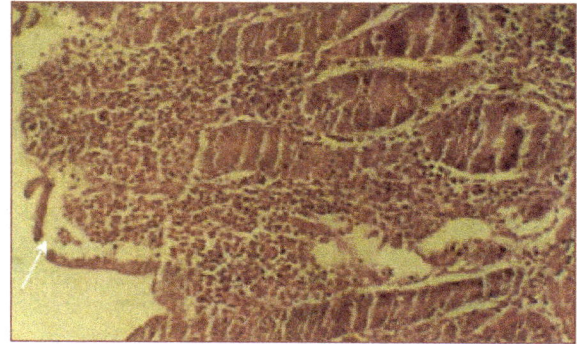

Figure 5. Small Intestine: note atrophic villi, denudation of lining epithelium and intense infiltration of mononuclear cells. H and E X10.

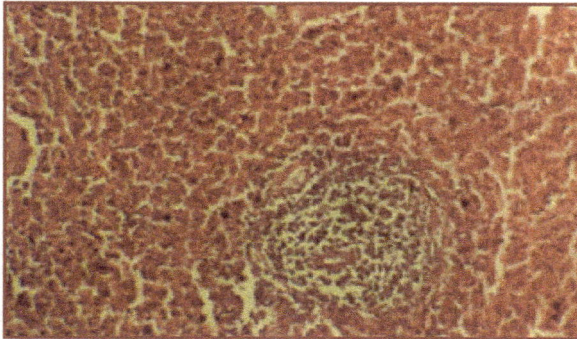

Figure 4a. Spleen: note accumulation of extravagated erythrocytes. H and E X10.

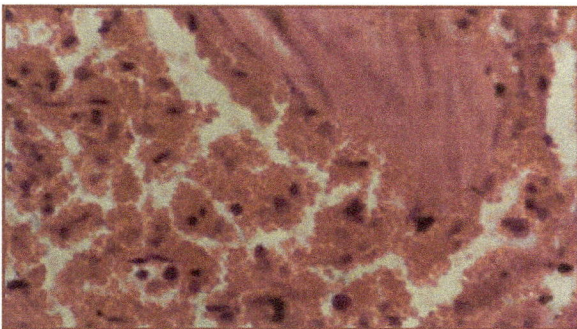

Figure 4b. Spleen: note accumulation of extravasated erythrocytes. H and E X40

DISCUSSION

The examination of nasal and lacrymal swabs from goats experimentally inoculated with PPRV by HA test resulted in agglutination of RBCs. Chicken RBCs were used for detection of PPRV in swabs depending on the highest sensitivity of the RBCs of this species upon the others. The HA test of swab samples resulted in slightly low titres ranging from 4 to 16. This result was in agree-ment with Wosu (1991) who documented that the HA titre was not a reflection of the concentration of the virus in secretions, but rather a reflection of the degree of dilution of the virus in the secretion with the diluents. The obtained HA titres were indication of the shedding of PPR virus in nasal and lachrymal swabs. This result was similar to that mentioned earlier by Abegunde and Adu (1977) whom detected the existence of PPRV in nasal and conjunctival secretions. Virus was confirmed in ocular and nasal swabs at the onset of clinical signs which is the most important epidemiological aspect in spread of the disease.

Of the 16 infected goats, 8 were inoculated with tissue culture propagated PPRV and the other 8 were inoculated with infected tissue suspension. The experimental infection of goats with PPRV revealed observable clinical signs and postmortem lesions in inoculated animal. This infection was accompanied with morbidity and mortality rates reaching 100% and 12.5% respectively. Serological studies indicated the presence of detectable antibodies against PPRV. Following challenge of inoculated animals, with virulent PPRV which was carried out at day 21 p.i., goats did not show rise in body temperature or any signs of the disease. From this result it was obvious that experimental infection with PPRV resulted in high percentage of morbidity but low percentage of mortality in contrast to natural PPRV infection. This is in agreement with Elhag Ali (1973) and Mann et al. (1974) whom found that animals experimentally infected with PPRV develop mild form of the disease. Although they demonstrated the appearance of the disease in experimentally infected animals, Mann et al. (1974) suggested that the successful transmission of the acute disease may require more than one challenge. Moreover some authors suggested that a more severe disease results from mixed infection of bacteria and viruses than a single infection. This substantiated the result of Onoviran et al. (1984) who reported that a combined infection of *Mycoplasma capri* and PPR was found to be much more severe in goats than infection by a single agent. Nutritional and environmental factors have an important effect on the appearance of the disease in a flock of animals. On the other hand, Saliki (1998) previously reported that poor nutrition status, stress of movement and concurrent parasitic and bacterial infections enhance the severity of clinical signs.

Examination of serum samples from goats infected with PPRV showed detectable antibodies at the 7th day and the 10th day p.i. with increase in the antibody titre at the following days. The detectable PPR antibodies were indicative of the humoral immune response due to the exposure of animals to the virus. This observation was in concordance with those reported by Taylor (1979).

Conclusion

The observed clinical signs, post mortem lesions and the detectable antibodies indicated that PPRV in the form of tissue culture propagated viruses or infected tissue homogenate were proved effective for initiation of infection.

REFERENCES

Abegunde AA, Adu FD (1977). Excretion of the virus of peste des petits ruminants by goats. Bull. Anim. Health Prod. Afr. 25(3): 307-311.

Abu-Elzein EME, Hassanien MM, Al-Afaleq AI, Abd-Elhadi MA, Housain FMI (1990). Isolation of peste des petits ruminants virus from goats in Saudi Arabia. Vet. Rec. 127: 309–310.

Biological Diagnostic Supplies LTD (BDSL). Flow Laboratories and Institute for Animal Health, Pirbright, Surrey, England (2000). PPR competitive ELISA Kit. Competitive ELISA for detection of antibodies to P.P.R. virus. Developed in Collaboration with: The Animal Production and Health Section joint FAO/IAEA Division, IAEA, and the FAO/IAEA Central Laboratory for ELISA and molecular techniques in diagnosis of animal diseases, Vienna, Austria.

Bundza A, Afshar A, Dukes TW, Myers DJ, Dulac GC, Becker SAWE (1988). Experimental peste des petits ruminants (Goat Plague) in goats and sheep. Can. J. Vet. Res. 52: 46-52.

Carleton HM (1967). Histological Technique, 4th edition. New York, London and Toronto. Oxford University Press. pp. 48-58.

Durojaiye OA (1980). Brief notes on history, epizootiology and the economic importance of PPR in Nigeria. In: Proceedings of the International Workshop on Peste des Petits Ruminants, IITA, Ibadan, Nigeria, 24-26 September, pp. 24-27.

Durtnell DR (1972). A disease of sokoto goats resembling "peste des petits ruminants". Trop. Anim. Health Prod. 4: 162-164.

Elhag Ali B (1973). A natural outbreak of rinderpest involving sheep, goats and cattle in Sudan. Bull. Epizoot. Dis. Afr. 21: 421-428.

Ezeokoli CD, Taylor WP, Diallo A (1986). Clinical and epidemiological features of Peste des petits ruminants in Sokoto red goats. Rev. Elev. Med. Vet. Pays Trop. 39(3-4): 219–273.

Ismail TM, Yamanaka MK, Saliki JT, EL-Kholy A, Mebus C, Yilma T (1995). Cloning and expression of the nucleoprotein of peste des petits ruminants virus in baculovirus for use in serological diagnosis. Virology. 20: 776–778

Jones L, Giavedoni L, Saliki JT, Brown C, Mebus C, Yilma T (1993). Protection of goats against peste des petits ruminants with a vaccinia virus double recombinant expressing the F and H genes of rinderpest virus. Vaccine. 11: 961–964.

Karber G (1931). Beitrag zur kollektiven behandlung pharmakologischer reihenversuche. Arch. Exp. Pathol. Pharmakol. 162: 480-483.

Lefevre PC, Diallo A (1990). Peste des petits ruminants. Rev. Sci. et Tech. de l'Office Intl des Epiz. 9:951-965.

Mann E, Isoun TT, Fabiyi A, Odegbo-Olukoya OO (1974). Experimental transmission of the stomatitis peumoenteritis complex to sheep and goats. Bull. Epizoot. Dis. Afr. 22(2): 99-102.

Nussieba AO, Mahasin EA, Ali AS, Fadol MA (2008). Rapid Detection of Peste des Petits Ruminants (PPR) Virus Antigen in Sudan by Agar Gel Precipitation (AGPT) and Haemagglutination (HA) Tests. Trop. Anim. Health Prod. 40(5): 363-368.

Onoviran O, Majiyagbe KA, Molokwu JU, Chima JC, Adegboye DS (1984). Experimental infection of goats with Mycoplasma capri and "peste des petits ruminants" virus. Rev. Elev. Med. Vet. Pays Trop. 37(1): 16-18.

Plowright W, Ferris RD (1959). Studies with rinderpest virus in tissue culture. I. Growth and cytopathogenicity. J. Comp. Pathol. 69(2): 152-172.

Plowright W, Ferris RD (1962). Studies with rinderpest virus in tissue culture. A technique for the detection and titration of virulent virus in cattle. Res. Vet. Sci. 3: 94-103.

Saliki JT (1998). Peste des petits ruminants. In: Foreign Animal Diseases: The Gray Book. 6th edition. Part IV. Richmond, VA: US Animal Health Association, Committee on Foreign Animal Diseases. pp. 344-352.

Spearman C (1908). The method of right and wrong cases (constant stimuli) without Gauss's formulae. Brit. J. Psychol., 2: 227-242.

Taylor WP (1979). Serological studies with the virus of peste des petits ruminants in Nigeria. Res. Vet. Sci. 26,: 236–242.

White G (1958). A specific diffusible antigen of rinderpest virus demonstrated by the agar double-diffusion precipitation reaction. Nature, London, 181: 1409.

Wosu LO (1991). Haemagglutination test for diagnosis of peste des petits ruminants disease in goats with samples from live animals. Small. Rumin. Res. 5: 169-172.

Wosu LO (1994). Current status of peste des petits ruminants (PPR) disease in small ruminants-a review article. Stud. Res. Vet. Med. 2:83-90.

17

Prevalence of Hepatitis B surface antigen and Hepatitis C virus antibodies among pre-surgery screened patients in Khartoum, Central Sudan

Emad-Aldin Ibrahim Osman[1,2]*, Nagwa Ahmed Abdulrahman[2], Osman Abbass[2], Waleed Hussein Omer[3], Hafi Anwer Saad[4] and Muzamil Mahdi Abdel Hamid[5]

[1]Department of Haematology, Faculty of Medical Laboratory Sciences, Elrazi College of Medical and Technological Sciences, Khartoum, Sudan.
[2]Department of Clinical Laboratories, Al-Shaab Teaching Hospital, Federal Ministry of Health, Khartoum, Sudan.
[3]Al-Neelain Medical Research Center, Faculty of Medicine and Health sciences, Al-Neelain University, Khartoum, Sudan.
[4]Department of Community Medicine, Faculty of Medicine, University of Shendi, Shendi, Sudan.
[5]Department of Molecular Biology, Institute of Endemic Diseases, University of Khartoum, P. O. Box 102, Khartoum, Sudan.

The prevalence of Hepatitis B and C infection varies between different regions of Sudan according to several published reports. The present study is a descriptive hospital-based study aimed to estimate prevalence of Hepatitis B surface antigen and anti-Hepatitis C virus antibodies among 3172 patients undergoing surgery at Al-Shaab Teaching Hospital in Khartoum from April 2008 to April 2011. All patients were screened by rapid chromatography immunoassay for qualitative detection of Hepatitis B surface antigen and anti-Hepatitis C virus antibodies. The mean age of the studied subjects was 44 years; 61.1% of them were males and 38.9% were females. Hepatitis B surface antigen was detected in 156 patients (4.91%), while anti-Hepatitis C virus antibodies were detected in 58 patients (1.82%). The prevalence of Hepatitis B surface antigen is slightly higher in males (5.46%), than females (4.04%); however, it is statistically insignificant ($P= 0.08$). In conclusion, the present study reported a prevalence of Hepatitis B surface antigen which is lower than what has previously been reported in Sudan.

Key words: Hepatitis B virus (HBV), hepatitis C virus (HCV), prevalence, Khartoum, Sudan.

INTRODUCTION

Hepatitis B virus (HBV) and Hepatitis C virus (HCV) are common causes of liver disease globally. The HBV is a major public health challenge due to its worldwide distribution, chronic persistence and complications. Its endemicity ranges from high (≥8%) to moderate (2-7%) and low (<2%) (Margolis et al., 1991; Maynard et al, 1989). The HCV was first identified in 1989 (Houghton, 2009). Infections with HCV are pandemic and the World Health Organization (WHO) estimates a world-wide prevalence of 3%, most of these cases occur in Africa (Madhava et al., 2002; WHO, 1999). Infection with HBV and HCV is primarily blood borne or parenterally transmitted. Routes of parenteral transmission include contaminated blood and blood products, needle sharing, reuse of contaminated razors by barbers, tattooing devices, haemo-dialysis, acupuncture needles and contaminated medical devices. Other modes of transmission include sexual promiscuity and vertical transmission in the early childhood (Alain et al., 2002; Henderson, 2003; Levine et al., 1994; Ranger-Rogez et al., 2002; Stauber, 2000).

In health care settings, occupational risk for

---

*Corresponding author. E-mail: emad.ibrahim.osman@gmail.com.

Table 1. Prevalence of HBsAg and Anti-HCV among the study group.

Parameter	Study group n (%)	HBsAg n (%)	P value	HCV n (%)	P value
Total patients	3172	156 (4.9)		58 (1.8)	
Male	1939 (61.1)	106 (5.5)	0.08*	35 (1.8)	0.99
Female	1233 (38.9)	50 (4.0)		23 (1.9)	

n= number, *=chi square test.

transmission of HBV and HCV might exist. Transmission could occur from infected patients to staff, patient to patient or infected providers to patients (Alain et al., 2002). In recent years, considerable advances have been made in diagnostic testing for HBV and HCV. These enable rapid blood samples screening for patients and health care staff to minimize the risk of infection.

Previous studies of HBV and HCV epidemiology in Sudan showed different prevalence, from as high as 23% to as low as 0.6% for HCV and a wide range of HBV prevalence were reported (5.6 to 68%) (El-Amin et al., 2007; Elsheikh et al., 2007; McCarthy et al., 1989a). However, most of the reported prevalence in Khartoum State was among pregnant women, haemo-dialysis or patients with liver diseases (El-Amin et al., 2007; Elsheikh et al., 2007; Mudawi et al., 2007a).

The aim of this study was to estimate the prevalence of Hepatitis B surface antigen (HBsAg) and anti-HCV antibodies among patients who undergo different surgical interventions in Al-Shaab Teaching Hospital in Khartoum State. The results could serve as guidelines to future healthcare strategies and vaccination campaigns in Sudan.

MATERIALS AND METHODS

The study was conducted at the department of clinical laboratories, Al-Shaab Teaching Hospital which is tertiary hospital, located in the centre of Khartoum, Sudan. The study involved 3172 patients from all ages and both sexes who were undergoing different surgical interventions from April 2008 to 2011. Ethical approval was obtained from the Ethical Research Committee, Institute of Endemic Diseases, University of Khartoum and a separate permission was taken from the hospital.

All patients were informed about the test. Blood sampling collection was performed using a standard procedure and plasma was used for HBV and anti-HCV antibodies screening. Each patient was screened by rapid chromatography immunoassay for qualitative detection of HBsAg and anti-HCV antibodies to find the carrier status of patients before surgery. The hospital routinely uses commercial HBsAg and HCV test kits, from Standard Diagnostics (SD-South Korea Inc). Tests were performed in accordance with the manufacturer's instructions. The HBsAg SD BIOLINE kit has both sensitivity and specificity of 99%, while the HCV SD BIOLINE kit has a sensitivity of 100% and specificity of 99.4%. Moreover, samples which were positive for either HBsAg or anti-HCV antibodies had been re-tested for second time by the same method. Samples repeatedly reactive for HBsAg or anti- HCV antibodies were considered positive.

Statistical analysis was done using SPSS for Windows version 11.5. The difference between categorical variables was tested using Chi square test. The P value was considered significant when it is less than or equal to 0.05.

RESULTS

The present study included a total of 3172 patients undergoing surgical intervention at Al-Shaab Teaching Hospital in Khartoum; the mean age was 44 years with an age range of 1 month – 92 years. Out of the studied group, 1939 (61.1%) were males and 1233 (38.9%) were females. The HBsAg was detected in 156 patients (4.91%, 95% CI 4.2%-5.7%), while anti-HCV antibodies were detected in 58 patients (1.82%, 95% CI 1.4%-2.4%). The prevalence of HBsAg is slightly higher in males 106 (5.5%, 95% CI 4.5%-5.6%), than females 50 (4.0%, 95% CI 3.1%- 5.3%); however, it is statistically insignificant (P = 0.08). The estimated prevalence of HBsAg is higher than that of anti-HCV antibodies. Table 1 shows the prevalence of HBsAg and Anti-HCV in males and females of the study group.

DISCUSSION

Most descriptions of HBV and HCV epidemiology rely mainly upon HBsAg and HCV sero-prevalence studies. These studies are typically cross-sectional in design and are done in selected populations e.g. haemo-dialysis patients, blood donors or patients with chronic liver disease which are not representative of the whole community in which they reside. Population-based studies representative of an entire community are far more informative, but this kind of study is not feasible in most parts of the world (Shepard et al., 2005). Previous studies of HBsAg and HCV sero-prevalence in Khartoum State were merely done in selected subjects such as pregnant women representing adult females in child bearing age. Other studies were done in haemo-dialysis patients who have an increased risk to blood-borne infections due to the frequent dialysis. The present study estimates the prevalence of HBsAg and HCV among 3172 patients from all ages and both sexes who underwent different surgical interventions at Al-Shaab Teaching Hospital in Khartoum State. The results showed a moderate prevalence of HBsAg (4.91%), which could be compared with other studies from Sudan. Elsheikh et al. (2007) reported a 5.6% HBsAg positivity in 728 pregnant women in Khartoum State, which is similar but slightly higher than our result. On the other hand, some other studies in Sudan reported a higher prevalence of HBV than this study. Recent study in Nyala, South Dar Fur region revealed 6.2% HBsAg prevalence in blood donors (Abou et al., 2009). A survey in Gezira State of Central Sudan found 6.9% HBsAg positivity in the general population (Mudawi et al., 2007b). Moreover, another study reported 7.0% prevalence in healthy controls from Gezira and North Kordofan States (Omer et al., 2001). Earlier reports showed a high prevalence of HBV infection in Sudan. In 1989 a study reported prevalence of 18.7% HBsAg positivity after a survey involved two rural villages in Gezira State (Hyams et al., 1989). Furthermore, in 1992 another study in the Gezira State reported HBsAg positivity of 17.3 and 12.1% in blood donors and laboratory technical staff respectively (Elshafie, 1992). In 1989 McCarthy et al. (1989a, 1989b) reported two studies, one detected HBV markers in 68% of sexually active heterosexuals on the coast of Sudan (Port Sudan and Suakin), the other study found that 78% of male soldiers had serological evidence of past hepatitis B infection. Moreover, McCarthy et al. (1994) reported that HBsAg was detected in 26% of out-patients in Juba city, southern Sudan. The probable explanation for the lower prevalence of HBsAg in this study in comparison to the previous reports in Khartoum; was the improvement of blood screening procedures and the introduction of the HBV vaccination program.

The prevalence of HCV infection was lower in various regions of Sudan than that of HBV infection. The present study reports a 1.82% anti-HCV sero-positivity which is closer to other previous studies in Sudan. A study by Mudawi et al. (2007c) detected a 2.2% prevalence of anti-HCV antibodies in Gezira State. Other studies reported a prevalence of 3 and 1.5% in southern and two different regions of Sudan respectively (McCarthy et al., 1994; Omer et al., 2001). However, a study in Khartoum State reported a lower prevalence of 0.6% of anti-HCV among pregnant women (Elsheikh et al., 2007). Another study in Dar Fur, western Sudan also reported a low prevalence of 0.65% (Abou et al., 2009). Higher prevalence, 4.5 and 23.7% were reported in Khartoum among patients with hepatosplenic schistosomiasis and haemo-dialysis patients respectively (El-Amin et al., 2007; Mudawi et al., 2007a). Similar prevalence to this study was reported in other countries, 1.7% in Yemen, 1.7% in Ethiopia and 1.8% in USA (Alter et al., 1999; Bajubair et al., 2008; Gelaw and Mengistu, 2007). Egypt the northern neighbouring country to Sudan reported the highest HCV seroprevalence in the world, 12 to 31%, transmission has been attributed to contaminated glass syringes used in nationwide schistosomiasis treatment campaigns from 1960 to 1987 (Frank et al., 2000; Lehman and Wilson, 2009;

Mohamed et al., 1996; Youssef et al., 2009). However, in Sudan no association was found between HCV infection and schistosomiasis or with parenteral antischistosomal therapy (Mudawi et al., 2007a).

In conclusion, this study shows the prevalence of HBsAg and anti-HCV antibodies in 3172 patients who underwent different surgical interventions, which is more representative for the population of Khartoum State than the previous studies. This study reports a lower HBsAg prevalence.

ACKNOWLEDGEMENTS

Authors thank all patients who participated in this study. We are grateful to the laboratory staff at Al-Shaab Teaching Hospital for their excellent technical assistance.

REFERENCES

Abou MA, Eltahir YM, Ali AS (2009). Seroprevalence of hepatitis B virus and hepatitis C virus among blood donors in Nyala, South Dar Fur, Sudan. Virol. J. 6:146.

Alain S, Loustaud-Ratti V, Dubois F, Bret MD, Rogez S, Vidal E, Denis F (2002). Seroreversion from hepatitis C after needlestick injury. Clin. Infect. Dis. 34:717-719.

Alter MJ, Kruszon-Moran D, Nainan OV, McQuillan GM, Gao F, Moyer LA, Kaslow RA, Margolis HS (1999). The prevalence of hepatitis C virus infection in the United States, 1988 through 1994. N. Engl. J. Med. 341:556-562.

Bajubair MA, Elrub AA, Bather G (2008). Hepatic viral infections in Yemen between 2000-2005. Saudi. Med. J. 29:871-874.

El-Amin HH, Osman EM, Mekki MO, Abdelraheem MB, Ismail MO, Yousif ME, Abass AM, El-haj HS, Ammar HK (2007). Hepatitis C virus infection in haemodialysis patients in Sudan: two centers' report. Saudi J. Kidney Dis. Transpl. 18:101-106.

Elshafie SS (1992). The prevalence of hepatitis B surface antigen in the Gezira (Sudan). Afr. J. Med. Med. Sci. 21:61-63.

Elsheikh RM, Daak AA, Elsheikh MA, Karsany MS, Adam I (2007). Hepatitis B virus and hepatitis C virus in pregnant Sudanese women. Virol. J. 4:104.

Frank C, Mohamed MK, Strickland GT, Lavanchy D, Arthur RR, Magder LS, El Khoby T, Abdel-Wahab Y, AlyOhn ES, Anwar W, others (2000). The role of parenteral antischistosomal therapy in the spread of hepatitis C virus in Egypt. Lancet 355:887-891.

Gelaw B, Mengistu Y (2007). The prevalence of HBV, HCV and malaria parasites among blood donors in Amhara and Tigray regional States. Ethiop. J. Health Dev. 22(1):3-7.

Henderson DK (2003). Managing occupational risks for hepatitis C transmission in the health care setting. Clin. Microbiol. Rev. 16:546-568.

Houghton M (2009). The long and winding road leading to the identification of the hepatitis C virus. J. Hepatol. 51:939-948.

Hyams KC, al-Arabi MA, al-Tagani AA, Messiter JF, al-Gaali AA, George JF (1989). Epidemiology of hepatitis B in the Gezira region of Sudan. Am. J. Trop. Med. Hyg. 40:200-206.

Lehman EM, Wilson ML (2009). Epidemic hepatitis C virus infection in Egypt: estimates of past incidence and future morbidity and mortality. J. Viral Hepat. 16:650-658.

Levine OS, Vlahov D, Nelson KE (1994). Epidemiology of hepatitis B virus infections among injecting drug users: seroprevalence, risk factors and viral infections. Epidemiol. Rev. 16:418-436.

Madhava V, Burgess C, Drucker E (2002). Epidemiology of chronic hepatitis C virus infection in sub-Saharan Africa. Lancet Infect. Dis. 2:293-302.

Margolis HS, Alter MJ, Hadler SC (1991). Hepatitis B: evolving epidemiology and implications for control. Semin. Liver Dis. 11:84-92.

Maynard JE, Kane MA, Hadler SC (1989). Global control of hepatitis B through vaccination: role of hepatitis B vaccine in the Expanded Programme on Immunization. Rev. Infect. Dis. 11 Suppl., 3: S574-578.

McCarthy MC, Burans JP, Constantine NT, el-Hag AA, el-Tayeb ME, el-Dabi MA, Fahkry JG, Woody JN, Hyams KC (1989a). Hepatitis B and HIV in Sudan: a serosurvey for hepatitis B and human immunodeficiency virus antibodies among sexually active heterosexuals. Am. J. Trop. Med. Hyg. 41:726-731.

McCarthy MC, el-Tigani A, Khalid IO, Hyams KC (1994). Hepatitis B and C in Juba, Southern Sudan: results of a serosurvey. Trans. R. Soc. Trop. Med. Hyg. 88:534-536.

McCarthy MC, Hyams KC, el-Tigani el-Hag A , el-Dabi MA, el-Sadig el-Tayeb M, Khalid IO, George JF, Constantine NT and Woody JN (1989b). HIV-1 and hepatitis B transmission in Sudan. AIDS 3:725-729.

Mohamed MK, Hussein MH, Massoud AA, Rakhaa MM, Shoeir S, Aoun AA, AboulNaser M (1996). Study of the risk factors for viral hepatitis C infection among Egyptians applying for work abroad. J. Egypt Public Health Assoc. 71(1-2):113-47.

Mudawi HM, Smith HM, Fletcher IA, Fedail SS (2007a). Prevalence and common genotypes of HCV infection in Sudanese patients with hepatosplenic schistosomiasis. J. Med. Virol. 79:1322-1324.

Mudawi HM, Smith HM, Rahoud SA, Fletcher IA, Saeed OK, Fedail SS (2007b). Prevalence of hepatitis B virus infection in the Gezira State of central Sudan. Saudi. J. Gastroenterol. 13:81-83.

Mudawi HM, Smith HM, Rahoud SA, Fletcher IA, Babikir AM, Saeed OK, Fedail SS (2007c). Epidemiology of HCV infection in Gezira State of central Sudan. J. Med. Virol. 79:383-385.

Omer RE, Van't Veer P, Kadaru AM, Kampman E, el Khidir IM, Fedail SS, Kok FJ (2001). The role of hepatitis B and hepatitis C viral infections in the incidence of hepatocellular carcinoma in Sudan. Trans. R. Soc. Trop. Med. Hyg. 95:487-491.

Ranger-Rogez S, Alain S, Denis F (2002). [Hepatitis viruses: mother to child transmission]. Pathol. Biol. (Paris) 50:568-575.

Shepard CW, Finelli L, Alter MJ (2005). Global epidemiology of hepatitis C virus infection. Lancet Infect. Dis. 5:558-567.

Stauber R (2000). Epidemiology and transmission of hepatitic C. Wien Med. Wochenschr, 150:460-462.

WHO (1999). Global surveillance and control of hepatitis C. Report of a WHO Consultation organized in collaboration with the Viral Hepatitis Prevention Board, Antwerp, Belgium. J. Viral Hepat. 6:35-47.

Youssef A, Yano Y, Utsumi T, abd El-alah EM, abd El-HameedAel E, SerwahAel H, Hayashi Y (2009). Molecular epidemiological study of hepatitis viruses in Ismailia, Egypt. Intervirology 52:123-131.

18

Rapid progression to human immunodeficiency virus infection / acquired immunodeficiency syndrome (HIV/AIDS) correlates with variation in viral 'tat' sequences

Mary Bridget Nanteza[2], David Yirrell[1,5], Benon Biryahwaho[2], Natasha Larke[3], Emily Webb[3], Frances Gotch[4] and Pontiano Kaleebu[1,2]

[1]MRC/UVRI Uganda Research Unit on AIDS, c/o Uganda Virus Research Institute, Plot 51-59 Nakiwogo Road, P. O. Box 49, Entebbe, Uganda.
[2]Uganda Virus Research Institute, Plot 51-59 Nakiwogo Road, P. O. Box 49, Entebbe, Uganda.
[3]MRC Tropical Epidemiology group, London School of Hygiene and Tropical Medicine, Keppel Street, London WC1E 7HT, United Kingdom.
[4]Department of Immunology, Imperial College, Chelsea and Westminster Hospital 369 Fulham Rd, London SW10 9NH, United Kingdom.
[5]Department of Medical Microbiology, Ninewells Hospital, Dundee, DD1 9SY United Kingdom.

Gene sequence diversity plays an important function in determining survival of micro-organisms. Pathogenicity of HIV is correlated to host as well as viral factors. We aimed to identify sequence variations in *tat, nef* and the membrane-proximal *gp41*. These genes regulate important viral functions: *tat* for trans-activation, *nef* for enhancing infectivity and the membrane-proximal gp41 for fusion which could correlate with HIV disease progression. We studied HIV sequences from ART naïve adult Ugandans. Sequence diversity was analysed for 19 rapid progressors and 22 long-term survivors, Rapid progressors were individuals who progressed to a CD4 count of <200 cells/µl in a median time of 3.7 (range 1.3 to 4.9) years. The median time is calculated as being from mid-way between the last HIV sero-negative result and the index HIV sero-positive result, to the time of obtaining the study blood sample. Long-term survivors were individuals who had a CD4 count of >500 cells/µl after a median time of 8.8 (range 7.5 to 9.3) years, measured from the time of the index HIV sero-positive result to the time of obtaining the study blood sample. Amplification of DNA by polymerase chain reaction (PCR) and subsequent sequencing of *tat, nef* and membrane-proximal gp41 was performed starting from viral DNA directly obtained from frozen uncultured peripheral blood mononuclear cells. A 'long' *tat* protein was only observed in rapid progressors (RPs). The 'long' *tat* appears to predict rapid disease progression and could be relevant for designing an HIV-1 prognostic assay.

Key words: HIV-1, progression, *tat*, *nef*, gp41.

INTRODUCTION

Disease progression among people living with human immunodeficiency virus infection/acquired immunodeficiency syndrome (HIV/AIDS) seems to be strongly associated with multiplicity of factors such as host HLA alleles and HIV-1 subtype. Genetic variation in individual viral sequence such as an insertion of two amino acids in the C-terminus of exon 2 of *tat* (Tzitzivacos et al., 2009) and deletions in *nef* were in some studies shown to correlate with slow disease progression. In gp41, low levels of antibodies to the epitope '*ELDKWA*' have been associated with advanced HIV-1 disease (Srisurapanon et al., 2005).

The *tat* protein trans-activates transcription by attaching to the trans-activating *responsive (TAR)* element of the 5' long terminal repeat (*LTR*). Studies have been performed to delineate the functional mechanisms of *tat* (Kuppuswamy et al., 1989, LeGuern et al., 1993). Peloponese et al.

(1999) showed that HIV strains circulating in Africa were more virulent than strains in Europe and America and was attributed to variations in the *tat* protein. Furthermore, Niyasom et al. (2009) also showed that subtype B *tat* activity was associated with reduced disease progression. Humoral and cytotoxic T-cell responses to *tat* have also shown inverse correlation with slow and non- progressive HIV-1 disease (van Baalen et al., 1997, Zagury et al., 1998, Gupta and Mitra, 2007). However this association was not replicated in a study conducted among Ugandans (Senkaali et al., 2008). The trans-activating and immuno-responsive functions of *tat* have been attributed to the cysteine-rich region (region III) and the basic region (region IV) of *tat* exon 1. *Tat* has also been shown to enhance HIV-1 replication.

With regard to *nef,* several studies have demonstrated that *nef* deleted mutants of HIV-1 and simian immuno-deficiency virus (SIV) were associated with diminished viral virulence and attenuated infection however some other work did not confirm this observation (Hofmann-Lehmann et al., 2003, Chakrabarti et al., 2003). The *nef* gene has been shown to exert its effect through acceleration of HIV-1 activation from latency and enhancement of viral replication. There are indications that *nef* is implicated in the downregulation of CD4 and MHC class I molecules (Jin et al., 2008), and Geriach et al., 2010) thus disabling the humoral and cytotoxic responses.

Apobec3 cytidine deaminases are antiviral proteins that inhibit the replication of HIV-1. The 'YXXL' motif in the membrane-proximal cytoplasmic gp41 has been identified to mediate the binding to the human Apobec3 (Pery et al., 2009). It has been hypothesised that sequence variations in the 'YXXL' motif could interfere in the binding of Apobec3 and result in up regulation of viral replication. Such events could subsequently result in rapid disease progression. We report on DNA sequences of *tat, nef and membrane-proximal gp41* and attempts to correlate the variation observed to HIV-1 disease progression.

METHODOLOGY

Study subjects

This was a retrospective cross sectional study conducted among HIV-infected adults from a natural history population-based cohort maintained by the MRC/UVRI Uganda Research Unit on AIDS in Uganda (Morgan et al., 1997). Rapid progressors were individuals who progressed to a CD4 count of <200 cells/µl (median 173) in a

---

*Corresponding author. E-mail: marybridgetnanteza@yahoo.com.

Abbreviations: RP, Rapid progressors, **LTS,** long term survivors.

median time of 3.7 (range 1.3 to 4.9) years. This median time was calculted from mid-way between the last HIV sero-negative result and the index HIV sero-positive result, to the time of obtaining the study blood sample. Long-term- survivors were individuals who had a CD4 count of >500 cells/µl (median 689) in a median time of 8.8 (range 7.5 to 9.3) years. The median time was measured from the time of the index HIV sero-positive result to the time of obtaining the study blood sample. 64% of the long-term survivors were prevalent cases with no prior documentation of a negative result. The remainder were incident cases where the true length of infection could be documented (data not shown). The blood samples were obtained before anti-retroviral therapy (ART) was widely implemented in Uganda and participants were selected on the basis of having no previous exposure to anti-retroviral drugs. The Uganda Virus Research Institute Scientific and Ethical Committee approved the study.

CD4/CD8 count estimation

CD4/CD8 lymphocytes were quantified from 50 µl of fresh ethylene diaminetetraacetic acid (EDTA) blood using flowcytometry on a fluorescence activated cell sorting (FACS) count according to the manufacturer's instructions (Becton Dickinson International, Belgium).

DNA extraction

DNA was extracted from 300 µl frozen uncultured PBMC using the Puregene kit (Gentra Systems Inc., North Carolina, USA) according to the manufacturer's protocol.

DNA amplification

We designed *tat* and *gp41* primers and the *nef* primers were adapted from (Jubier- Maurin et al., 1999) to suit HIV-1 subtype A and D that were dominantly present in Uganda. All the *tat, gp41* and *nef* primers were synthesized by Oswel DNA, Southampton, UK. For the first round PCR, 5 µl of the DNA extract (~1.0 µg DNA) was added to a 15 µl reaction containing x1 PCR buffer; 200 µM-of dCTP, dATP, dTTP and dGTP [Sigma, USA]; 0.2 pmoles outer primer pairs (*tat-1* and *tat-2*) or (*nef-1* and *nef-2*) or (*env-7* and *env-8*); 1.5 mM $MgCl_2$ for (*tat-1* and *tat-2*) and (*nef-1* and *nef-2*); and 1.4 mM $MgCl_2$ for (*env-7* and *env-8*). Finally, 0.05 U of Taq DNA polymerase was added. DNA samples were cycled: (i) 94°C (1 min), 55°C (1 min) and 72°C (1 min); for three cycles; (ii) 94°C (30 s), 55°C (45 s), 72°C (1min); for 30 cycles; and (iii) 72°C (5 min). Two microlitres of the first round PCR product was transferred to an 18 µl reaction mixture containing x1 PCR buffer; 1.4 mM $MgCl_2$; 200 µMof dCTP, dATP, dTTP and dGTP; 0.2 pmoles of the inner primer pairs (*tat-3* and *tat-4*) or (*nef-3* and *nef-4*) or (*env-5* and *env-6*) or (*env-5* and *env-4*); and 0.05U of Taq DNA polymerase. Amplification was performed using the cycling conditions stated above. The details of the primers used are shown in Table 1.

Sequencing

The template for sequencing was generated from a 120 µl secondary PCR reaction containing 3 µl of the corresponding primary PCR product. The generated product was cleaned using the QIAquick PCR Purification kit [QIAGEN, UK]. A sequencing PCR reaction was carried out in a volume of 10 µl which consisted of 1 µl of 3.2 pmole/µl single secondary primer; 4 µl dRhodamine deoxy terminator mix [Applied Biosystems, Warrington, UK]; and 5 µl of cleaned PCR product. The mixture was subjected to thermal cycling

Table 1. Details of primers used.

Primer name	Primer location on HXB2	Primer sequence
*Tat*1	5711-5730 outer	5'GGATACYTGGGMAGGAGTTG 3'
*Tat*2	6227 -6207 ,,	5'CATTGCCACTGTCTTCTGCTC 3'
*Tat*3	5775-5795 inner	5'CAGAATTGGGTGYCWACATAG 3'
*Tat*4	6137-6116 ,,	5'C*TAT*RGTCCACACAAC*TAT*TGC 3'
EnvVII	7932-7952 outer	5'GTCTGGGGCATTAAACAGCTC 3'
EnvVIII	8782-8761 ,,	5'CTTTCTAAGCCCTGTCTGATTC 3'
EnvV	8004-8032 inner	5'GGAATTTGGGGCTGCTCTGG 3'
EnvVI	8707-8686 ,,	5'C*TAT*CTRTCCMCYCAGCTACTG 3'
EnvIV	8537-8516 ,,	5'CAAKYGGTGGTAGCTGAAGAGG 3'
*Nef*1	8513-8533 outer	5'GTGCCTCTTCAGCTACCACCG 3'
*Nef*2	9508-9488 ,,	5'AGCATCTGAGGGYTAGCCACT 3'
*Nef*3	8696-8717 inner	5'GKGGAYAGATAGGGY*TAT*AGAA 3'
*Nef*4	9467-9448 ,,	5'CRCCTCCCCTGGAAAGTCCCC 3'

The second exon of *tat* from a different reading frame in gp41 was added to the first exon.

at 90°C (30 s), 50°C (15 s), and 60°C (4 min) for 25 cycles. The product was precipitated with ethanol and sequenced on an ABI 373A automated sequencer according to the manufacturer's instructions.

Sequence analysis

Sequences from each region were separately aligned with homologous regions from consensus HIV-1 strains D and A obtained from the Los Alamos database [http://hiv-web.lanl.gov] using version 2.2 of the Genetic Data Environment [GDE] package (Smith et al., 1994). Neighbour joining phylogenetic trees for each region were generated using the PHYLIP set of computer programs (Felsenstein et al., 2003) implemented through the Treecon package (Van de Peer et al., 1994) employing a Kimura distance matrix (Kimura, 1980). The nucleotide sequences were submitted to the Genbank; *tat* exon 1 AF425936 – AF425974, *nef* AF425870 – AF425900 and gp41 AF425901 – AF425935].

Statistical analysis

The frequencies of residues at each amino acid position were compared between RPs and long term survivors (LTSs) in an alignment with the corresponding consensus subtypes D and A using a Chi squared test. The crude association between presence/absence of the long *tat* and clinical outcome (RP vs. LTS) was evaluated using a Chi square test. To allow for the large number of statistical tests and correlations between positions, p-values adjusted for multiple testing were also calculated using an empirical permutation procedure with 100,000 iterations (Churchill and Doerge, 1994). To investigate the possibility that predictor residues could be specific for subtype, frequencies of residues at each position were examined separately by subtype; however formal statistical analyses were not undertaken because this was among a small sample size. Nevertheless several crude p values were found and evidence of an association at p<0.05 was provided, many of the adjusted p values were considerably larger, this was as a consequence of adjusting for the multiple statistical tests conducted in a small sample size. All analyses were carried out in Stata 10.

RESULTS

We found major and minor sequence variations affecting the three target HIV-1 genes.

The 'long' *tat* was only found in RPs;

Tat sequence variations were observed in *tat* exons 1 and 2. A major variation was found in exon 2; the 'long' *tat* (115+ amino acids) was only found in RPs (seven out of 18 RPs, 36.8% compared to 0 out of 14LTSs, 0% crude p = 0.002, adjusted p=0.008 chi square test (Figure 1). Furthermore, there was some suggestion that positions 75, 79, and 86 of *tat* exon 2 seem to be associated with rapid disease progression (crude p= 0.032, adjusted p = 0.301) (Figure 2A). Position 82 also seems to be associated with rapid disease progression (crude p = 0.036, adjusted p=0.356). Whilst these adjusted p values do not provide strong evidence (p > 0.05), this was among a small sample size. Sequence variations were observed among "long" *tat* subtype A sequences as 28% sequences had proline (P) at position 75, glutamine (Q) at position 79 and glutamic acid (E) at position 86. Position 79 constitutes part of the cellular adhesion region of *tat* and variations at this position could play a role in HIV pathogenesis. In *tat* exon 1, lysine (K) at position 63 appears to be associated with rapid disease progression (crude p=0.011, adjusted p = 0.073) and was also associated with the "long" *tat* in subtype A sequences. Position 8 in *tat* exon 1 was conserved in LTSs; leucine (L) was exclusively found in viruses isolated from LTSs whereas both leucine (79%) and isoleucine (I) (21%) were found in RPs (crude p = 0.034 adjusted p = 0.351)

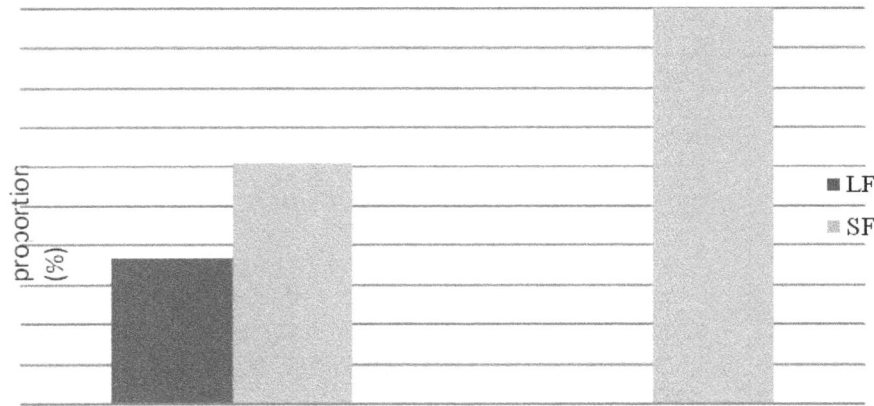

Figure 1. *Tat* sequence length variation and disease progression. LF, Long form; SF, short form; The figure shows the proportion of participants with 'long' *tat* and 'short' *tat* expressed as a percentage of total RPs and LTSs.

(Figures 2A and B). Among the "short" forms of *tat,* serine (S) at position 75, arginine (R) at position 79, and lysine (K) at position 86 were perfectly conserved in all LTSs and the amino acid residue at these positions was the same as the consensus subtype D sequence (Figure 2B). A neighbour joining phylogenetic tree for the *tat* exon 1 sequences is shown in Figure 3. Eight of the study isolates clustered with reference strains subtype A. Six were RPs and two were LTSs. Twenty six (26) clustered with reference strains subtype D. Nine were RPs and 17 were LTSs. Five study isolates did not identify with specific reference strains. These were 19RP, 04RP, 12RP, 14RP and 15LTS.

Nef sequence variation

The *nef* region exhibited a minor sequence variation at the protein kinase C binding site (data not shown). The alanine residue in the protein kinase C binding site of *nef* 102'PMTYK<u>A</u>A'108 was more common in RPs (56%) than LTSs (15%) whereas glycine (G) at the same site was more common in LTSs (85%) than RPs (44%) (crude p=0.023, adjusted p=0.381). A similar pattern was shown by Walker et al. (2007).

Gp41 sequence variation;

The epitope '55NWF<u>S</u>I'---'LW64' of the membrane-proximal external region *(MPER)* for gp41 neutralizing antibody '4E10' showed a minor sequence variation (data not shown). The serine residue within the epitope was more common in LTSs (76.5%) than RPs (27.8%) whereas aspartic acid (D) was more common in RPs (38.9%) than LTSs (17.6%) (crude p = 0.038, adjusted p = 0.502). The '96YXXL99' of the membrane-proximal cytoplasmic gp41 was conserved in the RPs and LTSs.

DISCUSSION

Campbell et al. (2004) performed a functional study on two subtype D sequences from our study population 05RP and 11LTS. A short alpha helix was observed in *tat* 05RP. *Tat* from 05RP was more efficient than *tat* from 11LTS in the trans-activation function. The differences between 05RP and 11LTS were the minor sequence variations at positions 8 and 63 of *tat* exon 1. Position 8 in the acidic region of *tat* exon 1 was quite conserved among the subtype A and D isolates however it exhibited sequence variation with the isoleucine residue among the rapid progressors. The acidic region is a domain for the neutralization antibody epitope of *tat* (Sneham et al., 2012). Sequence variation at position 8 might represent neutralization escape mutants for the subtype A isolates. Position 63 is located in the glutamine rich region. The glutamic rich region is a vital domain and plays a role in *tat* mediated apoptosis of the T-cells. Sequence variation in this domain may affect the rate of disease progression. In this study, the proline residue at position 63 was found among the RPs. This might play a role in enhancing *tat* mediated apoptosis of the T-cells and thus disease progression.

Position 79 of *tat* exon 2 is part of the '79RGD80' motif of *tat* exon 2 that is involved in cellular adhesion, uptake of extracellular *tat,* apoptosis and enhancing HIV replication. The 'RGD' motif enhances the adhesion of the extracellular *tat* via the $\alpha_v\beta$ and $\alpha5T\beta$ integrins. Extracellular *tat* stimulates the HIV LTR and results in the up regulation of the transcriptional process. Sequence variations at this site might therefore have various effects. Variations could down regulate or up regulate the transcription process resulting in slow or rapid disease progression. The subtype A isolates exhibited sequence variation at position 79 *tat* exon 2. Position 79 contained the glutamine residue '79Q/R' among the RPs. The variation could not

Tat Amino acid sequence for subtype A study isolates in RPs and LTSs

A

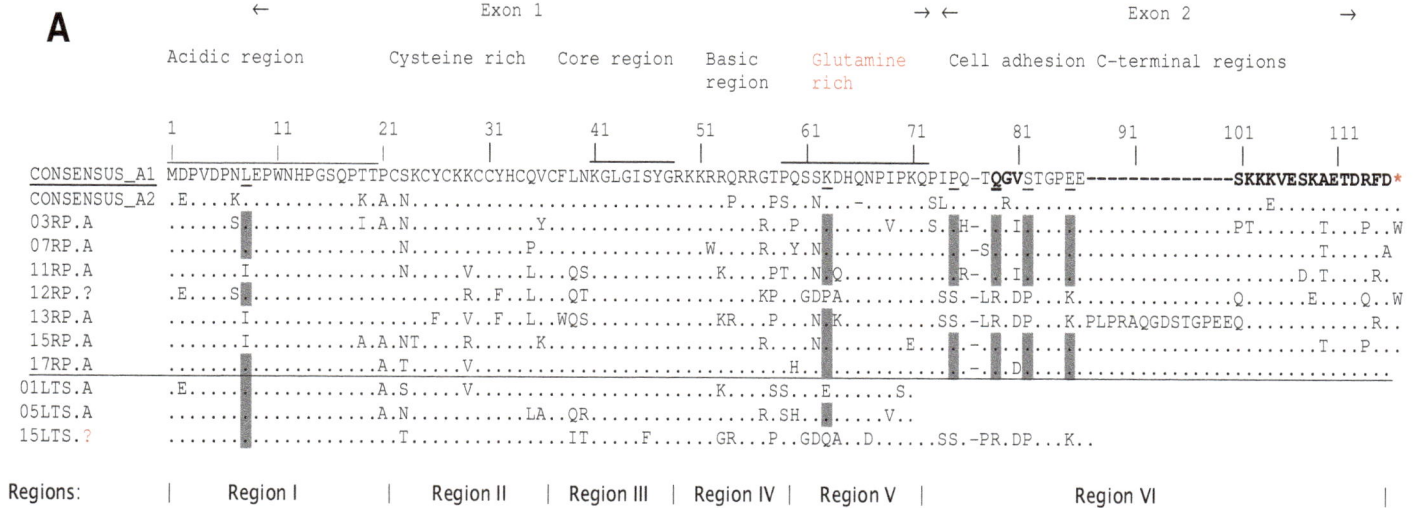

Tat Amino acid sequence for subtype D study isolates in RPs and LTSs

B

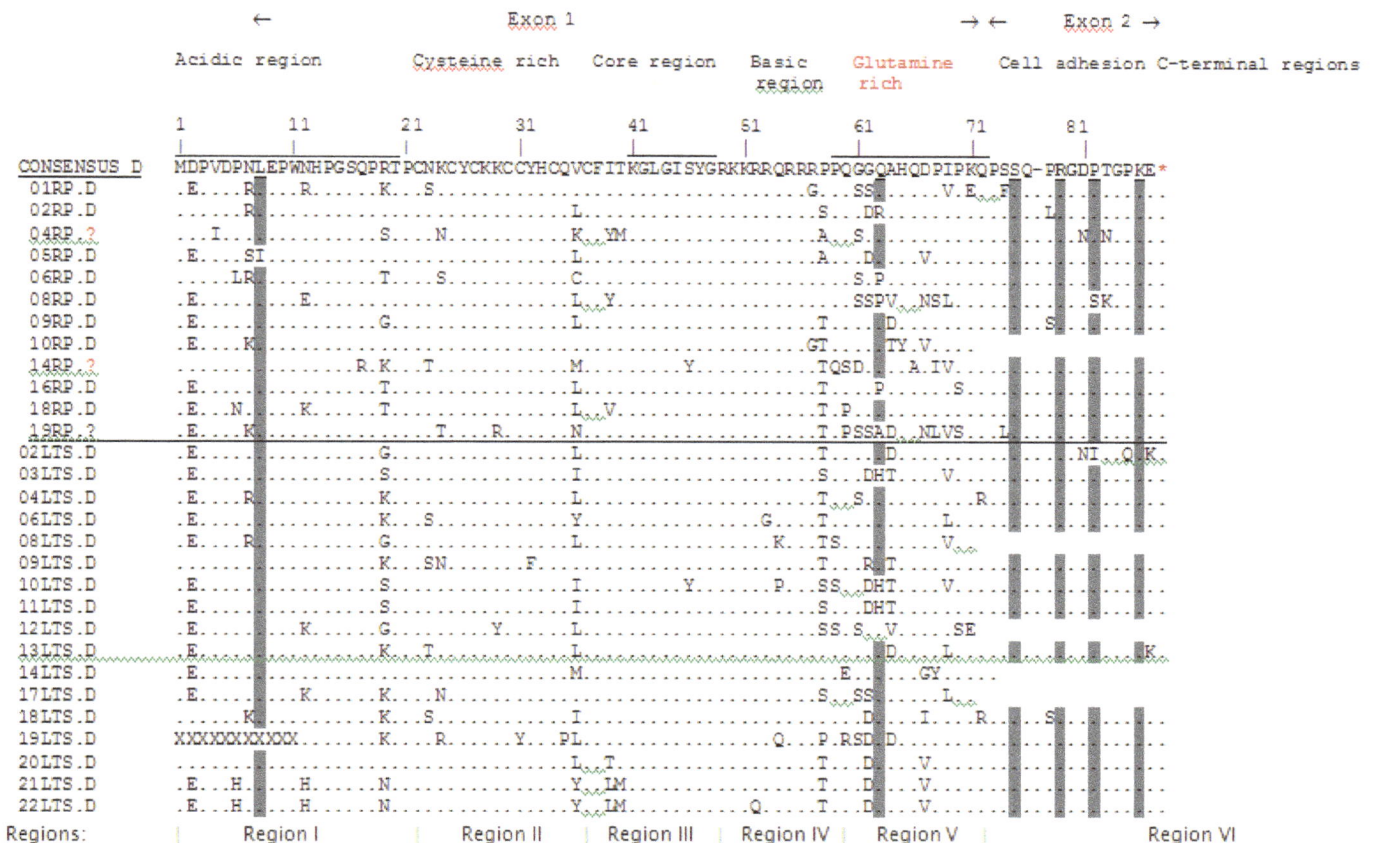

Figure 2. The amino acid alignments of *tat* sequences from RPs and LTSs of HIV-1 infection The three published consensus sequences A1, A2 and D were obtained from the Los Alamos National Laboratory. The (.), denotes same identity with the respective consensus sequence; (-), deletion; (x), sequence residue was not clear; the letters underlined represent residue positions of potential importance to HIV-1 disease progression; letters in bold within the consensus represent the amino acid sequences of the cell adhesion signal motif of *tat* and the 'long' *tat*; (?), sequence that could not be classified into a subtype and (*), is a stop codon.

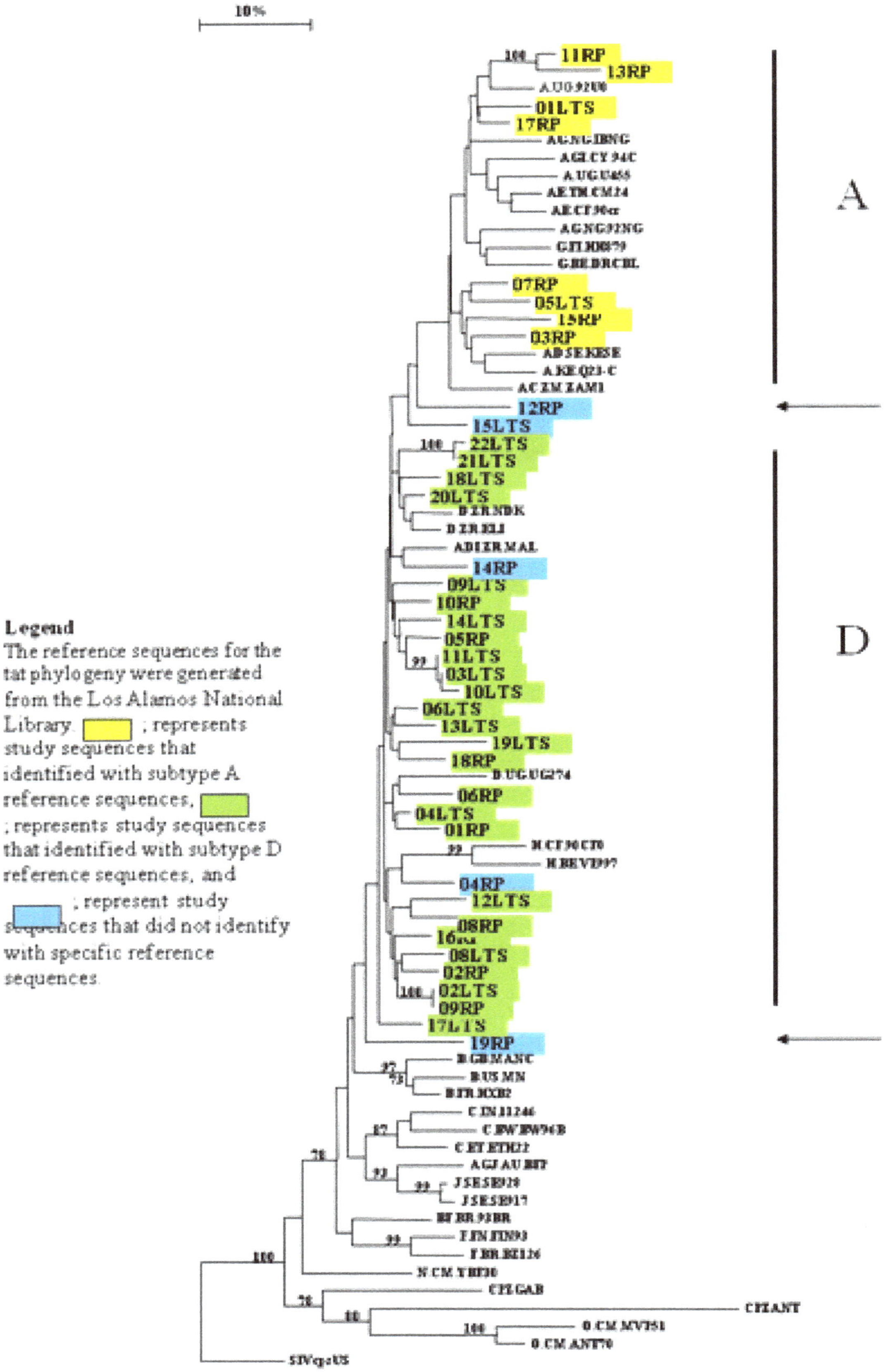

Figure 3. The phylogenetic tree of the first exon of TAT.

be studied in LTS because of the small sample size and short sequences. The 'long' tat was exclusively found in RPs in subtype A sequences. There was some suggestion that those residues with higher frequency in the sequences of RPs were likely to coexist with the 'long' tat in subtype A although this could not be formally evaluated due to the small study numbers. Tat exon 2 has been shown to improve the trans-activation process and induce HIV-1 pathogenic events (Lopez-Huertas et al., 2010). Thus variations within tat exon 2 could improve trans-activation function and contribute to disease progression. Although the cysteine-rich and basic regions have been reported to be critical for tat function data presented here in shows that tat exon 2 could also be critical for tat function. Functional studies of the 'long' and 'short' tat are required to understand better the role of the length of tat in disease progression. Future comparative studies using samples from European and American subjects could also give more insight on tat length and disease progression although the wide spread use of antiretroviral therapy makes this difficult.

CONCLUSION

In this study, the 'long' tat was associated with rapid HIV-1 disease progression. The 'long' tat could be a template for developing a prognostic screening assay as well as a therapeutic target for HIV disease control.

ACKNOWLEDGEMENTS

The study was funded by Rogers Research Fellowship through MRC/UVRI Uganda Research Unit on AIDS in Uganda. We are very grateful to Dr. Nicaise Ndembi for his useful comments, the MRC/UVRI Uganda Research Unit on AIDS in Uganda for its full support, and the 'Study Participants' who willingly donated blood and made it possible to perform this study.

CONFLICT OF INTERESTS: The authors have declared that no conflict of interest exists.

REFERENCES

Campbell GR, Pasquier E, Watkins J, Bourgarel-Rey V, Peyrot V, Esquieu D, Barbier P, de Mareuil J, Braguer D, Kaleebu P, Yirrell DL, Loret EP (2004). The glutamine-rich region of the HIV-1 Tat protein is involved in T-cell apoptosis. J. Biol. Chem. 279:48197-48204.

Chakrabarti LA, Metzner KJ, Ivanovic T, Cheng H, Louis-Virelizier J, Connor RI Cheng-Mayer C(2003). A truncated form of nef selected during pathogenic reversion of Simian Immunodeficiency Virus SIVmac239Δnef increases viral replication. J. Virol. 77:1245-1256.

Churchill GA, Doerge RW (1994). Empirical threshold values for quantitative trait mapping. Genetics.138:963-971.

Felsenstein J (1993). Phylip Manual Version 3.52c. Berkeley University Herbarium, University of Calfornia Berkeley.

Geriach H, Laumann V, Martens S, Becker CF, Goody RS, Geyer M (2010). HIV-1 Nef membrane association depends on charge, curvature, composition and sequence. Nat. Chem. Biol. 6:46-53.

Gupta S, Mitra D (2007). Human immunodeficiency virus-1 Tat protein: immunological facets of a transcriptional activator. Indian J. Biochem. Biophys. 44:269-275.

Hofmann-Lehmann R, Vlasak J, Williams AL, Chenine AL, McCure HM, Anderson DC, O'Neil S, Ruprecht RM (2003). Live attenuated, nef deleted SIV is pathogenic in most adult macaques after prolonged observation. AIDS 17:157-166.

Jin YJ, Cai CY, Zhang X, Burakoff SJ(2008). Lysine 144, a ubiquitin attachment site in HIV-1 Nef, is required for Nef-mediated CD4 down-regulation.J. Immunol.180:7878-7886.

Jubier-Maurin V, Saragosti S, Perret JL, Mpoudi E, Esu-Williams E, Mulanga C, Liegeois F, Ekwalanga M, Delaporte E, Peeters M(1999). Genetic characterization of the nef gene for human immunodeficiency virus type 1: Group M strains representing genetic subtypes A, B, C, E, F, G, and H. AIDS Res. Hum. Retroviruses 15:23-32.

Kimura M (1980). A simple method for estimating evolutionary rates of base substitutions through comparative studies of nucleotide sequences. J. Mol. Evol.16:111-120.

Kuppuswamy M, Subramanian T, Srinivasan A, Chinnadurai G (1989). Multiple functional domains of Tat, the trans-activator of HIV-1, defined by mutational analysis. Nucleic Acids Res. 17:3551-3561.

LeGuern M, Shioda T, Levy JA, Cheng-Mayer C (1993). Single amino acid change in Tat determines the different rates of replication of two sequential HIV-1 isolates. Virology 195:441-447.

Lopez-Huertas MR, Callejas S, Abia D, Mateos E, Dopazo A, Alcami J, Coiras M (2010). Modifications in host cell cytoskeleton structure and function mediated by intracellular HIV-1 Tat protein are greatly dependent on the second coding exon. Nucleic Acids Res. 38:3287-3307.

Morgan D, Maude GH, Malamba SS, Okongo MJ, Wagner HU, Mulder DW, Whitworth JA (1997). HIV-1 disease progression and AIDS-defining disorders in rural Uganda. Lancet 350:245-250

Niyasom C, Horthongkham N, Sreephiang A, Kantakamafakul W, Louisirirofchanakuf S, Chuenchire T, Sufhent R (2009). HIV-1 subtype B Tat gene activities and disease progression in HIV-1 CRF01_AE infection. Southeast Asian J. Trop. Med. Public Health. 40:748-758.

Peloponese JM Jr, Collette Y, Gregoire C, Bailly C, Campese D, Meurs EF, Olive D, Loret EP(1999). Full peptide synthesis, purification, and characterization of six Tat variants. Differences observed between HIV-1 isolates from Africa and other continents. J. Biol. Chem. 274:11473-11478.

Pery E, Rajendran KS, Brazier AJ, Gabuzda D (2009). Regulation of APOBEC3 proteins by a novel YXXL motif in human immunodeficiency virus type 1 Vif and simian immunodeficiency virus SIVagm Vif. J. Virol. 83:2374-2381.

Senkaali D, Kebba A, Shafer LA, Campell GR, Loret EP, Van Der Paal L, Grosskurth H, Yirrell D, Kaleebu P (2008). Tat-specific binding IgG and disease progression in HIV type 1-infected Ugandans. AIDS Res. Hum. Retroviruses 24:587-594.

Smith SM, Pentlicky S, Klase Z, Singh M, Neuveut C, Lu CY, Reitz MSJr, Yarchoan R, Marx PA, Jeang KT (2003). An in vivo replication-important function in the second coding exon of Tat is constrained against mutation despite cytotoxic T lymphocyte selection. J. Biol. Chem. 278:44816-44825.

Sneham T, Madhavan PN Nair, Shailendra KS (2012). Rendezvous with Tat: Transactivator of Transcription during Human Immunodeficiency Virus Pathogenesis. American Journal of Infectious Diseases 8(2):79-91.

Srisurapanon S, Louisirirotchanakul S, Sumransurp K, Ratanasrithong M, Chuenchirta T, Jintakatkom S, Wasi C (2005). Binding antibody to neutralizing epitope gp41 in HIV-1 subtype CRF 01_AE infection related to stage of disease. Southeast Asian J. Trop. Med. Public Health 36:221-227

Tzitzivacos DB, Tiemessen CT, Stevens WS, Papathanasopoulos MA (2009). Viral genetic determinants of nonprogressive HIV type 1 subtype C infection in antiretroviral drug-naive children. AIDS Res. Hum. Retroviruses 25:1141-1148

van Baalen CA, Pontesilli O, Huisman RC, Garetti AM, Klein MR, Miedema F, Gruterset RA, Ostahaus AD(1997). Human immunodeficiency virus type 1 Rev- and Tat-specific cytotoxic T lymphocyte frequencies inversely correlate with rapid progression to AIDS. J. Gen. Virol. 78:1913-1918.

Van de Peer Y, De Wachter R(1994). TREECON for Windows: a software package for the construction and drawing of evolutionary trees for the Microsoft Windows environment. Comput. Appl. Biosci.10:569-570.

Walker PR, Ketunuti M, Choge IA, Meyers T, Gray G, Holmes EC, Morris L (2007). Polymorphism in *nef* associated with different clinical outcomes in HIV type 1 subtype C-infected children. AIDS Res. Hum. Retroviruses 23:204-215.

Zagury JF, Sill A, Blattner W, Lachgar A, Le Buanec H, Richardson M, Rappaport J, Hendel H, Bizzini B, Gringeri A, Carcagno M, Criscuolo M, Burny A, Gallo RC, Zagury D(1998). Antibodies to the HIV-1 *Tat* protein correlated with nonprogression to AIDS: a rationale for the use of *Tat* toxoid as an HIV-1 vaccine. J. Hum. Virol. 1:282-292.

19

Molecular characterization of adenovirus causing acute respiratory disease in Malaysia from 2003 to 2011

Mohd Apandi Y.*, Zarina M. Z., Khairul Azuan O., Nur Ismawati A. R. and Zainah S.

Virology Unit, Infectious Disease Research Centre, Institute for Medical Research, Jalan Pahang, 50588 Kuala Lumpur, Malaysia.

Human Adenoviruses (HAdVs) are human pathogens that normally are associated with acute respiratory disease (ARD). Here, in this study, we characterized human adenovirus isolated from respiratory specimens collected from 2003 to 2011. Positive adenovirus were confirmed by cell culture and indirect immunoflourescence (IF) techniques. The isolates and positive clinical samples were subjected to adenovirus polymerase chain reaction, followed by DNA sequencing. BLAST searches and phylogenetic analysis revealed that only 2 species of HAdV, species C at 64% (47/73) and species B at 36% (26/73) were found circulating amongst ARD cases in Malaysia. Four types of HAdV species C were detected namely: HAdV-2 (66%), HAdV-1(17%), HAdV-5(13%) and HAdV-6(4%). For HAdV species B, only 2 types were detected with HAdV-3 being the highest at 58% and HAdV-7 at 42%. HadV-7 was associated with severe infections.

Key words: Human adenoviruses, acute respiratory disease, molecular characterization, gene sequences.

INTRODUCTION

Human Adenoviruses (HAdVs) were first isolated from human adenoids and identified as respiratory pathogens in 1953 (Rowe et al., 1953; Hilleman et al., 1954) and now HAdVs have been recognized as aetiological agent for variety of diseases. It is a common infectious agents in children less than 5 years and responsible for 5 to 10% of all lower respiratory tract infections in infants and children (Choi et al., 2006). They are ubiquitous, non-enveloped, double stranded DNA in the family of *Adenoviridae*. Diseases in human such as pharyngitis, pneumonia, gastroenteritis, haemorrhagic cystitis and keratoconjunctivitis have been associated with HAdVs infections (Mandell et al., 2009). The severity of the disease also ranges from mild or inapparent clinical syndromes to severe life threatening disease in immuno-compromised individuals (Mandell et al., 2009). Infections with HAdVs can occur sporadically or in epidemics; and the virus can be isolated and identified throughout the year (Hong et al., 2001). HAdVs are also responsible for outbreaks in settings that have close living condition such as in military barracks (Dudding et al., 1972), hospital wards (Straube et al., 1983), chronic care facilities (Finn et al., 1988) and police academy (Apandi et al., 2012). Previously, HAdVs were classified by haemagglutination and serum neutralization (Seto et al., 2011). Since genomic data and bioinformatics became available, new HAdVs have been identified including several emerging and recombinant viruses (Walsh et al., 2011) and today there are 57 recognised serotypes in 7 species, HAdV-A to HAdV-G (Martin et al., 2007). The hexon protein with serotype specific encoded by seven hyper variable regions (Crawford and Schnurr, 1996) is the most important components for serotype identification of adenoviruses (Takeuchi et al., 1999).

Epidemiologic characteristics of the HAdV vary by type. All are transmitted by direct contact, faecal oral transmission and occasionally waterborne transmission. Depending on the species, these viruses may infect respiratory, conjunctival, gastrointestinal and genitourinary sites and specific types have been linked to distinct clinical syndromes. Recently, HAdV have emerged as life-threatening opportunistic agents in severely immune-suppressed patients, such as stem cell recipients (Ganzenmueller et al., 2011). The outbreak of adenoviruses associated with respiratory disease has been reported worldwide (Schmitz et al., 1983; Zhu et al., 2009; Trei et al., 2010). The serotypes that most frequently associated with acute respiratory disease (ARD) were from species B and C. HAdV-3 and HAdV-7 in species B were reported as a common cause of severe pneumonia in neonates and young children (Carballal et al., 2002; Tsolia et al., 2004); whereas, HAdV-11 and HAdV-14 have been reported in febrile respiratory disease outbreaks in all ages (Kajon et al., 2010; Gu et al., 2012). In the species C, serotypes HAdV-1, HAdV- 2, HAdV-5 and HAdV-6 were endemic in parts of the world (CDC, 2007) and cause febrile respiratory illnesses in children and young adults (Metzgar et al., 2005; Abd-Jamil et al., 2010). HAdV-4 was the only serotype in species C that involved in ARD (Rubin, 1993; Kandel et al., 2010).

The objectives of the study were to determine the type of HAdV circulating from 2003 to 2011 and the association of HAdV type with severe acute respiratory disease. Therefore, we characterized all HAdVs associated with respiratory illness from samples received by the Institute for Medical Research (IMR) from patients who sought treatment for respiratory infection in the government hospitals in Malaysia.

MATERIALS AND METHODS

Virus

From 2003 to 2011, IMR received samples from 10,972 patients who seek treatment at 140 government hospitals for respiratory illness such as acute respiratory distress, pneumonia, bronchiolitis and bronchopneumonia. All samples were screened for respiratory viruses including adenovirus by indirect immunoflourescence assay (IFA respiratory panel 1 Kit, Millipore, UK) and cell culture technique. In the IFA technique, samples were considered positive for adenovirus when fluorescence was present in the nucleus or cytoplasm of the cells; whereas in tissue culture solely based on the appearance of the cytopathic effect (CPE). Positive samples by cell culture were recultured in Vero cells and harvested when cells showed CPE.

DNA extraction and PCR

Viral nucleic acid was extracted from the samples by using Roche high pure viral nucleic acid kit (Roche Applied Science, Mannheim, Germany). Briefly, 200 μl of sample was added to binding buffer supplemented with poly (A) and proteinase K in a microcentrifuge tube and incubated at 72°C for 10 min. Later, binding buffer was added and the mixture was transferred to high pure filter tube for centrifugation. Inhibitor removal buffer was used to remove PCR inhibitors and wash buffer for removal of residual impurities. Finally, elution buffer was added to elute viral nucleic acid in a clean micro-

centrifuge tube. Nested PCR was used to amplify the HAdV hexon gene. First PCR primer pair, forward primer HEXB-1 (5'-AAC ATG ACC AAR GAC TGG TT-3') and reverse primer HEXB-2 (5'-GCC GAG AAS GGY GTR CGC AGG TA -3') (IHCM, 2003) in a one tube reaction (20 μl) containing 10 μl DNA, 1 μl of 20 μM of each primer, 2.0 μl of 25 mM MgCl₂ with 10X PCR buffer, 0.6 μl of 10 mM dNTP, 0.5 μl of 5 U/μl Taq DNA polymerase and nuclease free water. Samples were subjected to 35 PCR cycles, denaturation at 95°C for 30 s, annealing at 60°C for 30 s and extension at 72°C for 30 s; and final extension at 72°C for 5 min. For the second PCR, forward primer HEXB-3 (5'-TTC AGA AAC TTC AGC CCY ATG AG-3') and reverse primer HEXB-4 (5'-TCC ATG GGA TCC ACC TCA AAR GTC AT-3') (IHCM, 2003) were used and 5 μl of first PCR product was used as a template and PCR was performed using the same conditions as the first PCR.

Expected PCR products (360 bp) were examined by gel electrophoresis and QIAquick gel extraction kits (QIAGEN Inc, Valencia, CA) was used to extract the DNA from the gel.

Nucleotide sequencing and phylogenetic analysis

The partial hexon gene amplicons were sequenced on both strands by using primers HEXB-3 and HEXB-4. Sequencing was performed by using the Big Dye Cycle Sequencing kit version 3.0 and an ABI377 automated DNA sequencer (Applied Biosystems, Foster City, CA, USA). The SeqMan software module in the Lasergene suite of programs (DNASTAR, Madison, WI, USA) was used to format the nucleotide sequences. Alignment of the gene sequences was undertaken by using the MegAlign software module in the Lasergene suite of programs (DNASTAR, Madison, WI, USA). Phylogenetic tree was constructed by using the neighbour-joining method from the Software MEGA4 (Tamura et al., 2007). The duck adenovirus type 1 was used as an outgroup for phylogenetic analysis together with different species of adenovirus obtained from GenBank for the purpose of generating dendograms. All sequences were submitted to GenBank with accession number JX182290-JX183062.

RESULTS

More than 76% specimens found positive for adenoviruses were derived mainly from nasopharyngeal aspirate (NPA) from children less than 5 years of age diagnosed with ARD (Table 1). All 73 positive adenovirus samples, 30 positive by tissue culture and 43 IFA positive were subjected to adenovirus polymerase chain reaction (PCR), and followed by DNA sequencing. Type of samples received, year of isolation, diagnosis and types of adenovirus isolated are shown in Table 1. All positive samples were sequenced, and basic local alignment search tool (BLAST) sequencing analysis (http://blast.ncbi.nlm.nih.gov/Blast.cgi) showed that the sequences belonged to HAdV-1, HAdV-2, HAdV-3, HAdV-5, HAdV-6 and HAdV-7. Phylogenetic tree, constructed on the basis of partial hexon gene (348 bp) nucleotides of all adenoviruses isolated from 2003 to 2011 together with adenovirus isolates retrieved from GenBank is shown in Figure 1. In 2003, only 2 HAdV-7 were detected from ARD specimens and no adenoviruses were detected in 2004. The following year, in 2005, 2 HAdVs were detected and were identified as HAdV-1 and HAdV-2. HAdV detection rate started to increase in 2006, with 10 HAdV-2 and 1 HAdV-1. The number increased to

Table 1. Summary of patients' gender, age, sample type, clinical diagnosis, HAdV typing result and GenBank accession number.

ID	Accession No.	Sample type	Sex/age	Clinical diagnosis	Isolation year	Adenovirus type
RV0227/03	JX182990	TA (i)	M/1.3	Respiratory distress	2003	Ad7
RV0259/03	JX182991	NPA (i)	M/1.5	Pneumonia	2003	Ad7
RV0092/05	JX182992	NPA (i)	M/2.5	Pneumonia	2005	Ad1
RV0198/05	JX182993	Sputum (i)	M/6	Chronic lung Disease	2005	Ad2
RV0700/06	JX182994	NPA (i)	M/0.3	Bronchopneumonia	2006	Ad1
RV1007/06	JX183003	NPA (p)	M/0.3	Pneumonia	2006	Ad2
RV1009/06	JX182995	NPA (p)	M/0.3	Bronchopneumonia	2006	Ad2
RV1016/06	JX182996	NPA (p)	M/0.6	Bronchopneumonia	2006	Ad2
RV1027/06	JX182997	NPA (p)	F/1.0	Severe pneumonia	2006	Ad2
RV1067/06	JX183002	NPA (p)	F/2.4	Bronchopneumonia	2006	Ad2
RV1083/06	JX182998	NPA (p)	F/0.1	Bronchopneumonia	2006	Ad2
RV1093/06	JX183004	NPA (p)	M/0.3	Bronchiolitis	2006	Ad2
RV1095/06	JX182999	NPA (p)	F/1.1	Pneumonia	2006	Ad2
RV1111/06	JX183000	NPA (p)	M/0.3	Pneumonia	2006	Ad2
RV1115/06	JX183001	NPA (p)	M/0.1	Pneumonia	2006	Ad2
RV0004/07	JX183005	NPA (p)	M/1.0	Severe bronchopneumonia	2007	Ad3
RV0037/07	JX183006	NPA (p)	M/0.1	Pneumonia	2007	Ad2
RV0062/07	JX183007	NPA (p)	M/0.5	Pneumonia	2007	Ad2
RV0114/07	JX183014	TA (p)	M/7.0	Bronchopneumonia	2007	Ad2
RV0353/07	JX183015	NPA (p)	F/1.5	Pneumonia	2007	Ad5
RV0440/07	JX183019	NPA (p)	M/1.0	Pneumonia	2007	Ad1
RV0447/07	JX183008	NPA (p)	M/0.1	pneumonia	2007	Ad2
RV0533/07	JX183009	NPA (p)	M/0.8	pneumonia	2007	Ad3
RV0603/07	JX183017	NPA (p)	F/6.4	Pneumonia	2007	Ad2
RV0619/07	JX183013	NPA (p)	M0.7	Bronchopneumonia	2007	Ad1
RV1031/07	JX183010	NPA (p)	M/0.8	Bronchopneumonia	2007	Ad1
RV1312/07	JX183018	NPA (p)	M/0.5	Bronchopneumonia	2007	Ad2
RV1426/07	JX183011	NPA (i)	F/1.0	pneumonia	2007	Ad3
RV1601/07	JX183020	Sputum (p)	M/30	AGE with sepsis	2007	Ad2
RV1694/07	JX183012	NPA (p)	F/2.6	Acute pharyngotonsilitis	2007	Ad2
RV1816/07	JX183016	NPA (p)	M/0.8	pneumonia	2007	Ad2
RV0018/08	JX183022	NPA (i)	M/0.6	Acute bronchiolitis	2008	Ad5
RV1293/08	JX183024	NPA (p)	M/0.5	Bronchopneumonia	2008	Ad3
RV1907/08	JX183023	ETS (i)	F/0.5	Nosocomial pneumonia	2008	Ad2
RV1987/08	JX183021	NPA (p)	M/0.5	pneumonia	2008	Ad3
RV0087/09*	JX183032	Lung tissue (i)	M/52	Pneumonia	2009	Ad3
RV0224/09	JX183025	NPA (p)	M/1.2	Pneumonia	2009	Ad7
RV0229/09	JX183033	NPA (i)	M/1.0	Bronchopneumonia	2009	Ad3
RV0272/09	JX183026	NPA (i)	M/0.8	Pneumonia	2009	Ad2
RV0322/09	JX183027	NPA (i)	M/0.8	Pneumonia	2009	Ad2
RV0475/09	JX183028	NPA (i)	F/0.4	Nosocomial pneumonia	2009	Ad5
RV0482/09	JX183029	NPA (i)	F/0.5	Pneumonia	2009	Ad1
RV0577/09	JX183030	NPA (i)	F/0.4	Pneumonia	2009	Ad1
RV0725/09	JX183031	NPA (i)	F/0.8	Pneumonia	2009	Ad2
RP0297/10	JX183036	NPA (p)	M/1	Bronchopneumonia	2010	Ad6
RP0320/10	JX183034	ETT (i)	F/1	Severe pneumonia	2010	Ad5
RP0393/10	JX183035	TS (i)	M/5	DHF with pneumonia	2010	Ad3
RP0263/11	JX183054	NPA(p)	M/1	pneumonia	2011	Ad3
RP0301/11*	JX183037	TA (p)	F/22	CAP with respiratory failure	2011	Ad7

Table 1. Contd.

RP0302/11*	JX183038	TA, lung, spleen (p)	F/25	CAP with respiratory failure	2011	Ad7
RP0320/11	JX183058	Stool (i)	M/20	CAP	2011	Ad5
RP0360/11	JX183060	NPA (p)	M/1	Acute bronchiolitis	2011	Ad1
RP0404/11	JX183039	NPA (p)	F/3	Bronchopneumonia	2011	Ad7
RP0423/11	JX183061	NPA (i)	M/1.5	pneumonia	2011	Ad2
RP0460/11	JX183059	NPA (p)	F/0.8	pneumonia	2011	Ad7
RP0485/11	JX183040	TS (i)	M/24	CAP	2011	Ad7
RP0507/11	JX183052	NPA (i)	M/0.8	pneumonia	2011	Ad3
RP0510/11	JX183055	NPA (p)	M/17	Severe pneumonia	2011	Ad7
RP0536/11	JX183062	NPA (p)	M/0.3	pneumonia	2011	Ad5
RP0540/11	JX183056	NPA (p)	F/0.7	Severe pneumonia	2011	Ad7
RP0568/11	JX183041	NPA (i)	M/4.3	pneumonia	2011	Ad2
RP0583/11	JX183042	NPA (p)	F/5	pneumonia	2011	Ad3
RP0592/11	JX183043	NPA (p)	F/0.5	Acute bronchiolitis	2011	Ad7
RP0612/11	JX183044	NPA (p)	M/1.3	pneumonia	2011	Ad3
RP0622/11	JX183045	NPA (i)	F/1.8	Severe pneumonia	2011	Ad3
RP0630/11	JX183046	NPA (i)	F/0.5	pneumonia	2011	Ad3
RP0689/11	JX183051	NPA (i)	M/1	bronchopneumonia	2011	Ad2
RP0695/11	JX183047	ETS (p)	F/0.1	Respiratory distress	2011	Ad6
RP0745/11	JX183048	NPA (p)	M/0.1	pneumonia	2011	Ad3
RP1081/11	JX183049	TS (i)	M/23	CAP	2011	Ad2
RP1082/11	JX183050	TS (i)	M/22	CAP	2011	Ad2
RP1095/11*	JX183053	TS (i)	M/12	Meningoencephalitis	2011	Ad2
RP1098/11	JX183057	TS (i)	M/25	ILI	2011	Ad2

M, Male; F, female; TA, trachea aspirate; NPA, nasopharyngeal aspirate; ETS, endotracheal secretion; TS, throat swab; (i), tissue culture isolate; (p), primary clinical samples; AGE , acute gastroenteritis; DHF , dengue haemorrhagic fever; CAP , community acquired pneumonia; ILI, influenza like illness; *, denotes fatal case

16 in 2007, and 15 (94%) of the positive PCR samples came from primary clinical specimens. Of the 16 positive samples, 15 were HAdV species C consisting of 9 HAdV-2, 3 HAdV-1 and 1 HAdV-5. HAdV-3 from Species B was identified from the remaining 2 HAdVs positive samples. Four HAdV were isolated in 2008, 2 were from primary clinical specimens and were identified as HAdV-3, and 2 from cell culture were identified as HAdV-5 and HAdV-2.

In 2009, from 9 samples found positive for HAdV, 5 different types were identified namely: HAdV-1, HAdV-2, HAdV-3, HAdV-5 and HAdV-7. In 2010, we detected HAdV-6 in species C for the first time from nasopharynx-geal aspirate; together with HAdV-3 and HAdV-5. Most of the positive samples for this study (~35%) were received in 2011 where HAdV-7 was the dominant type followed by HAdV-2, HAdV-3, HAdV-5, HAdV-6 and HAdV-1 (Table 1).

DISCUSSION

HAdV has been recognised as a cause of ARD, gastro-intestinal infection and other simple febrile illness (Horwitz, 2001). Specific HAdV types normally present with specific manifestations and severity (Wold and Horwitz, 2007),

although there are variations in different parts of the world (Baum, 2005). Young children and immunocompromised patients are the most vulnerable to severe complication of HAdV infections (Kojaoghlanian et al., 2003; Walls et al., 2003). In this study, partial hexon gene sequences were used to characterize the HAdV isolates. This gene region contains the hypervariable region and is the most important components for serotype identification of adenoviruses (Takeuchi et al., 1999). The finding showed that 2 species of HAdV; 64% of species C and 36% of species B were found circulating amongst ARD cases in Malaysia from 2003 to 2011. Overrepresentation of HAdV-C at 64% from total adenovirus detected in the past 9 years could suggest a high prevalence of the virus in the community. This finding was consistent with the report from previous study in Malaysia by Abd-Jamil et al. (2010). She found that 70% of the HAdV infection from paediatric patients who sought treatment for respiratory tract infection in UMMC, Malaysia from 1999 to 2005 was species C with HAdV-1 and HAdV-2 were the commonest. These were similar to the findings reported by Garcia et al. (2009) in South America involving 231 characterized adenoviruses collected from influenza like-illness during 2006 to 2008 showed that 161 (76%) adenoviruses belong to species C, 45 (21%) to species B

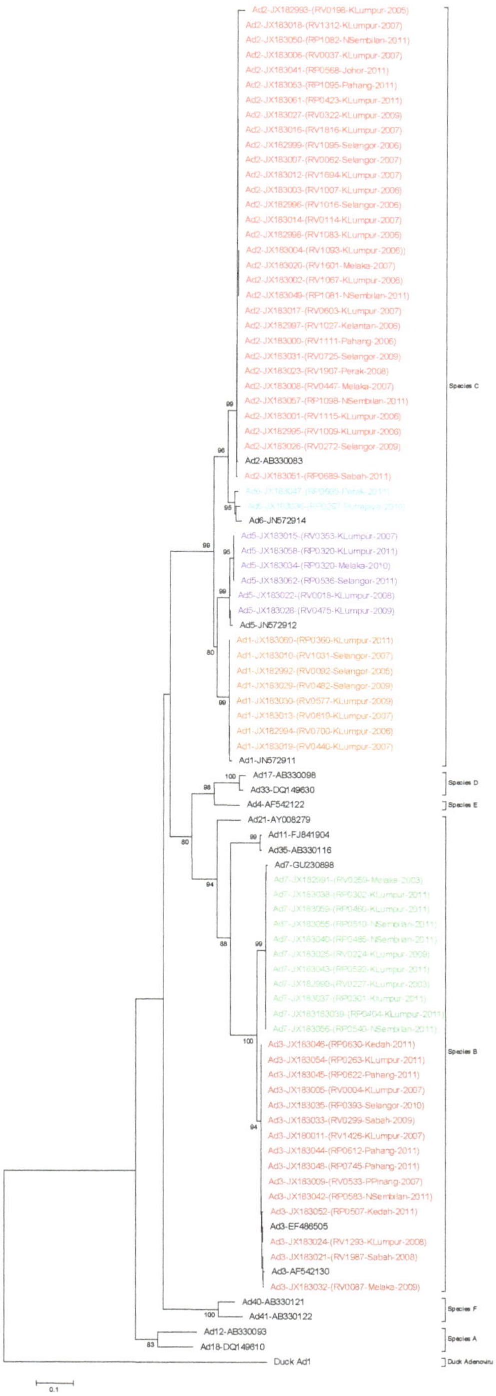

Figure 1. Phylogenetic tree of partial hexon gene sequences (348 bp) of human adenovirus inferred by using the neighbor-joining method from the MEGA4 software (www.megasoftware.net). The evolutionary distances were computed by using the maximum composite likelihood method. Species A to F are indicated by square brackets with duck adenovirus A as an outgroup. Seventy three human adenovirus from acute respiratory disease in Malaysia from 2003 to 2011 are indicated. Representative strains of each species obtained from GenBank are labeled by using the adenovirus species and accession number. Bootstrap values (>75%) for 1,000 pseudoreplicate datasets are indicated at branch nodes. Scale bar indicates nucleotide substitutions per site.

and 7 (3%) to species E. However, a study in Korea reported HAdV species B especially HAdV-3 and HAdV-7, was the predominant serotype among Korean children with respiratory tract infection (Lee et al., 2010).

Similar findings were also reported in Canada (Yeung et al., 2009) and in the military camp in USA (Kajon et al., 2007) where HAdV- B was the major species of HAdV identified from respiratory diseases. The possibility of high prevalence of species C especially HAdV-1, HAdV-2 and HAdV-5, could possibly be due to the virus ability to persist and cause latent infection in tonsils and adenoids of human, which at times cause ARD in young children (Pereira, 1972). Therefore, the prolonged presence of the virus in infected children increases its transmissibility and this could contribute to the persistence presence of HAdVs in young children in Malaysia. HAdV-C species have also been reported not only in respiratory diseases but also in digestive tract infection, regardless of the immune status of the patients and was also found in local ecology in France (Berciaud et al., 2012). In our study, 66% of HAdV species C detected was HAdV-2 and majority was isolated from children less than 5 years presenting with lower respiratory tract infection such as pneumonia, bronchopneumonia and bronchiolitis (Table 1). Our study showed that only 2 cases of HAdV-2 were related to upper respiratory tract infection (URTI); one isolated from a 25 year-old adult with URTI and another was from a 2.5 year-old child with acute pharyngotonsilitis. Report of fatal outcome associated with HAdV-2 is very rare. However, in this study we found a case of a 12 year-old boy that was admitted for meningoencephalitis, ventilated and died.

No other pathogen such as enterovirus, influenza virus and bacteria were detected except HAdV-2 was isolated from his throat swab. Although, considered as an isolated case, but it could suggest that this type could possibly cause severe infection in patient having underlying disease such as CNS infection. 17% of the species C detected was HAdV-1 and it was found in children less than 3 years old diagnosed with lower respiratory tract infections (Table 1). This finding is similar to findings from Sevaraju et al. (2011) who reported that HAdV-1 affecting mainly infants and young children less than 2 years of age and caused both upper and lower respiratory infections; and also from Casas et al. (2005) who implicated HAdV-1 with bronchiolitis in children less than 5 years of age. Others adenovirus in this species, HAdV-5 and HAdV-6 were also detected in small numbers in our study and showed to be low prevalence in Malaysia. For HAdV species B, HAdV-3 was more dominant than HAdV-7 at 58 and 42% detection rate, respectively. Majority of patients with HAdV-3 infection were children less than 5 years old presenting either with pneumonia or severe bronchopneumonia. HAdV-3 was also detected in lung tissue from a fatal case (Table 1), suggesting that type 3 could be associated with severe illnesses and fatal cases. A World Health Organization (WHO) survey on respiratory viral infections reported that HAdV-3 had a worldwide distribution and accounted for most of the HAdV associated infections.

Similar situation was also observed in Connecticut (Landry et al., 2009), where an increased incidence of HAdV-3 infections associated with a new variant, HAdV-3a51 that caused mostly mild infections; however, one fatality involving a patient with underlying disease was reported. The same scenario was also observed in Taiwan where HAdV-3 was the commonest HAdV detected among children with RTI from 1999 to 2000 (Hsieh et al., 2009). In our study, HAdV-7 has been found circulating at a low prevalence since 2003 but increased drastically in 2011 with more than 50% of HAdVs isolated were HAdV-7. The frequency of detection increased as a result of the outbreak of HAdV-7 reported in Malaysia in 2011 where there was an increase in both prevalence and disease severity (Apandi et al., 2012). This was similar to the report from Kansas City, where infections with HAdV-7 increased to 25 from 5.6% in the previous two years of surveillance and were associated with severe illness such as acute bronchiolitis and pneumonia (Gray et al., 2007). HAdV-7 was also known to cause outbreaks amongst hospitalised children (Choi et al., 2006; Selvaraju et al., 2011), police/army recruit camps (Apandi et al., 2012), and had the potential for morbidity and mortality. HAdV-7 is a well-known pathogen causing epidemics of severe lower respiratory tract infections in children, with a high mortality rate (Baum, 2005). Global survey has shown approximately one-fifth of all HAdV infections reported to World Health Organization (WHO) were attributed to HAdV-7 (Smith et al., 1983; Erdman et al., 2002). All of our HAdV-7 isolates from 2003, 2009 and 2011 belong to HAdV-7d2 which were very similar to the isolate CQ1198 isolated in China in 2010 from children with severe respiratory infection (Ni et al., 2012 unpublished). The same findings were reported by Selvaraju et al. (2011) where HAdV-7 strains 7d2 were responsible for severe lower respiratory tract infection in children in USA and by Tang et al. (2011), where it was associated with infants' pneumonia in China.

In Malaysia, we reported the first HAdV-7 outbreak in 2011 and it was associated with more severe disease and even fatal cases. Most of HAdV-7 isolated in the study was from patients with respiratory distress, severe pneumonia, bronchiolitis and fatal cases due to respiratory failure. These 2 fatal cases were patients related to the outbreak of HAdV-7 in Police Training Centre as reported by Apandi et al. (2012). HAdV-7 has been documented in outbreaks either among infants, older children or adults with serious outcomes (Wadell et al., 1980). The HAdV-7d2 strain was first reported in 1998 by Azar et al. (1998) and since then became the prevalent dominant strain of HAdV-7. In China, HAdV-7d was dominant during 1980 to 1994 and was a representative genome type in Asian until 1998 (Zhang et al., 1986) and associated with higher fatality rate than HAdV-3 (Li et al., 1996) and Erdman et al. (2002) reported 2

emergent genome types of HAdV-7 and both were associated with epidemic, severe illness and death. Other serotypes from species B, HAdV-11 and HAdV-14 which previously have been associated with acute respiratory disease with fatal outcome (Zhu et al., 2009; Ou et al., 2008), involved in outbreak associated pneumonia (Esposito et al., 2010) and fatal pneumonia (Hong et al., 2001; Tate et al., 2009) were not found in the study.

Conclusion

This study showed that only 2 species of adenovirus were found circulating among ARD cases in Malaysia in the 9 years period from 2003 to 2011. Majority was HAdV species C at 64% and the remaining 36% (26/73) was HAdV species B. The predominant type was HAdV type 2, followed by HAdV-3, HAdV-7, HAdV-1, HadV-5 and HAdV-6. HadV-7 was found to be associated with severe clinical presentations.

ACKNOWLEDGEMENTS

The authors would like to thank the Director General of Health and the Director of the Institute for Medical Research for permission to publish this paper. We would like to acknowledge the staff of the Virology unit in IMR for their technical contributions and laboratory staff from the government hospitals in Malaysia for sending the specimens to IMR. This research project was funded under JPP 11-024.

REFERENCES

Abd-Jamil J, Teoh BT, Hassan EH, Nuruliza R, Sazaly AB (2010). Molecular identification of adenovirus causing respiratory tract infection in paediatricpatients at the University of Malaya Medical Center. BMC Paed. 10:46.

Apandi Y, Tengku RTAR, Thayan R, Khairul AO, Norhasnida AH, Norfaezah A, Zainah S (2012). Human adenovirus type 7 outbreak in Police Training Center, Malaysia, 2011. Emerg. Infec. Dis: 18(5): 852-854.

Azar R, Varsano M, Mileguir F, Mendelson E (1998). Molecular epidemiology of adenovirus type 7 in Israel: Identification of two new genome types, Ad7k and Ad7d2. J. Med. Virol. 54:291-299.

Baum SG (2005). Adenovirus. In: Mandell GL, Bennett JE, Dolin R, editors. Mandell, Douglas, and Bennett's principles and practice of infectious diseases. 6th ed. Philadelphia, PA: Elsevier Churchill Livingstone. pp. 1835–41.

Berciaud S, Rayne F, Kassab S, Jubert C, Faure-Della CM, Salin F, Wodrich H, Lafon ME (2012). Typadeno Study Members. Adenovirus infections in Bordeaux University Hospital 2008–2010: Clin. virol. features. J. Clin.Viro. 54: 302– 307.

Carballal G, Videla C, Misirlian A, Requeijo PV, Aguilar MC (2002). Adenovirus type 7 associated with severe and fatal acute lower respiratory infections in Argentine children. BMC Pediatr, 2: pg. 6.

Casas I, Avellon A, Mosquera M, Jabado O, Echevarria JE, Campos RH, Rewers M, Perez-Breña P, Lipkin WI, Palacios GI (2005). Molecular identification of adenoviruses in clinical samples by analyzing a partial hexon genomic region. J. Clin. Microbiol. Dec; 43(12):6176-82.

CDC (2007). Acute respiratory disease associated with adenovirus serotype 14-Four states, 2006-200. Morb. Mortal weekly Rep.

56:1181-4

Choi EH, Lee HJ, Kim SJ, Eun BW, Kim NH, Lee JA, Lee JH, Song EK, Kim SH. (2006). Ten year analysis of adenovirus type 7 molecular epidemiology in Korea, 1994-2004: Implication of fiber diversity. J. Clin. Virol. 35(4):388-393.

Crawford IL, Schnurr DP (1996). Analysis of 15 adenovirus hexon proteins reveals the location and structure of seven hypervariable regions containing serotype-specific residues. J. Virol. 70: 1836-1844.

Dudding BA, Wagner SC, Zellar SC, Gmelith JT, French GR, Top FH (1972). Fatal pneumonia associated with adenovirus type 7 in three military trainees. N Engl. J. Med. 286:1289-1292.

Erdman DD, Xu W, Gerber SI, Gray GC, Schnurr D, Kajon AE, Anderson LJ (2002). Molecular epidemiology of adenovirus type 7 in the United States, 1966-2000. Emerg. Infect. Dis. 8:269-277.

Esposito DH, Gardner TJ, Schneider E, Stockman LJ, Tate JE, Panozzo CA, Robbins CL, Jenkerson SA, Thomas L, Watson CM, Curns AT, Erdman DD, Lu X, Cromeans T, Westcott M, Humphries C, Ballantyne J, Fischer GE, McLaughlin JB, Armstrong G, Anderson LJ (2010). Outbreak of pneumonia associated with emergent human adenovirus serotype 14--Southeast Alaska, 2008. J. Infect. Dis. Jul 15; 202(2):214-22. doi: 10.1086/653498.

Finn A, Andy E, Talbot GH (1988). An epidemic of adenovirus 7a infection in a neonatal nursery:course, morbidity and management. Infect. Control Hosp. Epidemiol. 9:394-404.

Ganzenmueller T, Buchholz S, Harste G, Dammann E, Trenschel R, Heim A (2011). High lethality of human adenovirus disease in adult allogeneic stem cell transplant recipients with high adenoviral blood load. J. Clin. Virol. 2011;52:55–9.

Garcia J, Sovero M, Laguna-Torres VA, Gomez J, Chicaiza W, Barrantes M, Sanchez F, Jimenez M, Comach G, de Rivera IL, Agudo R, Arango AE, Barboza A, Aguayo N, Kochel TJ (2009). Molecular characterization of adenovirus circulating in Central and South America during the 2006-2008 period. Influenza Other Resp. Viruses. 3(6):327-30.

Gray GC, McCarthy T, Lebeck MG, Schnurr DP, Russell KL, Adriana KAE, Landry LM, Leland DS, Storch GA, Ginocchio CC, Robinson CC, Demmler GJ, Saubolle MA, Kehl SC, Selvarangan R, Miller MB, Chappell JD, Zerr DM, Kiska DL, Halstead DC, Capuano AW, Setterquist SF, Chorazy ML, Dawson JD, Erdman DD (2007). Genotype prevalence and risk factors for severe clinical adenovirus infection, United States 2004–2006. Clin. Infect. Dis. 45:1120–31.

Gu L, Liu Z, Li X, Qu J, Guan W, Liu Y, Song S, Yu X, Cao B (2012). Severe community-acquired pneumonia caused by adenovirus type 11 in immunocompetent adults in Beijing. J. Clini. Virol. 54(4): 295–301.

Hilleman MR, Werner JH (1954). Recovery of new agent from patients with acute respiratory illness. Proc. Soc. Exp. Biol. Med. 85:183–188.

Hong JY, Lee HJ, Piedra PA, Choi EH, Park KH, Koh YY, Kim WS (2001). Lower respiratory tract infections due to adenovirus in hospitalized Korean children: epidemiology, clinical features, and prognosis. Clin. Infect. Dis. 32: 1423–1429.

Horwitz MS (2001). Adenoviruses. In Fields Virology 4th edition. Edited by: Fields BN, Knipe DM, Howley PM, Griffin DE. Philadelphia: Lippincott Williams & Wilkins; 2001:2301-2326.

Hsieh WY, Chiu NC, Chi H, Huang FY, Hung CC (2009). Respiratory adenoviral infections in Taiwanese children: a hospital-based study. J. Microbiol. Immunol. Infect. 42(5): 371-377.

Institute of Health and Community Medicine (2003). Standard operating procedure for hand, foot and mouth disease outbreak in UNIMAS Sarawak.

Kajon AE, Dickson LM, Metzgar D, Houng HS, Lee V, Tan BH (2010). Outbreak of febrile repiratory illness associated with adenovirus 11a infection in a Singapore Military training Camp. J. Clin. Microbiol. 48(4): 438-1441.

Kajon AE, Moseley JM, Metzgar D, Huong HS, Wadleigh A, Ryan MA, Russell KL (2007). Molecular epidemiology of adenovirus type 4 infections in US military recruits in the postvaccination era (1997–2003). J. Infect. Dis. 196:67–75.

Kandel R, Srinivasan A, D'Agata EM, Lu X, Erdman D, Jhung M (2010). Outbreak of adenovirus type 4 infection in a long term care facility for the elderly. Infect. Cont. Hosp. Epidemiol. 31(7): 755-757.

Kojaoghlanian T, Flomenberg P, Horwitz MS (2003). The impact of adenovirus infection on the immunocompromised host. Rev. Med. Virol. 2003, 13:155-171.

Landry ML, Lebeck MG, Capuano AW, McCarthy T, Gray GC (2009). Adenovirus type 3 outbreak in Connecticut associated with a novel variant. J. Med. Virol. 81(8): 1380-1384.

Lee J, Choi EH, Lee HJ (2010). Comprehensive serotyping and epidemiology of human adenovirus isolated from the repiratory tract of Korean children over 17 consecutive years (1991-2007). J. Med. Virol. 82(4): 624-631.

Li QG, Zheng QJ, Liu YH, Wadell G (1996). Molecular epidemiology of adenovirus type 3 and 7 isolated from children with pneumonia in Beijing. J. Med. Virol. 49:170-177.

Mandell GL, Bennett JE, Dolin R (2009). Mandell, Douglas, and Bennett's principles and practice of infectious diseases, 7th ed Churchill Livingstone/Elsevier, Philadelphia, PA.

Martin MA, Knipe DM, Fields BN, Howley PM, Peter M, Griffin D, Lamb R (2007). Fields' virology. Philadelphia: Wolters Kluwer Health / Lippincott Williams & Wilkins. Pp. 2395.

Metzgar D, Osuna M, Yingst S, Rakha M, Earhart K, Elyan D, Esmat H, Saad MD, Kajon A, Wu J, Gray GC, Ryan MAK, Russell KL (2005). PCR analysis of Egyptian respiratory adenovirus isolates, including identification of species, serotypes, and coinfections. J. Clin. Microbiol. 43: 5743-5752.

Ou ZY, Zeng QY, Wang FH, Xia HM, Lu JP, Xia JQ, Gong ST, Deng L, Zhang JT, Zhou R (2008). Retrospective study of adenovirus in autopsied pulmonary tissue of pediatric fatal pneumonia in South China. BMC Infect. Dis. 8:122.

Pereira HG (1972). Persistent infection by adenoviruses. J. Clin. Pathol. Suppl. 6:39-42.

Rowe WP, Huebner RJ, Gilmore LK, Parrott RH, Ward TG (1953). Isolation of a cytopathogenic agent from human adenoids undergoing spontaneous degeneration in tissue culture. Proc. Soc. Exp. Biol. Med. 84: 570–573.

Rubin BA (1993). Clinical picture and epidemiology of adenovirus infections. Acta Microbiol Hung 1993, 40:303-323.

Schmitz H, Wigand R, Heinrich W (1983). Worlwide epidemiology of human adenovirus infections. Am. J. Epidemiol. 117: 455-466.

Selvaraju SB, Kovac M, Dickson LM, Kajon AE, Selvarangan R (2011). Molecular epidemiology and clinical presentation of human adenovirus infections in Kansas City Children. J. Clin. Virol. Doi: 10.1016/j.jcv.2011.02/014.

Seto D, Chodosh J, Brister JR, Jones MS (2011). Using the whole genome sequence to characterize and name human adenoviruses. J. Virol. 85:5701–5702.

Straube RC, Thompson MA, Van Dyke RB, Wadell G, Connor JD, Wingaed D, Spector A (1983). Adenovirus type 7b in children's hospital. J. Infect. Dis. 147:814-819.

Takeuchi K, Itoh N, Uchio E, Aoki K, Ohno S (1999). Serotyping of adenoviruses on conjunctival scrapings by PCR and sequence analysis. J. Clin. Microbiol. 37: 1839-1845.

Tamura K, Dudley J, Nei M, Kumar S (2007) MEGA4: Molecular Evolutionary Genetics Analysis (MEGA) software version 4.0. Mol. Biol. Evol. 24:1596-1599.

Tang L, Wang L, Tan X, Xu W (2011). Adenovirus serotype 7 associated with a severe lower respirartory tract disease outbreak in infants in Shaanxi Province, China. Virol. J. 8:23.

Tate JE, Bunning ML, Lott L, Lu X, Su J, Metzgar D, Brosch L, Panozzo CA, Marconi VC, Faix DJ, Prill M, Johnson B, Erdman DD, Fonseca V, Anderson LJ, Widdowson MA (2009). Outbreak of severe respiratory disease associated with emergent human adenovirus serotype 14 at a US air force training facility in 2007. J. Infect. Dis. 199 (10): 1419-26.

Trei JS, Johns NM, Garner JL, Noel LB, Ortman BV, Ensz KL, Johns MC, Bunning ML, Gaydos JC (2010). Spread of adenovirus to geographically dispersed military installations, May-October 2007. EID.www.cdc.gov/eid.16 (5);769-775.

Tsolia MN, Psarras S, Bossios A, Audi H, Paldanius M, Gourgiotis D, Kallergi K, Kafetzis DA, Constantopoulos A, Papadopoulos NG (2004). Etiology of community-acquired pneumonia in hospitalized school-age children: evidence for high prevalence of viral infections. Clin. Infect. Dis. 39(5): 681–686.

Wadell G, Versanyi TM, Lord A, Sutton RN (1980). Epidemic outbreaks of adenovirus 7 with special reference to the pathogenicity of adenovirus genome type 7b. Am. J. Epidemiol. 112:619-628.

Walls T, Shankar AG, Shingadia D (2003). Adenovirus: an increasingly important pathogen in paediatric bone marrow transplant patients. Lancet Infect. Dis. 3:79-86.

Walsh MP, Seto J, Liu EB, Dehghan S, Hudson NR, Lukashev AN, Ivanova O, Chodosh J, Dyer DW, Jones MS, Seto D (2011). Computational analysis of two species C human adenoviruses provides evidence of a novel virus. J. Clin. Microbiol. 49(10): 3482-90. doi: 10.1128/JCM.00156-11. Epub 2011 Aug 17.

Wold WS, Horwitz MS (2007). Adenoviruses. In: Knipe DM, Howley PM, editors. Fields virology. 5th ed. Philadelphia, PA: Lippincott Williams &Wilkins, Inc. 2395–436.

Yeung R, Eshaghi AR, Lombos E, Blair J, Mazzulli T, Burton L, Drews SJ (2009). Characterization of culture-positive adenovirus serotypes from respiratory specimens in Toronto, Ontario, Canada. Virol. J. 6(11): 1-3.

Zhang ZJ, Wang ZL, Cao YP, Zhu ZH, Liu YL, Lin LM, Gao X (1986). Acute respiratory infections in childhood in Beijing: An aetiological study of pneumonia and bronchiolitis. Clin. Med. J. 99:695-702.

Zhu Z, Zhang Y, Xu S, Yu P, Tian X, Wang L, Liu Z, Tang L, Mao N, Ji Y, Li C, Yang Z, Wang S, Wang J, Li D, Xu W (2009). Outbreak of acute respiratory disease in China caused by B2 species of adenovirus type 11. J. Clin. Microbiol. 47(3): 697-703.

20

High prevalence of anti-hepatitis E virus among Egyptian blood donors

Endale Tadesse[1], Lobna Metwally [2]and Alaa EL- Din Saad Abd-El Hamid [3]

[1]Department of Medical Laboratory Science, College of Medicine and   Health Sciences, Hawassa University, P.O. Box 1560, Hawassa, Ethiopia.
[2]The Department of Medical Microbiology and Immunology, Faculty of Medicine, Suez Canal University, Egypt.
[3]The Department of Clinical Pathology, Faculty of Medicine, Suez Canal University, Egypt.

This study evaluated the seroprevalence of hepatitis E virus among blood donors attending blood transfusion Center of Suez Canal University Hospital from March to September, 2010. Four hundred eighty eight (488) subjects which consisted of 137 Anti- Hepatitis C virus positive donors, 35 Hepatitis B surface antigens positive donors and 316 blood donors who were negative Hepatitis B surface antigen, Anti- Hepatitis C virus and HIV were included in this study. Anti-hepatitis E virus (IgG and IgM) was detected in 17.7, 28.57, and 26.28% of blood donors negative for Hepatitis B surface antigen (HBsAg) and Anti- Hepatitis C virus, Hepatitis B surface antigen (HBsAg) positive and Anti- Hepatitis C virus positive donors, respectively. No significant (P > 0.05) association was found between anti- Hepatitis E virus positivity and Hepatitis B surface antigen (HBsAg) positivity and anti-Hepatitis C virus) positivity subjects. The overall prevalence of anti- Hepatitis E virus antibodies (IgG and IgM)   was 20.9% (102/488). Seroprevalence increased significantly with age; from 8.3% in subjects below 20 years of age, 16.94% in 20-34 years of age, 34.5% in 35-49 years of age and a slight decline of 33.3% over those of 50 years of age. All anti-HEV antibodies samples were negative for Hepatitis E virus RNA by reverse transcriptase polymerase chain reaction (RT-PCR) method. Even though, seroprevalence of hepatitis E virus antibody among blood donors in our study in Ismailia, Egypt is high, transfusion-associated with hepatitis E infection still needs further investigation.

Key words: Hepatitis E virus, Blood donors, seroprevalence.

INTRODUCTION

Hepatitis E virus (HEV) is the etiological agent of acute hepatitis. It is a non-enveloped, positive-sense, single stranded RNA virus. Originally classified within the family of caliciviruses, HEV is now classified as the sole member of the genus Hepevirus in the family Hepeviridae (Emerson et al., 2005, Tam et al. 1991).

Even if hepatitis E is a self-limiting viral infection followed by recovery, occasionally, HEV infection leads, in 1 to 2% of cases, to lethal fulminant hepatitis, and this level reaches 20% in pregnant women infected during their third trimester in areas of endemicity (Singh et al., 2003).

Hepatitis E is endemic in many developing countries of Asia and Africa where sanitation is sub optimal (Dalton et al., 2008) and is also endemic in many industrialized countries including the United States, European countries, and Japan (Emerson and Purcell, 2004; Purcell and Emerson, 2001; Okamoto et al., 2003).

Antibody to HEV, which is indicative of past infection, has been detected in 5 to 60% of the general population in developing countries where the disease is endemic

(Purcell, 1996). In Egypt, recent studies reported very high levels of anti-HEV prevalence among healthy adults and pregnant females in rural areas; 67.7 and 84.3%, respectively (Stoszek et al., 2006).

The virus is excreted in feces and primarily transmitted via the fecal-oral route through contaminated water or food. In developing countries with poor sanitation conditions, rare outbreaks of acute hepatitis E in more explosive epidemic form are generally associated with fecal contamination of drinking water (Arankalle et al., 2001; Purcell and Emerson, 2001). In addition most of the HEV outbreaks have been observed during the rainy seasons or after floods (Uchida, 1992).

Alternatively, the detection of serum antibodies to HEV and its viral genome in various animal species; like, swine, rodents, chickens, dogs, cows, sheep and goats (Arankalle et al, 2001; Favorov et al., 1998; Tien et al.,1997) amplified the idea of a possibility of zoonotic transmission of the virus.

Although transmission of HEV is generally via the faecal–oral route, person-to-person transmission, and transmission via the parenteral route or blood transfusion have been suggested (Dawson et al., 1997; Schlauder et al., 1993).

Blood borne transmission of HEV had been investigated as indirect evidence implicating HEV as a potential transfusion risk by many investigators worldwide. It has been reported that a substantial proportion of blood donors (1.5%) were positive for HEV RNA and viraemic blood donors are potentially able to cause transfusion-associated hepatitis E in areas of high endemicity (Arankalle and Chobe, 1999; Arankalle and Chobe, 2000).

This work was carried out to determine the seroprevalence of HEV and attempt to gain insight into the possible blood-borne transmission of HEV among blood donor who attended blood transfusion center of Suez Canal University Hospital during the study period.

MATERIALS AND METHODS

Serum samples

Between March 1, 2010, and September 15, 2010, serum samples were collected from consecutive, voluntary, apparently health blood donors attending blood transfusion Center of Suez Canal University Hospital, Ismailia, Egypt. Demographic data were collected using a questionnaire. After routine blood screening for Hepatitis B surface Antigen (HBsAg), anti-HCV and human immunodeficiency virus (anti-HIV) markers and alanine transaminase (ALT) level, 488 serum samples consisted of 137 anti- HCV positive samples, 35 HBsAg positive samples, and 316 samples who were negative HBsAg, anti- HCV and anti-HIV were recruited and the serum were stored in duplicate at -20°C for anti-HEV antibodies detection and HEV RNA extraction.

Detecting anti-HEV antibodies

Anti-HEV IgG and IgM were detected using Third generation Enzyme Immuno Assay (EIA) according to the manufacturer's instructions (DIA.PRO, Milano, Italy). The Cut-off was calculated by

addition of 0.350 with mean optical density value of the Negative Control (NC) and samples were considered as positive when ratio of the test result of sample (od450nm) and the cut-off value was above 1 (or ≤1), according to the manufacturer's instruction. All positive samples were retested in duplicate with the same EIA assay to confirm the initial results.

Purification of viral RNA primer used

The RNA was extracted from 140 μm serum samples using the QIAamp® Viral RNA Mini kit (Germany) according to the manufacturer's instructions. The viral RNA was eluted from the spin column with 45 μL of the elution buffer and stored at -20°C. Two sets of primers for reverse transcriptase polymerase chain reaction (RT-PCR) and reverse transcriptase-nested polymerase chain reaction (RTnPCR) to amplify different region of the capsid gene were used (Huang et al., 2002).

RT-PCR and nested PCR

RT-PCR was performed by using a RevertAid TMFirst Stand cDNA Synthesis RT-PCR kit (Fermentas, Canada). The RT step was carried out in a 20 μl reaction mixture volume containing 1 μL (concentration) Primer S1HE-R (Metabion GmbH, Germany), 4 μL 5xbuffer, 1 μL (units/uL) RNase inhibiter, 2 μl reverse transcriptase (units/uL); 2 μL dNTPs mix (Fermentas, Canada). The mixture was incubated at 37°C for 1 h, heated for 5 min at 70°C, and placed immediately on ice. The first round of PCR was carried out in a total reaction mixture volume of 25 μl including 2.5 μL 5 x buffer, 0.5 μl dNTPs mix (Fermentas ,Canada), 0.5 μl Primer S1HE-R, 0.5 μl Primer S1HE-F, (Metabion, GmbH, Germany),0.25 μl BioReady rTaq polymerase (Hangzhou, China), 18.5 μl RNase-free water and 2.5 μl of cDNA. 35 cycles of PCR (94°C for 5 min, 94°C for 30 s, 50°C for 30 s and 72°C for 1min) were carried out.

For the second round of polymerase chain reaction (PCR), 2.5 μl of the first PCR product was amplified as described above (The first round of PCR), except that the primers used were Primer S2HE-R and Primer S2HE-F (Metabion GmbH, Germany). The expected size of the PCR product amplified with the nest of primers was 348 bp (Huang et al., 2002).

Statistical analysis

Data were entered and analysis was performed using the statistical package for the social sciences (SPSS V-16) for windows (Version 16, SPSS). Descriptive summaries were the study findings presented, and were explained in words and tables. Chi-square test (x2) was used to compare and assess difference between the proportions between groups. In all cases, P-value less than 0.05 was taken and considered as statistically significant.

The study was endorsed by the ethical committee of the faculty of Medicine, Suez Canal University. Participation was fully voluntary and consent was obtained from all participants. Any information obtained during the study was kept confidential.

RESULTS

A total of 488 samples which consisted of 137 anti-HCV positive blood donors, 35 HBsAg positive donors and 316 blood donors who were negative HBsAg, anti-HCV, and HIV were included in this study. The mean and SD of age in the study group was 29.90 ± 7.44 years. Overall seroprevalence of HEV was 20.90% (102/488). Seroprevalence

Table 1. Seroprevalence of HEV Ab in different age groups of blood donors.

Age (year)	HEV Ab		Total	P Value
	Positive	Negative		
< 20	1 (8.3%)	11 (91.7%)	12	
20- 34	61 (16.94%)	299 (83.06%)	360	0.0031*SA
35-49	39 (34.5%)	74 (65.5%)	113	
≥50	1 (33.3%)	2 (66.7%)	3	
Total	102 (20.90%)	386 (79.1%))	488(100%)	

HEV Ab, Hepatitis E virus Antibody; *Chi-square test (x^2); SA, significant association (p < 0.05) (Between ≤ 35 year old and > 35 years old).

Table 2. Frequency of anti-HEV in blood donor positive for HCV Ab and HBsAg and Blood donors negative for HCV Ab, HBsAgand HIV.

Parameter	Anti- HEV		Total	P value
	Test positive (%)	Test negative (%)		
HCV Ab	36 (26.28%)	101 (73.72%)	137	0.068 *NS
HBsAg	10 (28.57%)	25 (71.43%)	35	0.247 *NS
Blood donors negative for HCV Ab ,HBsAgand HIV	56 (17.7%)	260 (82.3%)	316	
Total	102 (20.9%)	386 (79.1%)	488 (100%)	

HBsAg, Hepatitis B surface Antigen; HCV Ab, Hepatitis C virus Antibody; HIV, Human immunodeficiency Virus; * Chi-square test (x^2); NS, no significant association (p > 0.05) of anti-HEV positivity between of HBsAg or anti -HCV and Blood donors negative for HCV Ab, HBsAg and HIV .

Table 3. Frequency of Anti- HEV among blood donors negative for HCV Ab, HBs Ag and HIV and its relation with ALT levels.

Anti- HEV	ALT≤40IU/L	>40IU/L	TOTAL	P value
Test positive (%)	43 (19.6%)	13 (13.4%)	56 (17.7%)	0.183*NS
Test Negative (%)	176 (80.4%)	84 (86.6%)	260 (82.3%)	
Total	219	97	316 (100%)	

ALT, Alanine aminotransferase, * Chi-square test (x^2); NP, No significant association (p > 0.05) of HEV seropositivity between specimens with ALT level ≤40 and >40.

in male and female was 20.96% (100/477) and 18.82 % (2/11), respectively.

Age-specific prevalence of anti-HEV increased significantly with age (Table 1), which was 8.33% in subjects under 20 years, 16.94% in individuals 20 to 34 years, 34.51% in individuals 35 to 49 years and 33.33% over 50 years old. There was significant difference (p < 0.05) of seropositivity between age groups of ≤ 35year old and > 35 years old.

The highest seroprevalence of HEV was detected in specimens positive for HBsAg (28.57%), followed by specimens positive for anti- HCV (26.28%) while blood donors negative for HBsAg, Anti HCVandHIV had the lowest seroprevalence (17.7%) (Table 2). There was no statistical significant difference (p > 0.05) in anti-HEV positivity between HBsAg or anti HCV and blood donors negative for HCV Ab, HBsAg and HIV.

The frequency of Anti- HEV among blood donors with elevated ALT (>40 IU/L) and normal ALT level (<40IU/L) who was negative for HCV Ab, HBs Ag and HIV was 19.6 and 13.4% respectively (Table 3). There was no meaningful difference (p > 0.05) in seropositivity between the ALT level ≤ 40 IU/L and > 40 IU/L.

From 102 ELISA sera, 92 samples were tested for RNA by RT-PCR and all were negative, the other 10 samples were not tested because of inadequate sample volume.

DISCUSSION

Hepatitis E is one of the important hygienic infectious problems of the world with a high incidence in developing countries, mainly in Asia and in Africa (Dalton et al., 2008). A research conducted in rural area of Egypt reported 67.7 and 84% of anti-HEV prevalence among healthy adults and pregnant women respectively (Stoszek et al., 2006). The authors concluded that both zoonotic and anthroponetic transmission of virulent HEV occurs widely in these rural villages and that the rate of positive antibodies increased with age.

In this study, the overall prevalence of anti-HEV antibodies among our blood donors was 20.9%, which is more than figures obtained from blood donors in Germany (5.94%) (Vollmer et al., 2012), Spain (2.8%) (Mateos, et al., 1999), Ghana (4.6% anti-HEV IgG and5.9% anti-HEV IgM) (Meldal et al., 2013) and in Saudi Arabia (16.4%), (Abdelaal et al., 1998) general population in Pakistan (17.5%) (Hamid et al., 2002) and healthy blood donors in Riyadh (8.37%) (Arif, 1996), but it is lower than what was reported in previous Egypt study which was 45.2% (43/95) in blood donors and 39.6% (38/96) in hemodialysis patients (AbdelHady et al., 1998). This can be explained by varying epidemiologic condition in different geographical area and difference in diagnostic techniques between studies. This study was not able to assess the sex association of anti-HEV due to low flow of female donors to the center.

In the current study, seroprevalence increased significantly with age; from 8.3% in subjects below 20 years of age, 16.94% in 20 to 34 years of age, 34.5% in 35 to 49 years of age and a slight decline of 33.3% over those 50 years of age. The slight decrease of anti -HEV in age group over 50 may be the small study subjects in this age group. Similar finding of sero activity associated with increasing age was also reported in other studies among persons living in HEV endemic (Arankalle et al., 1995; Cheng, 2012) and non-endemic regions (Bernal et al, 1996).

In contrast to the above study, in the present work, all of 92 anti-HEV study subjects were negative for polymerase chain reaction PCR test, which point to the absence of viremia in the blood donors. A study conducted in Iran also showed that out of the 33 ELISA positive sera, only one was for RT-PCR, which is nearly similar to other finding (Keyvani et al., 2009). Another study conducted in Ghanaian blood donors also showed that all anti HEV IgG and anti- HEV IgM positive sera were HEV-RNA negative by RT-qPCR (Meldal et al., 2013).

The present study also searched the seroprevalence of HEV antibodies among HBsAg and HCV antibodies. Out of 137 individual positive for Anti- HCV, 36 (26.28%) were positive for anti-HEV and out of 36 individuals positive for HBsAg, 10/36 35(28.57%) individuals were positive for anti-HEV antibodies. In 316 individuals negative for anti-HCV, HBsAg and anti-HIV as control group, anti-HEV positivity was 56/316 (17.7%). The prevalence of anti-HEV in HBsAg and anti-HCV positive individuals was slightly higher than from individual without anti-HCV, HBsAg and anti-HIV. This study observed no statistical association between anti-HEV and anti-HCV or HBsAg positivity. One study conducted in Egypt has shown a striking association between HCV and HEV, pointing to similar or overlapping routes of transmission (AbdelHady

et al., 1998).

From 97 individual without anti-HCV, anti-HIV and HBsAg, only 13 individual were positive for anti-HEV with elevated ALT (>40). Out of this positive individual 9 were with ALT level range of 41-60 IU/L, 3 were with 61-80 IU/L and only one was with ALT level >120IU/L. This study shows that, the increase of ALT level has no correlation with anti-HEV prevalence, so the cause for elevated ALT level might be other factors.

In conclusion, Seroprevalence of HEV-antibody among blood donors in our study in Ismailia, Egypt is high, but we cannot recommend screening of all blood donors for HEV until more data becomes available and until more is known about the parenteral route of transmission. A careful surveillance in the general population is required and further appropriate investigations are needed to identify the exact mode of transmission and risk groups in Egypt.

ACKNOWLEDGEMENTS

The authors would like to thank NPT of Netherlands and Hawassa University for funding the research and Dr. Nora Faham from the Faculty of Pharmacy for her support during the laboratory investigation.

REFERENCES

Abdelaal M, Zawawi TH, al Sobhi E, Jeje O, Gilpin C, Kinsara A, Osoba A, Oni GA (1998). Epidemiology of hepatitis E virus in male blood donors in Jeddah, Saudi Arabia. Ir. J. Med. Sci. 167: 94–96.

AbdelHady SI, El-Din M S, El-Din ME (1998). A high hepatitis E virus (HEV) seroprevalence among unpaid blood donors and hemodialysis patients in Egypt. J. Egypt Public Health Assoc. 73: 165-79.

Arankalle VA, Chobe LP (1999).Hepatitis E virus: can it be transmitted parenterally? J. Viral Hepat. 6:161–4.

Arankalle VA, Chobe LP (2000). Retrospective analysis of blood transfusion recipients: evidence for post-transfusion hepatitis E. Sang. 79: 72–4.

Arankalle VA, Joshi MV, Kulkarni AM, Gandhe SS, Chobe LP, Rautmare SS (2001). Prevalence of anti-HEV antibodies in different Indian animal species. J. Viral Hepat. 8:223–227.

Arankalle VA, Tsarev SA, Chadha MS, Ailing DW, Emerson SU, Banerjee K, Purcell RH (1995). Age-specific prevalence of antibodies to hepatitis A and E viruses in Pune, India, 1982 and 1992. J. Infect. Dis. 171: 447–450.

Arif M (1996). Enterically transmitted hepatitis in Saudi Arabia: an epidemiological study. Ann. Trop. Med. Parasitol. 90:197-201.

Bernal W, Smith HM, Williams W (1996). A community prevalence study of antibodies to hepatitis A and E in inner-city London. J. Med. Virol. 49:230–4.

Cheng X, Wen Y, Zhu M, Zhan S, Zheng J, Dong C, Xiang K , Xia X, Wang G, Han L (2012). Serological and molecular study of hepatitis E virus among illegal blood donors. World J. Gastroenterol. 18(9): 986-990

Dalton HR, Bendall R, Ijaz S, Banks M (2008). Hepatitis E: an emerging infection in developed countries. Lancet Infect. Dis. 8:698-709.

Dawson GJ, Mast EE, Krawczynski K, Balan V (1997). Acute hepatitis E by a new isolate acquired in the United States. Mayo. Clin. Proc. 72:1133–1136.

Emerson SU, Anderson D, Arankalle A, Meng XJ, Purdy M, Schlauder GG (2005). Hepevirus. In: Fauquet CM, Mayo MA, Maniloff J, Desselberger U, Ball LA, editors. Virus Taxonomy, VIIIth Report of the ICTV. London, Elsevier/ Academic Press. pp. 851–855.

Emerson SU, Purcell RH (2004). Running like water-the omnipresence of hepatitis E. N. Engl. J. Med. 351:2367–2368.

Favorov MO, Nazarova O, Margolis HS (1998). Is hepatitis E an emerging zoonotic disease? Am. J. Trop. Med. Hyg. 59:242.

Hamid SS, Atiq M, Shehzad F, Yasmeen A, Nissa T, Salam A, Siddiqui A, Jafri W (2002). Hepatitis E virus superinfection in patients with chronic liver disease. Hepatology 36: 474-78.

Huang FF, Haqshenas G, Guenette DK, Halbur PG, Schommer SK, Pierson FW, Toth TE, Meng XJ (2002). Detection by Revers transcription-PCR and Genetic Characterization of Field Isolates of swine Hepatitis E Virus from Pig in Different Geographic Region of the United States. J. Clin. Microbiol. 40(4).1326-1332.

Keyvani H, Shahrabadi SM, Najafifard SI, Hajibeigi Z, Fallahian F, Alavian S (2009). Seroprevalence of anti-HEV and HEV RNA among volunteer blood donors and Patients with Hepatitis B and C in Iran. Viral Hepatitis Foundation Bangladesh. 1:34-37.

Mateos ML, Camarero C, Lasa E, Teruel JL, Mir N, Baquero F(1999).Hepatitis E virus: relevance in blood donors and risk groups. Vox Sang. 76: 78-80.

Meldal BHM, Sarkodie F, Owusu-Ofori S, Allain JP (2013). Hepatitis E virus infection in Ghanaian blood donors – the importance of immunoassay selection and confirmation. Vox Sanguinis. 104: 30–36

Okamoto H, Takahashi M, Nishizawa T (2003). Features of hepatitis E virus infection in Japan. Intern. Med. 42:1065–71.

Purcell R, (1996). Hepatitis E virus. In: Fields BN, Knipe DM, Howley PM, Chanock RM, Melnick JL, Monath TP, Roizman B, Straus SE, editors. Fields Virology, 3rd Edn. Philadelphia, Lippincott-Raven. pp. 2831– 2843.

Purcell RH, Emerson SU (2001).Hepatitis E virus. In: Knipe DM, Howley PM, Griffin DE, Martin MA, Lamb RA, Roizman B, Straus SE, editors. Fields virology, 4th Edn. Philadelphia, Lippincott Williams and Wilkin. pp. 3051–3061.

Schlauder GG, Dawson GJ, Mushahwar IK, Ritter A, Sutherland R, Moaness A, Kamel MA (1993). Viraemia in Egyptian children with hepatitis E virus infection. Lancet 341:378.

Singh S, Mohanty A, Joshi YK, Deka D, Mohanty S, Panda Sk (2003). Mother-to- child transmission of hepatitis E virus infection. Indian J Pediat. 70: 37–39

Stoszek SK, Abdel-Hamid M, Saleh DA, El Kafrawy S, Narooz S, Hawash Y, Shebl FM, El Daly M, Said A, Kassem E, Mikhail N, Engle RE, Sayed M, Sharaf S, Fix AD, Emerson SU, Purcell RH, Strickland GT (2006). High prevalence of hepatitis E antibodies in pregnant Egyptian women. Trans. R. Soc. Trop. Med. Hyg. 100: 95-101.

Tam AW, Smith MM, Guerra ME, et al.(1991).Hepatitis E virus (HEV): molecular cloning and sequencing of the full- length viral genome. Virology 185:120–131.

Tien NT, Clayson HT, Khiem HB, Sac PK, Corwin AL, Myint KS, Vaughm DW (1997). Detection of immuno-globulin G to the hepatitis E virus among several animal species in Vietnam. Am. J. Trop. Med. Hyg. 57:211.

Uchida T (1992). Hepatitis E: review. Gastroenterol Jpn. 27: 687-96.

Vollmer T, Diekmann J, Johne R., Eberhardt M., Knabbe C, Dreier J (2012). A novel approach for the detection of Hepatitis E virus infection in German blood donors. J. Clin. Microbiol. doi: 10.1128/JCM.01119-12.

21

The potential of house fly, *Musca domestica* (L.) in the mechanical transmission of influenza A subtype H1N1 virus under laboratory conditions

Nazni W. A.[1], Apandi M. Y.[2] Eugene M.[1] Azahari A. H.[1] Shahar M. K.[1] Zainah S.[2] Vthylingam I.[3] and Lee H. L.[1]

[1]Medical Entomology Unit, Infectious Diseases Research Centre, Institute for Medical Research, Jalan Pahang, 50588 Kuala Lumpur, Malaysia.
[2]Virology Unit, Infectious Diseases Research Centre, Institute for Medical Research, Jalan Pahang, 50588 Kuala Lumpur, Malaysia.
[3]Department of Parasitology, Faculty of Medicine, University of Malaya, 50603 Kuala Lumpur, Malaysia.

A study on house flies was carried out to establish whether house flies can transmit the H1N1 virus mechanically due to their abundance, ability to transport pathogens and their behavioral traits of regurgitation and defecation. The objectives of this study were to examine the efficiency of house fly legs in picking up the influenza H1N1 virus particles, persistency of the virus particles on the legs at different time interval, viability of the virus dislodged from the legs and the presence of the virus in vomitus and fecal discharge of the house flies. The findings indicated that the persistency of H1N1 virus on fly legs could be detected up to 24 h in chilled and actively flying flies. Furthermore, the viability of virus was evidenced from immobilized flies exposed for 30 s. However, H1N1 virus was not detected in the vomitus and feces. Further, epidemiological studies are needed before the significance of house flies as transmitter of influenza virus can be determined.

Key words: H1N1 virus, epidemiological studies, house flies, influenza, transmit.

INTRODUCTION

The house fly *Musca domestica* (L.) is a cosmopolitan species and its synanthrophic behavior of breeding in animal manure; human excreta, garbage and animal bedding have evolved the house fly to live in association with man (Thaddeus et al., 2001). House flies are a nuisance and mechanically transport a host of diseases to human such as trachoma, cholera, anthrax, diphtheria, tuberculosis and leprosy by food contamination due to their behavioral patterns of ovipositing and feeding on decaying matter, human excrement and animal manure (Lane and Crosskey, 1993). Transmission of pathogens

by adult flies occurs through dislodgement from exoskeleton, fecal deposition and regurgitation (Greenberg, 1973). Several studies have indicated that house flies play an important role in transmission of viruses. Newcastle virus has been isolated from the surface of the fly body (Milushev et al., 1977). Medvecky et al. (1988) reported their experimental study on the role of house flies in transmission of pseudo rabies in pig, rabbit and lamb, while Emerson et al. (2000) and Forsey (2001) stated that house flies played an important role in transmission of eye disease among children. Satoshi et al. (2003)

conducted a laboratory study on a colony of 210 house flies fed on the feces of a pig infected with Porcine Reproductive Respiratory Syndrome Virus (PRRSV). All the samples from the pigs and house flies were tested positive for PRRSV nucleic acid by reverse transcriptase polymerase chain reaction, demonstrating that infectious PRRSV could remain viable in the intestinal visceral of a house fly up to 12 h following feeding on the faeces of a viraemic pig. These findings together with detection of infection from exterior surface indicated the ability of a house fly to transmit PRRSV mechanically.

Highly pathogenic avian influenza, or "fowl plaque" was identified as an infectious disease of birds and chicken in Italy in 1878. Avian influenza A (H5N1) virus strains that emerged in Asia in 2003 continued to evolve (Lupiani et al., 2009). There were 341 human fatalities due to Avian Influenza A/(H5N1) reported worldwide (Globalhealthfacts.org, 2012). Normally, influenza is not transmitted by insects, but since the virus is relatively environmentally stable, mechanical transmission by insects such as house fly could be possible. As the virus can survive long period in chicken feces and feed, flies may pick it up when landing on feces or infected dead bird and then carry it to other animals. Fomite can also be important in transmission of viruses and flies may act as a mechanical vector (Beard, 1998; USDA APHIS, 2002; WHO, 2004). Bean et al. (1985) isolated influenza virus of low virulence from chicken and the virus subsequently became virulent which was later identified as influenza H5N1 serotype. Subsequently, Sawabe et al. (2006) detected and isolated a highly pathogenic H5N1 virus from blow flies from an infected poultry farm in Kyoto, Japan. However, to date, no studies have been reported on the possibility of house flies in the transmission of Influenza A subtype H1N1 virus.

The objectives of our study were to examine the efficiency of house fly legs as a model in picking up the influenza H1N1 virus particles, persistency of the virus particles on the legs, viability of the virus dislodged from the legs and the presence of the virus in vomitus and fecal discharge. Our experimental work focused on the house flies because of their abundance, ability to transport pathogens and behavioral traits of regurgitation and defecation (Greenberg, 1973).

MATERIALS AND METHODS

House fly (*M. domestica*)

The laboratory- bred WHO IJ2 strain of house fly (F202) used in this study was colonized and maintained in the ACL-2 (Arthropod Containment Level-2) insectarium of Medical Entomology Unit, Infectious Disease Research Centre, Institute for Medical Research, Kuala Lumpur. The house flies were reared in a wooden cage, on larval food consisting of damp ground mouse chow pellet with a photoperiod of 12:12, constant temperature of 26±2°C and relative humidity of 80±5%. The adults were fed on a diet of sugar and soaked cotton wool served as water source. Fermented mouse chow pellet served as breeding media for the adults. This colony was maintained in the insectarium without exposure to any pathogens and re-confirmed negative with influenza H1N1 virus infection by RT-PCR.

Virus

The influenza A virus, A/New Caledonia/20/99 strain, subtype H1N1 belonging to the family of Orthomyoxiviridae (Lamb and Kruger, 1996) used in this study was cultivated from a virus culture stock belonging to the Unit of Virology, Infectious Disease Research Centre, Institute for Medical Research, Kuala Lumpur.

Virus isolation in cell culture

An ampoule of Madin-Darby Canine Kidney (MDCK) cells stored in liquid nitrogen was transferred to a water bath until thawed completely. The content was pipetted slowly drop by drop into a 75 cm^2 tissue culture flask containing Dulbecco's Modified Eagle Medium (DMEM) with 10% fetal calf serum (GIBCO, Invitrogen, USA) with 50 U/ml benzyl penicillin and 50 µg/ml streptomycin sulphate (Sigma, St Louis, USA) and incubated overnight at 36°C. The degree of cell confluence was checked under an inverted microscope. Once confluence, the growth medium from the cell culture flask was removed and the cells were washed twice with sterile phosphate buffered saline (PBS) without calcium and magnesium. 0.25% of trypsin-EDTA (Biowest) was added to the monolayer cells and the flask was placed in a 36°C incubator for about 5 min until the cells were completely detached from the surface. The detached cells were re-suspended in growth medium (DMEM) and the suspension was gently aspirated a few times through a fine Pasteur pipette to break up cell clumps. The number of cells were counted using a haemacytometer and normally 1 × 10^5 cells/ml were enough to obtain a confluent and monolayer cells within 2 to 3 days after incubation at 36°C. 1 ml of cells was transferred to each culture tubes and placed in the 36°C incubator until nearly confluent. The growth medium in the culture tubes was removed and the test specimens were added to the cells and allowed adsorption for 30 min at 37°C. Later, 1 ml of complete medium containing trypsin solution without fetal bovine serum (FBS) (Biowest) was added to the tubes and incubated in a stationary sloped (5°) position at 37°C. Cultures were observed daily for cytopathic effect (CPE) in which the cells become rounded, refractile and ultimately shrunk before detaching from the tubes.

Cultures were harvested when more than 90% of the cell monolayer showed cytopathic effect. The virus was reconfirmed as Influenza type A virus by reverse transcriptase-polymerase chain reaction (RT-PCR).

Reverse transcription-polymerase chain reaction (RT-PCR)

RNA extraction was performed using EZ-10 Spin Column Total RNA Minipreps Super Kit™ from Bio Basic Inc (Canada). In brief, 200 µl of infected virus solution was mixed with 400 µl RLT solution and the manufacture's procedure was subsequently adhered to. The extracted RNA was stored at -70°C until further use. One step RT-PCR using cMaster RTplusPCR System™ and Cmaster RT Kit™ (Eppendorf™, Hamburg, Germany) was performed. RT-PCR was carried out in a 20 ul reaction mixtures containing 3.95 µl of RNase free water, 2.0 µl of RT-PCR buffer, 0.4 µl of dNTP mix, 0.2 µl RNAse inhibitor, 0.25 µl cMaster RT, 0.2 µl cMaster PCR and 10 µl of RNA template. 1.5 µl (20 pico mole) of forward primer NPF 5'-CAG-RTA-CTG-GGC-HAT-AAG-RAC-3' and reverse primer NPR 5'-GCA-TTG-TCT-CCG-AAG-AAA-TAA-G-3' (Lee et al., 2001) were used to amplify the 330 bp nucleoprotein gene of influenza virus. Reverse transcription was carried out at 42°C for 40 min followed by 45 PCR cycles, denaturation at 94°C for 1 min, annealing at 55°C for 1 min and extension at 72°C for 0.4 min. The PCR products were examined by gel electrophoresis (Promega™, USA) stained with ethidium bromide (GelStar®).

Experimental design

Four experiments were conducted to determine the potential of house fly to transmit the virus. In all experiments, house flies of both sex starved overnight and aged 3 to 4 days old were used. All the experiments were conducted in a BSL-2 containment laboratory in accordance to established biosafety protocols.

Experiment 1: Influenza A H1N1 virus load acquisition by house fly legs

The experiment was conducted to detect the presence of Influenza A H1N1 virus mixed with bovine serum albumin to enhance adherence of virus particle on fly legs. A serial dilution of virus concentration in 10% bovine serum albumin was prepared. The dilution ratios used were 1:10, 1:4, 1:3, 1:2, 10:1 and 1:1 in the ratio of Influenza A infected tissue culture fluid: Bovine serum albumin. 50 µl suspension of each dilution was pipetted onto a sterile plastic surface placed in an ice box and exposed to flies. The titre of virus used in all the experiments was 3×10^6 TCID$_{50}$/ml. Flies were immobilized for 2 min at -20°C and were individually held over the virus suspension using forceps so that their legs came into contact with the suspension in order to simulate a walking motion on the virus suspension for 10 s. Ten flies were used per dilution ratio. The fly legs were then immediately washed with 200 µl of RNAnase free water to dislodge the virus particles from their legs. RNA was extracted from the dislodged virus solution and PCR was conducted to detect the virus.

Experiment 2: Persistency of Influenza A H1N1 virus on house fly legs

The presence of virus in the dilution shown as the brightest band on gel electrophoresis from experiment 1 was used to determine the persistency of virus on house fly leg. House flies were chilled at -20°C and individually held over the selected dilution of virus suspension so that their legs came into contact with influenza A virus for 10 s. Exposed flies were tested for the presence of the virus at interval of 30 s, 1 min, 5 min, 15 min, 30 min, 1 h, 2 h and 24 h post-exposure to the virus suspension. Exposed flies were divided into 2 groups: those chilled at 0°C and the actively moving flies. The survival of the virus on chilled flies and active flies were compared. After the respective time interval, the fly legs from both groups were immediately washed with 200 µl of RNAnase free water to dislodge the virus particles from the legs. RNA was extracted from the dislodged virus solution and PCR was conducted. Two replicates were conducted with 10 flies per replicate.

Experiment 3: Viability of Influenza A H1N1 virus dislodged from house fly legs

The experiment was similar to experiment 2, except that the virus was isolated from flies sampled at 30 s and 5 min after a 10-s exposure. The fly legs were washed with 200 µl of RNAnase free water to dislodge the virus particles from both the immobilized and active flies. The solution was then filter sterilized through a 0.22 µm Spritzan Syringe Filter (TPP, Europe). The filtrate was inoculated into the MDCK cell line and observed daily for the presence of CPE before being harvested when more than 90% of the cell monolayer showed CPE. Two replicates were conducted with 10 flies per expe-

riment. In the absence of CPE, the presence of virus was detected using immunofluorescence. Viral screening and identifica-tion IFA Kit (Millipore™, CA, USA) was used to stain the virus infec-ted cells. The slide was viewed under an immunofluorescence microscope at 400 to 500 nm range (blue light). Two passages of the virus were conducted to re-confirm the test results.

Experiment 4: Influenza A H1N1 virus in fly vomitus and fecal discharge

A total of 50 house flies (<7 days old) were collected from the laboratory colony and starved for 24 h prior to feeding on the virus. The flies were then immobilized for 2 min at -20°C and introduced into a plastic container (15 × 8 × 3 cm) consisting of a sterilized glass slide coated with 200 µl of tissue culture medium containing Influenza A virus. The flies were allowed to feed on the medium at room temperature for 1 h. After the exposure period, the flies were immobilized and removed into a sterile centrifuged container consisting of 3 ml RNAnase free water and vortexed vigorously for 1 min to dislodge the virus particles picked up by the fly legs, mouthparts, wings and the hairy body structure while feeding. RNA was extracted from the dislodged virus solution and PCR was conducted. The feces and vomitus were collected separately with 500 µl of RNAnase free water from the glass slide and the spots of vomitus and feaces were picked from the plastic container which was observed to contain large amount of vomitus and fecal discharges.

RESULTS

Experiment 1: Influenza A virus load acquisition by house fly legs

The initial dilution of the virus in tissue culture fluid with the diluent bovine serum albumin indicated positive RT-PCR results in all the dilutions tested. A weak band was observed at 1:10 dilution (virus in tissue culture fluid: bovine serum albumin). The dilution of 1:2, gave a clear and sharp band at 330 bp, indicating the strong presence of Influenza A virus dislodged from the fly legs (Figure 1). Hence, the dilution of 1:2 was used in subsequent experiments.

Experiment 2: Persistency of Influenza A virus on house fly legs at different time intervals

Influenza A virus was detected by RT-PCR from the house fly leg that has been simulated to walking on infected virus medium. The virus was detected in all post-exposure intervals of 1 min, 5 min, 15 min, 30 min, 1 h, 2 h and 24 h in chilled and actively flying flies. However, as the bands detected were faint possibly due to low concentration of virus obtained from the single leg, only the results of the 24-h post-infection was shown (Figure 2). The results indicated that Influenza A nucleic acid can be detected up to 24 h on house flies legs.

Experiment 3: Viability of Influenza A virus dislodged from house fly legs

In the first passage, there was no cytopathic effect observed in all samples, but the virus was detected by immunoflou-

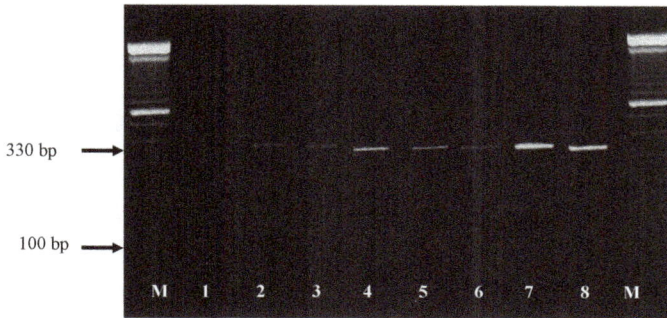

Figure 1. Agarose gel electrophoresis of RT-PCR amplified products for the detection of Influenza A virus at 330 bp. Lane M, size markers (100 bp ladder, Bioron, Germany); lanes 1 to 8 dilution of Influenza A infected tissue culture fluid: Bovine serum albumin; lane 1- 1:10; lane 2- 1:4; lane 3- 1:3; lane 4- 1:2; lane 5- 10:1; lane 6- 1:1; lane 7- Influenza A infected tissue culture fluid; lane 8- positive control (Influenza A virus); lane M– size markers (100 bp ladder, Invitrogen Life Technology).

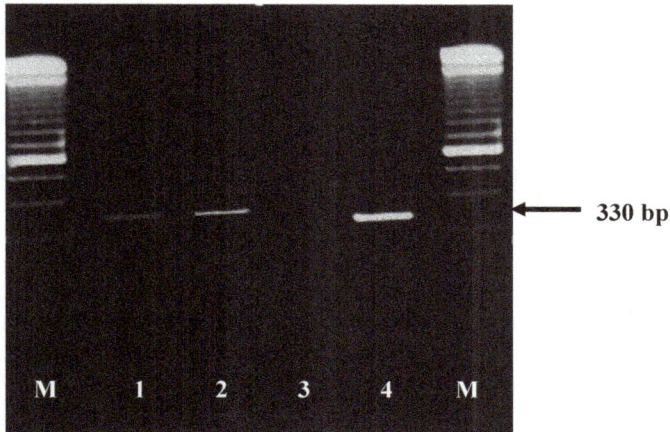

Figure 2. Agarose gel electrophoresis of RT-PCR amplified products for the detection of Influenza A virus at 330 bp after 24 h exposure. Lane M, size markers (100 bp ladder, Bioron, Germany). Lane 1- active flies after 24 h exposure to Influenza A virus; lane 2- chilled flies 24 h after exposure to Influenza A virus; lane 3- negative control (master = mix only); lane 4- positive control (Influenza A virus); lane M- size markers (100 bp ladder, Invitrogen Life Technology).

rescence for the 30 s exposure in both immobilized and active flies (Figure 3). In the second passage, the influenza virus was detected in both the 30 s immobilized and active flying flies using the immunofluorescence technique; however, cytopathic effect was only observed for flies exposed for 30 s in immobilized state (Table 1).

Experiment 4: Presence of Influenza A virus in fly vomitus and fecal discharge

The vomitus specks were pinkish in colour while the fecal specks were yellowish. Results of RT-PCR indicated that no PCR product was detected at 330 bp, indicating that

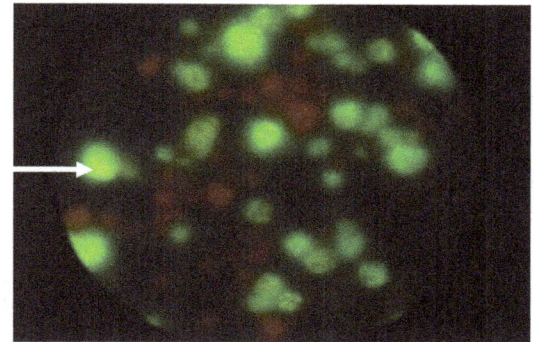

Figure 3. Detection of Influenza A virus using immunoflouscence technique.

the Influenza A virus was not found in the vomitus, feces and the external body of the flies.

DISCUSSION

Influenza A virus adherence to the fly legs persisted up to a period of 24 h, suggesting that the virus particles were trapped in the hairs and bristles on the legs of the house flies while the fly was in motion and therefore justifying the ability of house fly to transmit other viral infection like swine virus mechanically (Satoshi et al., 2003). In the experiment to detect the viability of virus at the different exposure periods in immobilized condition as well as in active state, virus in flies exposed for 5 min in both the immobilized and active flying state were not detected by immunofluorescence or tissue culture. The reason that CPE was not detected at 30 s in the actively flying insects could be due to the low viral load. as "However, the results showed that immunofluorescence was detected in both the actively flying and immobilized flies exposed for 30's in passages 1 and 2". "Furthermore, the results indicated that influenza A virus was still viable post exposure in immobilized flies suggesting that virus transmission through flies is possible". In the feeding experiment, while feeding, flies had the habit of regurgitating resulting in vomitus specks on the surface they explored. In addition, such surface also had fecal specks. It is noteworthy that the density of vomitus specks was higher than the fecal specks, indicating that the fly could probably spread infected organisms more often by vomitus rather than by feces. However, the current study demonstrated that Influenza A virus was not present in the vomitus and feces of the fly, probably due to the low viral load (<10 to 30 pico gram of RNA) detectable by RT-PCR.

Swabe et al. (2011) reported that crops and intestines dissected from fly at various times after virus exposure were used for virus isolation and titration. Virus was isolated from fly crops and intestines up to 24 h post-exposure and from feces and vomitus matter of 1 out of 3 blow flies at 48 h after exposure. At 14 days after exposure, no virus was isolated from any blow fly at 20 or 10°C. The H5N1 viral gene could be detected in blow flies up to 14

Table 1. Detection of Influenza A H1N1 virus using tissue culture and immunofluorescence techniques.

Exposure time	1st passage		2nd passage	
	T/ culture	IF	T/culture	IF
30 s ACT	No CPE	+	No CPE	+
30 s IMB	No CPE	+	CPE	+
5 min ACT	No CPE	-ve	No CPE	-ve
5 min IMB	No CPE	-ve	No CPE	-ve

ACT, Flies flying actively; IMB, immobilized flies; T/culture, tissue culture; IF, immunofluorescence technique; CPE, cytopathic effect.

days after exposure; no viable virus was detected 48 h post- exposure. These findings are important since the ability of influenza virus to reside within the body of a house fly may protect it against certain environmental factors known to be detrimental to influenza survivability outside the host such as ultraviolet light and drying. Furthermore, house flies frequently inhabit the interior of transport vehicles and livestock trailers. This may enhance exposure of the insects to influenza infected animal, and allow the movement of the insects to travel over greater geographical distances up to 7 km (Nazni et al., 2005). Hence, in the process of flying, house fly may serve as a mechanical vector of influenza virus. Because of the paucity of study on the potential of house fly to transmit Influenza A H1N1 virus, the findings from this study could only be compared with studies on other related viruses such as the H5N1 subtype transported by flies. Highly pathogenic H5N1 virus was detected and isolated from blow flies in an infected poultry farm in Kyoto, Japan (Sawabe et al., 2006). The virus genes were detected from the intestinal organs, crop and gut of two blow fly species, *Callliphora nigribarbis* and *Aldrichina grahami*, by RT-PCR. The authors suggested that it is possible that blow flies could be mechanical transmitters of H5N1 virus.

Earlier investigation in United States indicated that a virulent H5N1 virus was detected in house fly in a poultry farm (Bean et al., 1985). In a recent study, Wanaratana et al. (2010) showed that the virulent AI H5N1 virus consumed by house flies as food contaminated with this virus could carry the virus within their bodies and remained infective up to 72 h post infection and their RT-PCR was positive up to 96 h post infection. However, they showed that viruses could be detected in external surfaces of house flies for only up to 24 h post-exposure. Tyasasmaya et al. (2012) had also isolated highly pathogenic avian influenza virus H5N1 (AIV H5N1) from field collected house flies in Java, Indonesia. In a recent review by Sawabe (2011), H5N1 virus in blow flies could be detected up to 14 days after exposure, but no viable virus was detected at 48 h post- exposure. They further mentioned that contamination of the body surface of blow fly was much less compared to house flies where the body surface could be easily contaminated by the viruses. The author also stated that the transmission mechanism could be complicated due to the habit of blow flies which prefer to lick carcasses and droppings of chicken and pigs. These behavioural traits are not only present in blow flies but also in most flies including the house flies. Elsewhere, Sievert et al. (2006) reported the existence of avian influenza virus in the house flies, while studies in Malaysia showed that rotavirus can be mechanically transported by fly contaminated surfaces (Tan et al., 1997).

The prevalence of Influenza H1N1 is a serious public health problem in human. It is important to note that flies often occur in millions in human habitations, indicating that flies will definitely play an important role in transmission of virus particles. Field and laboratory studies on mechanical transmission of virus by flies would contribute greatly to the control of influenza outbreaks. Therefore, further studies on field-collected house flies are needed before a final conclusion can be drawn regarding the significance of house flies as a transmitter of influenza virus and impact on the human population.

ACKNOWLEDGMENTS

We thanked the Director General of Health, Malaysia, and the Director, Institute for Medical Research, for permission to publish this paper. This study was supported by a SEAMEO-TROPMED Grant-06.

REFERENCES

Bean WJ, Kawaoka, JM, Wood JE, Pearson RG, Webster (1985). Characterization of virulent and avirulent A/chicken/ Pennsylvania/83 influenza A viruses: potential role of defective interfering RNAs in nature. J. Virol. 54 (1):151-160

Beard CW (1998). Avian influenza. In foreign Animal Diseases, Richmond, V.A; United States Animal Health Association .pp. 71-80

Emerson PM, Bailey RL, Mahdi OS, Walraven O, Lindsay SW (2000). Transmission ecology of fly *Musca sorbens*, a punitive vector of trachoma. Transaction of Royal Society Trop. Med. Hyg. 94(1): 28-32

Forsey T, Drouugar S (2001) Transmission of chlamydia by the housefly. Brit. J. Ophtalmol. 65: 147-150

Globalhealthfacts.org. (2012). Avian Influenza A/(H5N1) cumulative number of confirmed human deaths. as of January 18, 2012, US Global Health www.globalhealthfacts.org/data/topic/map.aspx?ind=33

Greenberg B (1973) Flies and Diseases: Volume II; Biology and Disease transmission, Princeton University Press, Press, Princeton, New Jersey.

Lamb RA, Krug RM (1996). Orthomyxoviridae: the viruses and their replication. In: BN Fields, DM, Knipe PM, Howley RM, Chanock JL, Melnick TP, Momath, B Roizman (Eds.), Fields Virology, 3rd ed., Lippincott-Ravee, Philadelphia. PA.

Lane RP, Crosskey RW (1993). Houseflies, Blowflies and their allies (Calypterate Diptera). In Lane RW and Crosskey RW (eds) Medical insects and arachnids. Champman and Hall, London, pp. 403-428

Lee MS, Chang PC, Shien JH, Cheng MC, Shieh HK (2001). Identification and subtyping of avian influenza viruses by reverse transcription-PCR. J. Virol. Methods 97(1-2):13-22.

Lupiani B, Reddy MS (2009). The history of avian influenza. Comparative Immunology. Microbiol. Infectious Dis. 32: 311–323

Medvecky I, Kovacs L, Kavacs F, Papp L (1988). The role of the house flies Musca domestica in the spread of Aujesky's disease (pseudorabies). Med. Vet. Entomol. 2: 81-86

Milushev I, Greganov G, Shishkov N (1977). The role of the house flies in the epizootiology pseudorabies in birds. Vet. Med. Nausaki 4: 97-100

Nazni WA, Luke H, Wan Rozita WN, Abdullah AG, Sa'diyah I, Azahari AH, Zamree I, Tan SB, Lee HL, Sofian MA (2005). Determination of the flight range and dispersal of the housefly Musca domestica (L) using mark release and recapture methods. Trop. Biomed. 22(1): 53-61

Satoshi O, Scott A, Dee RD, Moon K, Rossow D, Carlos T, Mac DF, Carlos P (2003). Survival of porcine reproductive and respiratory syndrome virus in the house flies. Canadian J. Vet. Res. 67(3): 198-203

Sawabe K, Hoshino K, Sasaki T, Isawa H, Hayash, T, Tsuda Y, Kuruhashi H, Tanabayashi K, Hotta A, Saito T, Yamada A, Kobayashi M (2006). Detection and isolation of highly pathogenic H5N1 avian influenza A viruses from blow flies collected in the vicinity of an infected poultry farm in Kyoto, Japan. Amer. J. Trop. Med. Hyg. 75(2): 327-332.

Sawabe K, Hoshino K, Isawa H, Sasaki T, Kim KS, Hayashi T, Tsuda Y, Kurahashi H, Kobayashi M (2011). Blow flies were one of the possible candidates for transmission of highly pathogenic H5N1 Avian Influenza Virus during the 2004 outbreaks in Japan. Influenza Res. Treatment, Article ID 652652, p. 8.

Sievert K, Alvarez R, Cortada R, Valks M (2006). House flies carrying avian influenza virus (AIV). International Pest Control 48 (3): 114–116.

Tan SW, Yap KL, Lee HL (1997). Mechanical transport of rotavirus by the legs and wings of Musca domestica (Diptera: Muscidae). J. Med. Entomol. 34(5): 527-531

Thaddeus KG, Ronald K, Robert HG, Michael RC (2001). The role of non biting flies in the epidemiology of human infectious diseases. Microb. infect. 3: 231-235

Tyasasmaya T, Wuryastuty H, Wasito R, Sievert K (2012). Experimental evaluation of the house flies (Musca domestica spp.) as a possible vector for avian influenza virus H5N1. In: Current Status of Veterinary Biologicals and Opportunities and Challenges for the Future www.ivvdc2012.org/assets/IVVDC/ivvdc-abstracts.pdf

US Department of Agriculture, Animal (USDA), And Plants Health Inspection Services (APHIS), veterinary Vaccines for terrestrial animal. Highly pathogenic avian influenza. US poultry vaccines for terrestrial animal.(2002). Highly Pathogenic Avian Influenza, a threat to http://www.aphis.usda.gov.oa/pubs/avianflu.html.

Wanaratana S, Panyim S, Pakpinyo S (2011). The potential of house flies to act as a vector of avian influenza subtype H5N1 under experimental conditions. Medical and Veterinary Entomology 25 (1): 58-63

World Health Organization (WHO) (2004). Avian influenza fact sheet. http//www.who.int/crs/don/2004-01-15-en/.

22

Efficiency of cassava brown streak virus transmission by two whitefly species in coastal Kenya

B. Mware[1,2], R. Narla[1], R. Amata[2], F. Olubayo[1], J. Songa[2], S. Kyamanyua[4], and E. M. Ateka[3*]

[1]Department of Plant Science and Crop Protection, Faculty of Agriculture, University of Nairobi. P. O. Box 30197-00200 Nairobi, Kenya.
[2]Kenya Agricultural Research Institute/National Agricultural Research Laboratories, P. O. Box 14733, 00800 Nairobi, Kenya.
[3]Department of Horticulture, Jomo Kenyatta University of Agriculture and Technology. P. O. Box 62000-00200 Nairobi, Kenya.
[4]Makerere University Kampala-Uganda.

The efficiency of cassava brown streak virus (CBSV) transmission by *Bemisia tabaci* (Gennadius) (Hemiptera: *Aleyrodidae*) and spiraling whitefly (*Aleurodicus dispersus)* Russell (Hom, Aleyrodidae) was determined. The transmission utilized field collected adult whitefly populations fed on (allowed 48 h acquisition access feeding period (AAP)) on CBSD (cassava brown streak virus disease) symptomatic leaves before transfer onto clean recipient plants. In subsequent transmission experiments, adult whitefly numbers of each species were varied per plant to determine the effect of whitefly numbers on the rate of CBSV transmission. CBSV was transmitted by *B. tabaci* allowed 48 h AAP on CBSD infected cassava leaves at a higher rate of 40.7% compared to that of *A. dispersus* at 25.9%. This work reports for the first time the transmission of CBSV by *A. dispersus*. A likely biological property of CBSV reported here for the first time is its ability to be transmitted by two whitefly species belonging to two different genera (*Bemisia* and *Aleurodicus)*. Management of CBSD therefore needs to focus on the control of the two whitefly species to reduce the chances and rates of infection and disease spread.

Key words: *Aleurodicus dispersus, Bemisia tabaci*, CBSD, spiraling whitefly, transmission efficiency.

INTRODUCTION

Cassava brown streak disease (CBSD) first described in Tanzania (Storey, 1936) attacks cassava leading to root weight losses of up to 70% in susceptible cultivars (Hillocks et al., 2001). The disease is caused by cassava brown streak virus (CBSV), an *Ipomovirus* in the family Potyvirideae (Monger et al., 2001b) a virus that is graft-transmissible from cassava to cassava (Storey, 1936) and is mechanically transmitted from cassava to a number of herbaceous hosts (Lister, 1959). An earlier report pointed out that CBSV is insect-transmitted and that the most probable vector is the whitefly, *Bemisia tabaci* (Gennadius) (Hemiptera: Aleyrodidae) (Storey, 1939; Bock, 1994). More recent work by Maruthi et al.

(2004) reported *B. tabaci* exhibiting low CBSV transmission rates ranging from 20 - 22%. The low rates of natural spread (Storey, 1939; Bock, 1994; Maruthi et al., 2004 and (Mware et al., 2009) are inconsistent with the high incidences of CBSD observed in the field surveys of up to 64%, (Alicai et al., 2007). The high incidences observed in the field could be due to a wide range of vectors responsible for transmission and accumulation of the virus through continuous use of same planting material year in year out by farmers. Successful CBSV transmission by *B. tabci* has been reported (Maruthi et al., 2004 and Mware et al., 2009), but this does not preclude the possibility that under suitable conditions, *Aleurodicus dispersus* whose population has been direct-ly correlated with CBSD incidence may also transmit the virus (Mware et al., 2009). *A. dispersus* is an emerging pest infesting cassava in coastal Kenya and may also

*Corresponding author. E-mail: eateka@agr.jkuat.ac.ke.

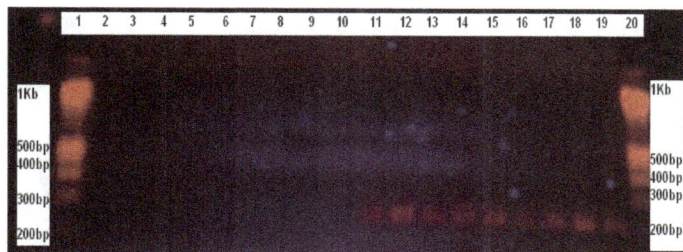

Figure 1. Shows diagnostic bands from CBSV infected plants and none for healthy plants used as controls in the transmission experiment. Lanes 2 - 8: healthy plants (controls), 9: negative control (water), 11, 13 - 17, 19: CBSV infected plants and 12, 18: positive control.

Plate 2. Showing symptoms in vector-inoculated plants of cultivar MM96/5280 (A, B, C) and MM96/4466 (D, E) Leaf inoculated with CBSV by *B. tabaci* showing CBSD early vein clearing and feathering after 26 days.

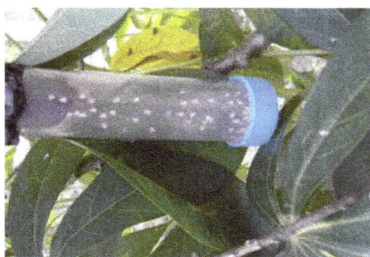

Plate 1. Spiraling whitefly feeding on CBSV-infected cassava leaves within a falcon tube to acquire CBSV before transfer onto recipient plants.

transmit CBSV (Mware et al., 2009). Prior to the transmission trials, *A. dispersus* populations were observed to be highest on lower mature CBSD symptomatic leaves during a whitefly collection survey in Kilifi, Malindi, Lunga lunga and Msambwueni within coastal cassava growing regions. Its population was mostly high on lower mature leaves (Mare et al., 2009).

Although *B. tabaci* is already reported to transmit CBSV, its efficiency to transmit the virus has not been determined. Furthermore, the fact that high populations of *A. dispersus* seemed to coincide with resurgence of CBSD in Coastal Kenya prompted transmission trials to elucidate this relationship. The objective of this study was to determine the transmission efficiency of the two whitefly species.

MATERIALS AND METHODS

Collection of cassava cultivars and whiteflies

CBSV susceptible cassava cultivars were identified during diagnostic surveys in Western Kenya (Mware et al., 2009). They included MM96/5280 and MM96/4466 which are most preferred by farmers due to their high yield, are early maturing (12 months), sweet, are consumed fresh and are resistant to cassava mosaic geminiviruses (CMGs). The cuttings of the cultivars were collected, established within an insect-proof glasshouse to ensure the absence of the CBSV. All plant materials were subjected to reverse

transcriptase polymerase chain reaction (RT-PCR) to confirm freedom from CBSV infection as described by Monger et al. (2001) (Figure 1). The materials were then kept in whitefly exclusion cages (0.3 mm mesh) prior to the start of the transmission experiments.

CBSV transmission tests

Virus transmission tests were done according to the protocols developed by Maruthi et al. (2004), with some modifications. Adult whiteflies were collected directly from CBSV infected cassava in the field then immediately transferred to recipient plants. However, in another treatment adult whitefly populations were allowed 48 h access acquisition feeding time under no choice confinement (in Falcon tube cages; plate 1) on CBSV infected cassava plants followed by an access inoculation feeding period of 48 h on recipient plants within 0.3 mm mesh cages.

The first transmission trial involved an adult whitefly population collected from CBSV infected cassava leaves in the field and then allowed a 48 h inoculation access feeding period within 1 x 1 x 1.5 m whitefly tight cages. Three cages (9 recipient plants per cage) were set up for each species in the first and second experiments (a total of 27 recipient plants were used in each experiment).

In the second trial, a mesh cage and modified falcon tube cages were used to confine CBSV cassava plants and leaf petioles, respectively for 48 h acquisition access feeding period (Plate 2). Specifically colonies of the adult whitefly species were allowed 48 h acquisition access feeding period on CBSV infected cassava leaves then transferred onto 9 recipient plants of cultivar MM96/5280 for 48 h inoculation access feeding period. The set up had 9 target plants replicated three times (27 recipient plants) for each whitefly species. Approximately 30 adult whiteflies were confined within the modified clip cages in which a single leaf was introduced with petiole undetached from the main recipient plant. The whiteflies were then eliminated by spraying with an insecticide (Brigade) and CBSD symptom development monitored specifically on the inoculated leaves and the entire recipient plant for 26 - 60 days. A control cage for each whitefly species was set up with 3 CBSV-free cassava plants not infested with whitefly since the population collected were from diseased plants in the field. The experiments were repeated three times for each whitefly species. The rate of transmission was determined as a proportion of infected target plants expressed as a percentage of the total number of plants tested.

Efficiency of transmission by whiteflies

To determine the transmission efficiency of the two whitefly spe-

Table 1. CBSV transmission probability by adult whitefly species allowed 48 h acquisition access feeding period and those not allowed the 48 acquisition access feeding period.

	Directly collected adult whiteflies without 48 h AAP								
Experiment	Bemisia tabaci				Aleurodicus dispersus				
Replicate	Recipient plants	Probability			Replicate	Recipient plants	Probability		
		Exp 1	Exp 2	Exp 3			Exp 1	Exp 2	Exp .
1	9	0.01	0	0.012	1	9	0.003	0	0.01
2	9	0.003	0.01	0	2	9	0	0.003	0
3	9	0.003	0	0.01	3	9	0.012	0	0
Probability		(0.005)	(0.002)	(0.006)			(0.005)	(0.001)	(0.001

	Adult whiteflies allowed 48 h AAP on CBSV- infected plants								
Experiment	Bemisia tabaci				Aleurodicus dispersus				
Replicate	Recipient plants	Probability			Replicate	Recipient plants	Probability		
		Exp 1	Exp 2	Exp 3			Exp 1	Exp 2	Exp :
1	9	0.012	0.033	0.012	1	9	0.01	0	0.012
2	9	0.01	0.01	0.024	2	9	0.003	0.003	0.017
3	9	0.024	0.012	0.017	3	9	0.012	0.017	0.024
Probability		(0.014)	(0.016)	(0.017)			(0.007)	(0.006)	(0.017

Note: Probability calculated as $P = 1-(1-I)^{1/k}$ where I is the proportion of CBSV infected recipient plants and K is the number of whitefly adults per plant

cies, viruliferous populations of each were introduced on leaves of several recipient CBSV free cassava plants. In addition, the effect of whitefly numbers on transmission efficiency was also assessed using 1, 5, 15 and 30 adult whiteflies of each species on target plants. The colonies of the two whitefly species were given 48 h acquisition feeding period on CBSD symptomatic leaves then transferred onto 4 recipient plants of cultivar MM96/5280 for an inoculation feeding period of 48 h within modified clip cages made from falcon tubes. The 1, 5, 15, 30 adult whiteflies were confined within the modified clip cages in which a single leaf was introduced with petioles undetached from the main recipient plant. Four cages were set up for the (1, 5, 15, 30) different whitefly populations of each whitefly species. The whiteflies were eliminated by spraying with an insecticide (Brigade) after inoculation. CBSD symptom development was monitored on inoculated leaves and on entire recipient plants. Virus transmission efficiency was calculated as the proportion (in percentage) of the total number of plants infested with viruliferous whitefly species that became infected. Comparisons of virus transmission efficiency were made using probability estimates of transmission by a single whitefly (Gibbs and Gower, 1960; Ng and Perry, 1999).

RESULTS

Transmission of CBSV by B. tabaci and A. dispersus

In the experiments with falcon caged B. tabaci (given 48 h AAP) in KARI-Mtwapa 33 of 81 test plants (40.7%) developed CBSD symptoms whereas the mass fed (collected and introduced on diseased plants within cages without 48 h AAP) had 17 of 72 test plants (23.6%)

developing CBSD symptoms. The control plants were not infected (Figure 1)

A total of 18 cassava plants var. MM96/5280 exposed to A. dispersus (7) and B. tabaci (11) collected from CBSV-infected cassava plants in the field without 48 h AAP, developed CBSD symptoms. CBSV was transmitted more efficiently when adult whiteflies were allowed 48h CBSV infected cassava plants then when not allowed 48 h acquisition access feeding. For instance, a transmission rate greater than 35% was achieved by B. tabaci population allowed a 48 h AAP as compared to a population not allowed 48 h AAP which was lower than 20% (Table 1).

Similarly A. dispersus also transmitted CBSV efficiently when given a 48 h AAP than when not allowed acquisition access feeding (Table 1). It is noteworthy here that the field collected adult whiteflies populations were directly transferred onto recipient plants immediately after collection without 48 h AAP (simulating a field scenario). Effectiveness of B. tabaci over A. dispersus as a vector was shown by higher rates of transmission by both populations allowed 48 h AAP and those not allowed 48 h AAP. CBSV was transmitted at higher rate by B. tabaci adult whiteflies (40.7%) fed on CBSV infected cassava leaves for 48 h whereas the spiraling whitefly had an overall CBSV transmission rate of 25.9%. Symptom took a long time (26 - 60 days) to appear in inoculated plants and none of the recipient plants showed symptoms until after 26 days.

Table 2. CBSV transmission rate by a single *B. tabaci* and *A. dispersus*.

Experiment	Adult whitefly species per pant								
		Bemisia tabaci				*Aleurodicus dispersus*			
Replicate	Recipient plants	1	5	15	30	Recipient plants	1	5	15	30
1	4	1(25.0)	2(50)	3(75)	4(100)	4	0	1(25)	2(50)	3(75)
2	4	0	1(25)	2(50)	3(75)	4	0	2(50)	2(50)	4(100)
3	4	1(25.0)	0	2(50)	3(75)	4	0	1(25)	1(25)	2(50)
4	4	0	2(50)	3 (75)	2(50)	4	0	0	1(25)	2(50)
		(12.5)	(31.2)	(62.5)	(75)		(0)	(25)	(37.5)	(68.8)

Figure 2. CBSV diagnostic bands following amplification of cDNA obtained from plants inoculated using *B. tabaci* (B and C) and *A. dispersus* (D and E). A is a DNA marker, whereas H and I are negative controls and G is a positive control.

Transmission efficiency

Efficiency of transmission differed among the two whitefly species examined in this study. The most efficient transmission was observed with *B. tabaci* (Table 2), following a 48 h AAP on CBSV-infected cassava plants and a 48 h IAP. CBSV was transmitted (Figure 2) at low rates by individual whiteflies of *B. tabaci* with 12.5% (2/16) transmission. In addition, *B. tabaci* transmitted at highest efficiency when 30 whiteflies per plant were used in transmission experiments compared to when 1, 5 or 15adults were used per plant. On the other hand *A. dispersus* was less efficient at transmitting CBSV (Fisher, P < 0.0001) (Table 2). However, unlike *B. tabaci*, transmission was not observed with individual spiraling whiteflies over the course of experiments due to adverse conditions that lead to high mortality. Non-choice feeding on CBSD symptomatic leaves within falcon cages led to greater efficiency of transmission by both species.

Figures in parenthesis are percentage infection per replicate.

Transmission was scored as the number of infected target plants over total number of plants tested. Experiments were repeated 4 times using 4 plants for each whitefly species.

DISCUSSION

Cassava brown streak virus was transmitted (Figure 1) with different efficiencies by both *B. tabaci* and *A. dispersus* with *B. tabaci* being more efficient (Table 1 and 2). However, the transmission rates achieved were low compared to the field recorded CBSD incidences of up to 64%, (Alicai et al., 2007). The low transmission rates of CBSV by the two whitefly species may be due to technical difficulties in the transmission protocols such as high temperatures within confined cages (mass mortality) and high humidity levels in falcon tubes. Environmental conditions may adversely affect transmission (Maruthi et al., 2004). For instance, high humidity within the clip cages lead to mass mortality of *A. dispersus* although *B. tabaci* was able to survive despite the humid conditions.

The feeding behaviour of adult *B. tabaci* on cassava plants seems to greatly influence CBSV transmission. More than 90% of adult *B. tabaci* feed on the top five leaves of cassava plants in the field (Maruthi et al., 2004b), whereas the most obvious CBSD symptoms and presumably higher virus titres develop in the lower leaves. The transmission mechanism employed here tried to overcome this challenge by allowing *B. tabaci* to feed on the most symptomatic leaves of field-grown cassava using clip cages (no choice feeding), thus providing ready access to virus for whiteflies. This resulted into higher efficiency of transmission than in mass feeding.

The transmission of CBSV by adult whitefly populations when both species were not allowed 48 h AAP, demonstrates the ability of the vectors to acquire the virus and naturally transmit it under field conditions. Up to 1.7% in a population of the adult *B. tabaci* whiteflies had been shown to be infective when collected in heavily infected cassava fields in Ivory Coast then transferred to young test seedlings of cassava (Fargette et al., 1990). During this trial the whiteflies were collected from CBSV-infected cassava and also from non-choice feeding then immediately transferred on to the recipient plants. In both cases, transmission occurred meaning that adults which had acquired the virus did not loose the ability to transmit it during the transfer, suggesting that both vectors may not require a latent period before they can transmit the virus after acquisition. Different modes of virus transmission have been characterized depending on the retention time, sites of retention, and internalization of virions by vectors (Andret-Link and Fuchs, 2005). Non-persistent viruses are retained by their vectors for less than a few hours whereas semi persistent viruses are retained for days, weeks, or even years. Viruses in these two categories are acquired from infected plants and inoculated within seconds or minutes to recipient plants. In addition, they do not require a latent period, e.g. time interval between acquisition and transmission, and do not replicate in the vector (Andret-Link P. and Fuchs M., 2005). Further work need therefore to focus on categorizing the mode of CBSV transmission by the vectors involved and the specificity of the transmission relationship.

Transmission of plant virus by a single *B. tabaci* has been reported previously such as for cotton leaf curl virus (Kirkpatrick, 1931), tomato yellow leaf curl virus (Mehta et al., 1994) and tobacco leaf curl virus (Aidawati et al., 2002). In most cases, the efficiency of transmission increased as the number of adult *B. tabaci* was increased. A similar result was achieved from this experiment when CBSV was transmitted by a single *B. tabaci* adult. The adult whiteflies per pant greatly influenced the transmission efficiencies achieved as it was observed that transmission rates increased with increase in whitefly numbers used. The ability of *B. tabaci* to transmit CBSV also seemed to be affected by the inoculation and acquisition feeding periods. When 48 AAP was allowed there was higher percent transmission rate achieved by both the adult whitefly species unlike when the whiteflies were not allowed the 48 AAP.

One remarkable biological property of CBSV is its ability to be transmitted by two different whitefly vectors belonging to two genera (*Bemisia* and *Aleurodicus*). This is however not very unusual for a whitefly-transmitted virus. Earlier studies have demonstrated that tomato chlorosis virus (ToCV) is transmitted with equal efficiency by both *Trialeurodes abutilonea* and *B. tabaci* biotype B, members of two different genera (*Trialeurodes* and *Bemisia*, respectively), and was achieved using individual whiteflies of either vector (Wintermantel and Wisler, 2006). Moreover, both *B. tabaci* biotype A and *Trialeurodes vaporariorum* can transmit ToCV, but single insect transmission was not observed with either of these vectors over five independent experiments (Wintermantel and Wisler, 2006).

These findings report for the first time the ability of spiraling whitefly to transmit CBSV and may explain its contribution in the spread of CBSD in cassava growing areas in coastal Kenya. High whitefly populations in the fields, comprising B. tabaci and A. dispersus may be correlated with the high CBSD incidences observed. The results of the investigations on ability of the insects collected from the infected cassava field to acquire and transmit the virus increases the understanding of the role the whitefly species play in the spread of CBSD.

Management options need to focus on the control of the vectors in addition to other control measures.

ACKNOWLEDGEMENT

We thank the Eastern African regional Research Network for Biotechnology, Biosafety and Biotechnology Policy Development BIO-EARN for financial support.

REFERENCES

Aidawati N, Hidayat SH, Suseno R, Sosromarsono S (2002). Transmission of an Indonesian isolate of *tobacco leaf curl virus* (Geminivirus) by *Bemisia tabaci* Genn. (Hemiptera: Aleyrodidae). Plant Path. J. 18: 231-236.

Alicai T, Omongo C A, Maruthi MN, Hillocks RJ, Baguma Y, Kawuki R, Bua A, Otim-Nape GW, Colvin J (2007). Re-emergence of Cassava Brown Streak Disease in Uganda. Plant Disease 91: 1-24.

Andret-Link P, Fuchs M (2005). Transmission Specificity of Plant Viruses by Vectors. J . Plant Path. 87: 153-165.

Bock KR (1994). Studies on cassava brown streak virus disease in Kenya. Trop. Sci. 34: 134–145.

Hillocks RJ, Raya MD, Mtunda K, Kiozia H (2001). Effects of brown streak virus disease on yield and quality of cassava in Tanzania. J. Phytopathol. 149: 389–394.

Lister RM (1959). Mechanical transmission of cassava brown streak Virus. Nat London 183: 1588–1589.

Maruthi MN, Hillocks RJ, Rekha AR, Colvin J (2004). Transmission of Cassava brown streak virus by whiteflies. In: Sixth International Scientic Meeting of the Cassava Biotechnology Network– Adding Value to a Small-Farmer Crop, 8-14 March 2004, CIAT, Cali, Colombia p. 80.

Mehta PJ, Wyman JA, Nakhla, MK Maxwell DP (1994). Transmission of *tomato yellow leaf curl geminivirus* by *Bemisia tabaci* (Homoptera: Aleyrodidae). J. Econ Entomology 87:1291-1297.

Monger WA, Seal S, Isaac AM, Foster GD (2001a). Molecular characterization of the Cassava brown streak virus coat protein. Plant Pathol. 50: 527–534.

Monger WA, Seal S, Cotton S, Foster GD (2001b). Identification of different isolates of cassava brown streak virus and development of a diagnostic test. Plant Pathol. 50: 768–775.

Mware BO, Ateka EM, Songa JM, Narla RD, Olubayo F, Amata R, (2009). Transmission and distribution of cassava brown streak virus disease in non coastal cassava growing areas of Kenya. J. App. Biosci. 16: 864– 870.

Storey HH (1936). Virus diseases of East African plants. VI-A progress report on studies of the disease of cassava. East Afr. Agric. J. 2: 34–39.

Storey HH (1939). Report of the Plant Pathologist. East Afr. Agric. Res. Station Rep.

Wintermantel WM, Wisler GC (2006). Vector specificity, host range, and genetic diversity of Tomato chlorosis virus Plant Dis. 90: 814-819.

William M. Wintermantel, United States Department of Agriculture–Agricultural Research Service, Salinas CA, Gail C, Wisler, Department of Plant Pathology, University of Florida, Gainesville.

23

Subgenogroup B5 maintains its supremacy over other human Enterovirus71 strains that circulated in Malaysia from 2010 to 2012

Mohd Apandi Yusof , Hariyati Md Ali, Hamadah Mohammad Shariff, Noor Khairunnisa Ramli, Zarina Mohd Zawawi, Syarifah Nur Aisyatun, Jasinta Anak Dennis and Zainah Saat

Virology Unit, Infectious Disease Research Centre, Institute for Medical Research, Kuala Lumpur, Malaysia.

Human enterovirus71 (HEV71) is responsible for hand, foot and mouth diseases (HFMD). Several outbreaks of HFMD were associated with severe neurological disease and deaths. In Malaysia, outbreaks normally occur periodically every two to three years but HEV71 were isolated throughout the year from HFMD cases. From 2010 to 2012, HEV71 strains were isolated from 37 children presented with typical HFMD. All isolates were sequenced and BLAST searched. A phylogenetic tree constructed based on the complete *VP1* gene showed that all isolates belonged to subgenogroup B5. This subgenogroup has been found dominant since 2003.

Key words: Human enterovirus71, molecular epidemiology, gene sequences, circulating subgenogroup.

INTRODUCTION

Human enterovirus71 (HEV71) is classified under the human enterovirus A species. It is a positive-sense RNA virus from enterovirus genus in the family picornaviridae. Together with Coxsackie A16 and Coxsackie A6, they cause major outbreaks of hand foot and mouth disease (HFMD) in children. During outbreaks in Sarawak, Malaysia in 1997 (Chan et al., 2000) and Taiwan in 1998 (Wang et al., 2002), HEV71 infections not only presented with typical HFMD but were also associated with neurological disorders such as encephalitis, aseptic meningitis and meningoencephalitis (McMinn et al., 2001; Kehle et al., 2003), and paralysis (Melnick, 1984) due to its affinity to anterior horn cell (Chumakov et al., 1979).

HFMD outbreaks associated with HEV71 had been reported globally. In the Asia Pacific region, large outbreaks have been recorded in Sarawak and Peninsular Malaysia in 1997, with 41 and 4 deaths, respectively (Chan et al., 2000). The largest outbreak so far, occurred in Taiwan in 1998 where an estimated 1.5 million people were infected with 405 children being hospitalised, of which 78 died (Wang et al., 2002). The recent epidemic in China in 2008, recorded 0.5 million cases with 126 deaths (Zhang et al., 2009, 2010). Many countries including Japan, Singapore, Sarawak, Vietnam and Peninsular Malaysia have ex-perienced cyclical epidemics that occur every 2-3 years (Fujimoto et al., 2002; Podin et al., 2006; Tu et al., 20077; Apandi et al., 2011).

Most of these outbreaks not only caused huge hospital admission but also resulted in fatality. One of the most important findings of these outbreaks was the emergence of different strains of HEV71.

Following the 1997 outbreak in Sarawak, the Malaysian Ministry of Health (MOH) made a ruling that all HFMD cases with or without neurological manifestations must be notified. Since then, a surveillance system for monitoring HFMD has been established.

MATERIALS AND METHODS

Clinical samples such as vesicle swabs, throat swabs, rectal swabs and stool from hospitalized HFMD patients were collected from 15 states in Malaysia. They were screened for enterovirus (EV) by RT-PCR (Perera et al., 2004). Positive samples were cultured in rhabdomyosarcoma (RD) cells and harvested once cytopathic effect was observed.

Viral RNA was extracted using the QIAamp® Viral RNA Mini Kit (Hilden, Germany). RT-PCR with specific HEV71 primers (Tu et al., 2007) was used for complete VP1 gene amplification. All isolates were sequenced using PCR primers as described by Tu et al.(2007) and internal primers VP1 Int F and VP1 Int R (Apandi et al., 2011). Sequencing was performed by using the Big Dye Cycle sequencing kit version 3.0 and an ABI377 automated DNA sequencer (Applied Biosystems, Forster City, USA). The SeqMan and Megalign software modules in the Lasergene Suite of programs (DNASTAR, Madison, WI, USA) were used to format the nucleotide sequences. All sequences were BLAST searched (http://blast.ncbi.nlm.nih.gov/Blast.cgi) and a phylogenetic tree was constructed by using the neighbour-joining method from the MEGA4 software (www.megasoftware.net).

RESULTS

In 2010, Institute for Medical Research (IMR) Kuala Lumpur, Malaysia received specimens from 634 hospitalised HFMD cases and EV was detected from 440 (69%) cases by RT-PCR. Of these, 36 isolates comprised of CA16, HEV71, CA5 and ECHO 4 were isolated.

In 2011, we received specimens from only 268 HFMD cases and 185 (69%) were found positive for EV and 36 isolates of CA16, HEV71, CA10 and ECHO9 were detected. There was an increase in number of cases in 2012, specimens from 583 hospitalised cases were analysed and EV was detected in 407 cases (70%). Eighty five isolates of CA16, HEV71 and CoxB3 were isolated.

Overall, a total of 37 HEV71 strains were isolated with 10 isolates in 2010, 5 in 2011 and 22 in 2012. Details of the isolates are shown in Table 1.

Based on the complete VP1 gene which consists of 891 nucleotides, the phylogenetic tree revealed that all HEV71 circulating in Malaysia from 2010 to 2012 in this study belonged to the HEV71 subgenogroup B5. The tree is shown in Figure 1.

DISCUSSION

Analyzing the VP1 gene is very important to determine the relationships between the genogroups (Ggs) and

pathogenicity. Based on this, HEV71 has been classified into three genogroups (Ggs) namely GgA, GgB and GgC (Brown et al., 1999). GgA was the prototype strain of HEV71, isolated in California, USA in 1970 (Schmidt et al., 1974) and has never been reported since. Another two Ggs, GgB evolved over the years into 5 subgenogroups: GgB1, GgB2, GgB3, GgB4 and GgB5; while GgC evolved into subgenogroup GgC1, GgC2, GgC3, GgC4 and GgC5. The genetic variation within genogroups was about 12% or was lesser than the variation between genogroups which was about 16.5 to 19.7% (Brown et al., 1999).

In Malaysia, HFMD cases are detected throughout the year. However, outbreaks of HFMD occur periodically nearly every t years. Many researchers have reported different strains of HEV71 in different outbreaks. The huge HFMD outbreak in Sarawak in 1997 (Cardosa et al., 2003) was the starting point of molecular epidemiology study of HEV71 in Malaysia. Several outbreaks in 2000 (8 fatalities) followed by outbreak in 2003 (no fatality) and in 2005 (two fatalities) (Chua and Kasri, 2011), had contributed more HEV71 genomic diversity.

GgB3 isolated in 1997, during the Sarawak outbreak, was documented as the first HEV71 strain circulating and associated with severe encephalitis and fatalities in Malaysia (Cardosa et al., 2003). To date, this strain has never been found circulating in peninsular Malaysia; indicating that GgB3 only appeared in very short period of time and confined to Sarawak State only for Malaysia. However, it has been isolated and reported in Japan at the same time (Fujimoto et al., 2002).

Between 1998 and 1999, HEV71 was still isolated from a few HFMD cases although there were no outbreaks reported. GgC1 was found to be circulating sporadically in 1998 in Sarawak (Cardosa et al., 2003) together with GgC2 in peninsular Malaysia (AbuBakar et al., 1999) from typical HFMD cases. However, GgC2 found circulating during the large HEV71 outbreak in Taiwan in 1998, was associated with fatal encephalitis (Wang et al., 2002).

HEV71 activity appeared low in 1999, and an analysis of 43 HEV71 isolates from peninsular Malaysia from 1997-2000 (Herrero et al., 2003) showed that GgB4 was the only predominant strain circulating in that year. Later in 2000, during the large outbreak of HFMD in peninsular Malaysia and Sarawak, GgB4 together with GgC1 emerged as the dominant strains (Chua and Kasri, 2011; Cardosa et al., 2003; Herrero et al., 2003). These two Ggs were also found widespread in the outbreak in Singapore and Taiwan, and continued to circulate in Malaysia in 2001 (Apandi et al., 2011), replacing GgB3.

GgC1 was found circulating in 2001, 2003, 2006 and 2007, mostly in peninsular Malaysia. From 2001 to 2009, 70 isolates from peninsular Malaysia and Sabah of HEV71 were analysed (Apandi et al., 2011). GgC1 was lastly detected in 2007 and it has neither been documented as the cause of large outbreak nor asso-

Table 1. Summary of patients' gender, age, sample type, clinical diagnosis and GenBank accession number.

Isolate	Accession No.	Sample type	Sex/age	Diagnosis	Isolation year
EV1004-Terengganu-11	KC894865	V/S	M/5	HFMD	2011
EV1056-Terengganu-11	KC894866	T/S	M/2	HFMD	2011
EV1268-Pahang-11	KC894867	V/S	F/0.8	HFMD	2011
EV0978-Sarawak-11	KC894868	R/S	F/1.5	HFMD	2011
EV0984-Sarawak-11	KC894869	V/S	M/1.3	HFMD	2011
EV0691-Terengganu-10	KC894872	V/S	M/1	HFMD	2010
EV0733-PPinang-10	KC894873	T/S	F/2	HFMD	2010
EV0744-Johor-10	KC894874	T/S	F/1	HFMD	2010
EV0994-Terengganu-10	KC894875	V/S	M/2	HFMD	2010
EV1233-Kedah-10	KC894876	V/S	F/4	HFMD	2010
EV1297-Melaka-10	KC894877	T/S	M/3	HFMD	2010
EV1299-Melaka-10	KC894878	T/S	M/8	HFMD	2010
EV1301-Melaka-10	KC894879	V/S	M/6	HFMD	2010
EV1312-Johor-10	KC894880	T/S	M/1	HFMD	2010
EV1389-KLumpur-10	KC894881	Stool	M/0.5	HFMD	2010
EV0615-Johor-12	KC894882	T/S	M/1.5	HFMD	2012
EV0616-Johor-12	KC894883	T/S	F/0.5	HFMD	2012
EV0655-Kedah-12	KC894884	V/S	M/4	HFMD	2012
EV0659-Pahang-12	KC894885	V/S, T/S	M/5	HFMD	2012
EV0665-Kelantan-12	KC894886	T/S	F/4	HFMD	2012
EV0673-Johor-12	KC894887	T/S	F/2.1	HFMD	2012
EV0710-Johor-12	KC894888	R/S	M/1.7	HFMD	2012
EV0769-Johor-12	KC894889	T/S	F/2	HFMD	2012
EV0775-Johor-12	KC894890	T/S	F/1.7	HFMD	2012
EV0779-Johor012	KC894891	V/S	F/7	HFMD	2012
EV0791-Johor-12	KC894892	R/S	F/1	HFMD	2012
EV0834-Johor-12	KC894893	T/S	M/2	HFMD	2012
EV0891-Johor-12	KC894894	T/S	F/1.2	HFMD	2012
EV0894-Kedah-12	KC894895	V/S	M/2.7	HFMD	2012
EV0896-Johor-12	KC894896	T/S	F/2	HFMD	2012
EV0953-Johor-12	KC894897	T/S	M/1.6	HFMD	2012
EV0961-Johor-12	KC894898	T/S	F/1.2	HFMD	2012
EV1002-Johor-12	KC894899	R/S	M/5	HFMD	2012
EV1003-Johor-12	KC894900	R/S	M/2.5	HFMD	2012
EV1170-Selangor-12	KC894901	T/S	M/5	HFMD	2012
EV1325-Johor-12	KC894902	T/S	F/1.5	HFMD	2012
EV0997-Pahang-12	KC894903	T/S	M/1.5	HFMD	2012

T/S = throat swab; R/S = rectal swab; V/S = vesicle swab; M = male; F = female; age in years.

ciated with encephalitis.

A new subgenogroup in GgC, named GgC4, first emerged in China in early 2000. It later became a predominant strain in Taiwan in 2004 (Lin et al., 2006) and was responsible for huge outbreaks of HFMD in Shandong and Fuyang, China in 2007-2008 (Zhang et al., 2009, 2010).

However in Malaysia, only one GgC4 was detected from HFMD cases in 2004 (Apandi et al., 2011) and other strains, GgC3 and GgC5, so far have never been reported (Apandi et al., 2011; AbuBakar et al., 1999;

Herrero et al., 2003). The GgC3 was found only during the major HEV71 outbreak in Korea in 2000 and GgC5 circulated widely in Southern Vietnam in 2005 (Tu et al., 2007).

Over the years, HEV71 in GgB has evolved. GgB3 that emerged in 1997 was displaced by GgB4 in 2000 and 2001. In 2003, the GB4 was displaced by GgB5 and since then, from 2004 to 2009, GgB5 became the only predominant strain in both peninsular Malaysia and Sarawak (Apandi et al., 2011). In this study, GgB5 was still found to be the dominant strain from 2010 to 2012

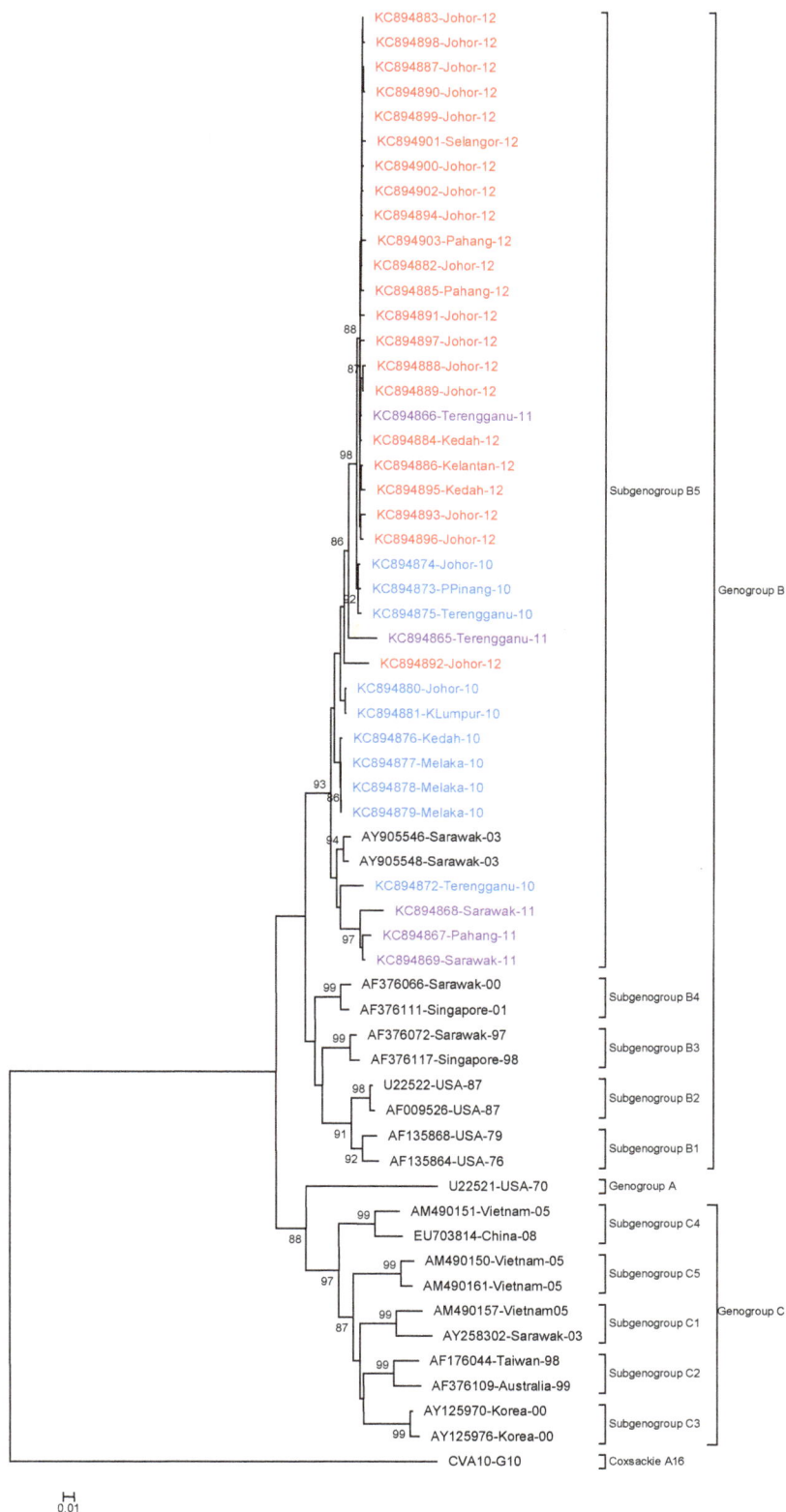

Figure 1. Phylogenetic tree of HEV71 based on the complete VP1 gene. Genogroups and subgenogroups are indicated by square brackets with CA16-G10 as an outgroup. HEV71 isolates in 2010, 2011 and 2012 are indicated in blue, purple and red, respectively, together with representative retrieved from GenBank.

and was the cause of the 2010 HEV71 outbreak in Malaysia.

Hence, for the past 25 years, 6 strains of HEV71, 3 from GgB and 3 from GgC were circulating in Malaysia. In GgB, GgB3 was isolated in 1997 and was followed by GgB4 in 2000-2001 which was later replaced by GgB5 from 2003 till now. In GgC, GgC1 was first detected in 1997 in Sarawak, continued to circulate in peninsular in 2001, 2003, 2006 and 2007, while GgC2 circulated in 1998 and GgC4 in 2007.

Conflict of Interests

The author(s) have not declared any conflict of interests.

ACKNOWLEDGEMENTS

We thank the Director General of Health and the Director of the Institute for Medical Research for permission to publish this article. This research project was funded under the Virology Operational Budget 2012.

REFERENCES

AbuBakar S, Chee HY, Al-Kobaisi MF, Xiaoshan J, Chua KB, Lam SK (1999). Identification of enterovirus 71 isolates from an outbreak of hand, foot and mouth disease with fatal cases of encephalomyelitis in Malaysia. Virus Res. 61:1-9.

Apandi MY, Fazilah R, Maizatul AA, Liyana AZ, Hariyati MA, Fauziah MK, Zainah S. (2011). Molecular epidemiology of human enterovirus71 (HEV71) strains isolated in Peninsular Malaysia and Sabah from year 2001 to 2009. J. General Mol. Virol. Vol.3(1), pp. 18-26, January 2011. http://www.academicjournals.org/JGMV

Brown BA, Oberste MS, Alexander JP, Kennett ML, Pallansch MA (1999). Molecular epidemiology and evolution of enterovirus 71 strains isolated from 1970 to 1998. J. Virol. 73:9969-9975.

Cardosa MJ, Perera D, Brown BA, Cheon D, Chan HM, Chan KP, Cho H, McMinn PC (2003). Molecular epidemiology of human enterovirus 71 strains and recent outbreaks in the Asia-Pacific region: Comparative analysis of the VP1 and VP4 genes. Emerg. Infect.Dis. 9(4):461-468.

Chan LG, Parashar UD, Lye MS, Ong FG, Zaki SR, Alexander JP, Ho KK, Han LL, Pallansch MA, Suleiman AB, Jegathesan M, Anderson LJ (2000). Deaths of children during an outbreak of hand, foot, and mouth disease in Sarawak,Malaysia: clinical and pathological characteristics of the disease. For the Outbreak Study Group. Clin. Infect. Dis. 31:678-683.

Chua KB, Kasri AR (2011). Hand foot and mouth disease due to enterovirus 71 in Malaysia. Virol Sin. 2011 Aug;26(4):221-8. doi: 10.1007/s12250-011-3195-8. Epub 2011 Aug 17.

Chumakov M, Voroshilova M, Shindarov L, Lavrova I, Gracheva L, Koroleva G, Vasilenko S, Brodvarova I, Nikolova M, Gyurova S, Gacheva M, Mitov G, Ninov N, Tsylka E, Robinson I, Frolova M, Bashkirtsev V, Martiyanova L, Rodin V (1979). Enterovirus 71 isolated from cases of epidemic poliomyelitis-like disease in Bulgaria. Arch. Virol. 60:329-340.

Fujimoto T, Chikahira M, Yoshida S, Ebira H, Hasegawa A, Totsuka A, Nishio O (2002). Outbreak of central nervous system disease associated with hand, foot, and mouth disease in Japan during the summer of 2000: detection and molecular epidemiology of enterovirus 71. Microbiol. Immunol. 46:621-627.

Herrero LJ, Lee CS, Hurrelbrink RJ, Chua BH, Chua KB, McMinn PC (2003). Molecular epidemiology of enterovirus 71 in peninsular Malaysia, 1997-2000. Arch. Virol. 148(7):1369-1385.

Kehle J, Roth B, Metzger C, Pfitzner A, Enders G (2003). Molecular characterization of an enterovirus 71 causing neurological disease in Germany. J. Neurovirol. 9:126-128.

Lin KH, Hwang KP, Ke GM, Wang CF, Ke LY, Hsu YT, Tung YC, Chu PY, Chen BH, Chen HL, Kao CL, Wang JR, Eng HL, Wang SY, Hsu LC, Chen HY (2006). Evolution of EV71 genogroup in Taiwan from 1998 to 2005: An emerging of subgenogroup C4 of EV71. J. Med. Virol. 78(2):254-262.

McMinn PC, Lindsay K, Perera D, Chan HM, Chan KP, Cardosa MJ (2001). Phylogenetic analysis of enterovirus 71 strains isolated during linked epidemics in Malaysia, Singapore and Western Australia. J. Virol. 75(16).

Melnick JL (1984). Enterovirus type 71 infections: a varied clinical pattern sometimes mimicking paralytic poliomyelitis. Rev. Infect. Dis. 6(2):S387-S390.

Perera D, Podin Y, Winnie A, Tan CS, Cardosa MJ (2004). Incorrect identification of recent Asian strains of Coxsackievirus A16 as human enterovirus 71: Improved primers for the specific detection of human enterovirus 71 by RT PCR. BMC Infect. Dis. 4:1-11.

Podin Y, Edna LMG, Ong F, Leong YW, Yee SF, Apandi Y, Perera D, Teo B, Wee TY, Yao SK, Kiyu A, Arif MT, Cardosa MJ (2006). Sentinel surveillance for human enterovirus 71 in Sarawak, Malaysia: lessons from the first 7 years. BMC Pub. Health. 6(180):7732-7738.

Schmidt NJ, Lennette EH, Ho HH (1974). An apparently new enterovirus isolated from patients with disease of the central nervous system. J. Infect. Dis. 129:304-309.

Tu PV, Thao NT, Perera D, Huu TK, Tien NT, Thuong TC, Ooi MS, Cardosa MJ, McMinn PC (2007). Epidemiologic and virologic investigation of hand, foot, and mouth disease, Southern Vietnam, 2005. Emerg Infect Dis.13:1733-1741.

Wang JR, Tuan YC, Tsai HP, Yan JJ, Liu CC, Su IJ (2002). Change of major genotype of enterovirus 71 in outbreaks of hand-foot-and-mouth disease in Taiwan between 1998 and 2000. J. Clin. Microbiol. 40:10-15.

Zhang Y, Tan XJ, Wang HY, Yan DM, Zhu SL, Wang DY, Ji F, Wang XJ, Gao YJ, Chena, Qiu An H, Xin Li D, Wang SW, Xu AQ,Wang ZJ, Xu WB (2009). An outbreak of hand, foot, and mouth disease associated with subgenotype C4 of human enterovirus 71 in Shandong, China. J. Clin. Virol. 44:262-267.

Zhang Y, Zhu Z, Yang W, Ren J, Tan X, Wang Y, Mao N, Xu S, Zhu S, Cui A, Zhang Y, Yan D, Li Q, Dong X, Zhang J, Zhao Y, Wan J, Feng Z, Sun J, Wang S, Li D, Xu W. (2010). Aneemerging recombinant human enterovirus 71 responsible for the 2008 outbreak of Hand Foot and Mouth Disease in Fuyang city of China. Virol. J. 7:94

24

Phenotypic and molecular screening of cassava (*Manihot esculentum* Crantz) genotypes for resistance to cassava mosaic disease

P. A. Asare[1], I. K. A. Galyuon[2], E. Asare-Bediako[1], J. K. Sarfo[3] and J. P. Tetteh[1]

[1]Department of Crop Science, University of Cape Coast, Cape Coast, Ghana.
[2]Department of Molecular Biology and Biotechnology, University of Cape Coast, Ghana.
[3]Department of Biochemistry, University of Cape Coast, Cape Coast, Ghana.

Cassava mosaic disease (CMD), caused by cassava mosaic geminivirus (CMG) is the most-important disease threatening production of cassava (*Manihot esculenta*) in Ghana. The disease is best managed through host-plant resistance. The study was conducted to assess resistance of 38 cassava genotypes to CMD, determine the associated resistance gene, and to identify the strains of CMG infecting cassava in Ghana. Both morphological and molecular markers were used to screen 38 cassava accessions against CMG infection. Morphological studies revealed one genotype (Capevars) as highly resistant whilst three others (Adehye, Nkabom and KW085) were tolerant, showing mild symptoms. PCR analyses using strain specific primers, however, detected the virus in all the three tolerant genotypes, but absent in Capevars. However, the dominant CMD resistance gene, *CMD2*, was detected in both the resistant and the tolerant genotypes. Apart from Capevars, the other 37 cassava genotypes were infected by, at least, one of the four ACMV variants of ACMV1, ACMV2, ACMV-AL and ACMV3. It is, therefore, concluded that field screening for CMD resistance, should integrate phenotypic evaluation and detection of the virus.

Key words: Cassava, African cassava mosaic virus, simple sequence repeats, resistance.

INTRODUCTION

Cassava (*Manihot esculenta* Crantz), an Euphorbiaceae (Webster, 1994), is the sixth world food crop for more than 500 million people in tropical and sub-tropical Africa, Asia and Latin America (FAO, 2008). Cassava is the number one staple food crop for majority of Ghanaians, with per capita consumption of 152.9 kg/head/year (MOFA, 2011) and has played a key role in food security in Ghana. It contributes 22% of Agricultural Gross

Domestic Product (AGDP) (FAO, 2014) and is also fast becoming an important crop for industries because of its high starch content. In Ghana, cassava is grown across all agro-ecological zones and ranks first in the area under cultivation (MOFA, 2011). However, the average yield of the crop in the country, which is 13.8 Mt ha^{-1}, is far below an achievable yield of 48.7 Mt ha^{-1} (MOFA, 2011). Pests and diseases are a major contributing factor to the low yield of the crop (Akinlosotu, 1985; Thresh et al., 1994). Major pests of cassava include the cassava mealybug (*Phenacoccus manihoti*), green spider mite (*Mononychellus tanajoa*) (Akinlosotu, 1985) and whitefly (*Bemisia tabaci*) (Perrings, 2001).

Cassava mosaic disease (CMD), caused by cassava mosaic geminiviruses of the family *Geminiviridae* and genus *Begomovirus* (Fauquet and Stanley, 2003; Fauquet et al., 2005), is the most important factor limiting cassava yields in many parts of Africa (Fauquet and Fargette, 1990; Legg and Fauquet, 2004). CMD is responsible for an estimated loss of yield of over 1.5 billion US dollars a year (Thresh et al., 1994). It is undoubtedly the most important constraint to the production of cassava in Ghana (Lamptey et al., 1998). The characteristic severe distortion and stunting of leaf and entire plant associated with the disease, especially on local genotypes, indicates how serious yields could be affected (Lamptey et al., 2000). ACMV has been reported to cause 80% yield loss in susceptible cultivars in Ghana (Moses et al., 2007). Losses due to ACMV disease reported elsewhere range from 20 to 95% (Fargette et al., 1988; Hahn et al., 1989; Terry and Hahn, 1990; Otim-Nape et al., 1994; Braima et al., 2000).

The mosaic virus spread is highly linked with its whitefly (*Bemisia tabaci*) vector (Fargette et al., 1985). The virus can also be transmitted from infected planting materials. Plants grown from infected cuttings are much more seriously affected than those infested later by the whitefly vector (*Bemisia tabaci*) and plants infected at a late stage of crop growth are almost unaffected (Thresh et al., 1994).

Nine distinct cassava mosaic viruses have been characterized worldwide from CMD-affected cassava plants and seven of them are from sub-Saharan Africa (Fauquet and Stanley 2003; Alabi et al., 2011). These viruses are African cassava mosaic virus (ACMV), East African cassava mosaic virus (EACMV), East African cassava mosaic Cameroon virus (EACMCV) (Fondong et al., 2000), East African cassava mosaic Kenya virus (EACMKV) (Bull et al., 2006), East African cassava mosaic Malawi virus (EACMMV) (Zhou, et al., 1998), East African cassava mosaic Zanzibar virus (EACMZV) (Maruthi et al., 2004) and South African cassava mosaic (SACMV) (Berrie et al., 1998). Two other viruses, Indian cassava mosaic virus (ICMV) (Matthew and Muniyappa, 1992; Saunders et al., 2002) and Sri Lankan cassava mosaic virus (SLCMV) (Saunders et al., 2002), were reported from the Indian sub-continent.

Cassava mosaic geminivirus (CMG) strains reported so far in Ghana are ACMV (Clerk, 1974; Lamptey et al., 1998) and EACMV (Offei et al., 1999). ACMD was first observed near Accra in 1926 (Doku, 1966) and its spread was more significant in the coastal areas of the country around 1930 (Leather, 1959; Clerk, 1974). At present, ACMD is widespread and found in all the agro-ecological zones in Ghana (Lamptey et al., 1998). The EACMV was first reported in Ghana in 1999 (Offei et al., 1999). The emergence of EACMV, which has its origin from East Africa but has been documented in Central and West Africa (Fondong et al., 1998; Offei et al., 1999; Ogbe et al., 1999), raises a lot of concern to cassava growers in the sub-region including Ghana.

Effective management of the CMD-pandemic in Ghana is quite important in order to improve yields. The most effective means of controlling CMD is by the deployment of resistant varieties (Thresh et al., 1997). CMD-resistant cassava had been developed through integration of resistance traits from *Manihot glaziovii* by interspecific hybridization (Nicholas, 1947), which has become the major source dominating CMD resistance in Africa (Fargette et al., 1996). Two CMD resistance genes *CMD1* (recessive gene) and *CMD2* (major dominant gene) have so far been placed on the map and important molecular markers associated with the *CMD2* gene have been identified (Fregene et al., 2001; Akano et al., 2002). Through cassava breeding programmes, these markers are very useful and hold great promise in fast-tracking the identification of CMD-resistant germplasms (Bi et al., 2010). Knowledge of genetic diversity or an under-standing of which viral strain, and strain combinations and how they are distributed, is important to such breeding programmes for resistance.

This work was, therefore, aimed at assessing the genetic diversity of ACMV currently infecting cassava in Ghana, identifying resistant cassava cultivars and determining the presence of the *CMD2* resistance gene using its associated simple sequence repeats (SSR) markers.

MATERIALS AND METHODS

Collection of cassava planting materials

Thirty-eight (38) distinct cassava genotypes were used for the study. Thirty (30) of them were obtained from the Plant Genetic Resources Research Institute (PGRRI), Bunso, Ghana and the remaining eight from the University of Cape Coast (U.C.C.) Teaching and Research Farm, Cape Coast, Ghana. Three of the materials (Capevars, Adehye, and Nkabom) have been released as cultivars for farmers.

Field experiment

Experimental site and field layout

The 38 cassava genotypes were evaluated in 2007/2008 and 2008/2009 growing seasons, on the Teaching and Research Farm,

Table 1. Disease rating and the corresponding symptom expression for cassava mosaic disease (CMD).

Rating	Symptom
1	No symptoms observed
2	Mild chlorotic pattern on entire leaflets or mild distortion at base of leaflets appearing green and healthy
3	Strong mosaic pattern on entire leaf, and narrowing cum distortion of lower one-thirds of leaflets
4	Severe mosaic distortion of two-thirds of leaflets and general reduction of leaf size
5	Severe mosaic distortion of four-fifths or more of leaflets, twisted and misshapen leaves.

U.C.C., Ghana. The location (5.1000° N, 1.2500° W) is a coastal savanna zone with a ferric luvisol soil type and is a high pressure (highly endemic) site for CMD. The soil has been described by Asamoa (1973) as Atabadze, equivalent to Ultisol in the United States Department of Agriculture, (USDA) classification. Cape Coast has a typical climate of the coastal savannah lowland characterized by an annual rainfall range of 800 to 1000 mm and mean monthly temperature of about 26.5°C.

A 380 m² land (38 × 10 m) was ploughed, harrowed and divided into 10-m rows with 1.0 m between rows in the 2007 and 2008 major planting seasons. A total of 38 cassava genotypes were planted in single rows in completely randomised plots. Ten 20 cm-long cuttings (bearing three to four nodes) were planted per genotype, in single rows at a spacing of 1 m within rows and 1 m between rows.

Cultural practices

The ploughed and harrowed field was lined and pegged before planting. The experiment was set out under rain-fed conditions and weeding was done manually using a hoe or cutlass when necessary.

Morphological screening of the cassava genotypes for CMD resistance

The 38 cassava genotypes were evaluated at 6, 12, 20 and 48 weeks after planting (WAP) in both 2007/8 and 2008/9 growing seasons to ascertain the resistance status of each genotype to CMD. Each plant was examined for symptom severity of the whole plant. Plants were assigned disease severity scores based on the standard 1-5 disease rating (Hahn, 1980; IITA, 1990; Ariyo et al., 2005), where 1 represents no disease symptom and 5 being the presence of the most severe symptoms, including severe chlorosis, leaf distortion and plant stunting (Table 1).

Five plants for each genotype were scored and the mean ordinal score determined. Plants with a mean CMD severity score of "1" were then classified as highly resistant (HR), those with a score of "2" were moderately resistant (MR), those with a score of "3" were classified as susceptible (S) and those with scores of "4" and "5" were classified as highly susceptible (HS), according to Lokko et al. (2005)

Determination of population of whitefly

Since whiteflies are the vectors of CMD, their population on cassava plants were determined in order to assess their relationship with the severity of the CMD disease infection. Direct counts of adult whiteflies on the crop were made as previously described (Hill, 1968; Fargette et al., 1985; Abdullahi et al., 2003).

Whitefly counting was usually done between 0600 and 0800 h when the environment was cooler and whiteflies were relatively immobile compared to later in the day as reported by Fauquet et al. (1987). Adult whitefly populations on the five topmost fully expanded leaves of the selected cassava cultivars were counted according to Otim-Nape et al. (2005) and Ariyo et al. (2005).

Whitefly count was often carried out on the five topmost fully expanded leaves. The counts were done one month after planting and were repeated at three and six months after planting. Five plants were randomly selected for each cassava genotype. On each plant, leaves were carefully turned over and the number of adult whiteflies on the abaxial leaf surfaces were counted and recorded. The mean number of whiteflies per 5 top leaves was then determined.

Screening for CMD resistance using molecular markers

Collection of cassava leaf samples

Young leaves from the 38 cassava genotypes were collected from both CMD-infected plants (symptomatic) and uninfected (non-symptomatic) plants at the experimental site.

DNA extraction and purification

Genomic DNA was extracted from the fresh samples, according to the method described by Dellaporta et al. (1983) with slight modifications. The leaf tissues were lysed using a lysis buffer, followed by extraction of DNA from the leaf tissues and DNA precipitation. DNA pellets from precipitation were washed with 700 µl of 80% ethanol, air-dried on tissue paper at room temperature (25-30°C) re-dissolved in 100 µl of 1x TE buffer and stored at -20°C until required.

PCR amplification

The ACMV strains or variants causing the mosaic symptoms in the 38 accessions were detected using the PCR method described by Zhou et al. (1997). The DNA samples of the cassava genotypes were tested for presence or absence of CMG using primers that could detect the four variants of ACMV (ACMV1, ACMV2, ACMV-AL and AVMV3). Four pairs of primer sequences designed by Zhou et al. (1997) were used (Table 2). The PCR reactions were conducted using Applied Biosystems® 2720 Thermal Cycler in 96-well plates (Life Technologies, New York, USA). The reaction mixture composed of 10 µl, which consists of AccuPower® PCR Premix (BIONEER Inc., Alameda, USA), genomic DNA, sterile distilled water (SDW) and primers. The PCR mixture contained 9 µl of PCR premix and primers and genomic DNA (10 ng µl⁻¹). The PCR programme consisted of an initial denaturation for 4 min at 94°C and then 35 cycles of denaturation for 30 s at 94°C, annealing for 30 s depending on the annealing temperature of the primer, and

Table 2. Primers for PCR amplification and strain differentiation of cassava mosaic virus diseases.

Virus strain	Name of primer	Primer sequence (5' - 3')	Reference
ACMV1	ACMV-F1	TTC AGT TAT CAG GGC TCG TAA (F)	Zhou et al. (1997)
	ACMV-R1	GAG TG AAG TTG ACT CAT GA (R)	Zhou et al. (1997)
ACMV2	ACMV-F2	GTG AGA AAG ACA TTC TTG GC (F)	Zhou et al. (1997)
	ACMV-R2	CCT GCA ATT ATA TAG TGG CC (R)	Zhou et al. (1997)
ACMV-AL	ACMV-AL1/F	GCG GAA TCC CTA ACA TAA TC (F)	Zhou et al. (1997)
	ACMV-ARO/R	GCT CGT ATG TAT CCT CTA AGG CCT (R)	Zhou et al. (1997)
ACMV3	ACMV-1	GCTC AAC TGG AGA CAC ACT TG (F)	Zhou et al. (1997)
	ACMV-2	CCT GCA ACA TAC TTA CGC TT (R)	Zhou et al. (1997)

extension at 72°C for 1 min and final extension of 5 min at 72°C. The PCR products were separated by electrophoresis in a 1% agarose gel at 100 V for 1.5 h. The gel was stained with ethidium bromide and viewed under UV light.

Detection of *CMD2* resistance gene in ACMD-resistant cassava genotypes

Plant DNA samples that did not show presence of any of the strains of cassava mosaic virus following PCR amplification with strain specific primers were further amplified with specific SSR markers (SSRY28, NS158, NS169 and RME1) associated with the *CMD2* gene, the dominant gene, which confers resistance to ACMD. PCR amplification and gel electrophoresis were carried out as described earlier.

Data analysis

Scatter plots showing the relationship between mean whitefly population and mean CMD severity scores during 2007 and 2008 crop seasons were drawn using MICROSOFT EXCEL (Microsoft Corporation, USA). The corresponding correlation coefficients were also determined using GenStat statistical software version 12 (Payne et al., 2009).

The relationships among cassava accessions, with respect to their susceptibility to the four ACMV strains were determined based on band patterns produced in the gel. Bands of alleles were scored as 1 for presence of virus or infection, and 0 as absence of alleles, denoting no infection or healthy, for various primers-cassava accessions combinations. The band scores were then used to calculate genetic distances (Nei, 1983) between pairs of cassava accessions. Then, using the unweighted pair-group mean average (UPGMA) cluster method of Nei's genetic distance (Sneath and Sokal, 1973), a dendrogram of genetic similarity was constructed using the Power Marker software version 3.5 (Liu and Muse, 2005).

RESULTS

Cassava mosaic disease (CMD) severity

The mean CMD severity scores recorded for the cassava genotypes planted during 2007 and 2008 growing seasons showed a varying and an interesting pattern (Table 3). At 6 weeks after planting (WAP) in 2007 the mean score for all the cassava genotypes on the field was 2.8, with a range score of 1-5.

With this range of scores, five accessions had a score of 1, 12 had a score of 2, 14 had a score of 3, nine were scored 4 while three accessions registered the highest score of 5. Thus, DMA 002, ADW 004 and OFF 029, which had the highest score of 5, were the most susceptible to ACMV infection at 6 WAP.

AT 12 WAP, four genotypes had a score of 1, twelve a score of 2, sixteen a score of 3, nine a score of 4 and three had a score of 5. The mean severity score was 2.9 for 2007. In 2008 the severity scores at 12 WAP were 1, 2, 3, 4 and 5 for four, nine, seven, twenty and four accessions, respectively, with a mean score of 3.3. This indicates that the severity of infection of the cassava genotypes by the ACMV was higher in 2008 than in 2007. This indicates that the cassava genotypes were more susceptible to the ACMV infection in 2008 than in 2007.

At 20 WAP in 2007, the mean score was 2.6 and that of 2008 was 3.4 with severity scores for both years ranging between 1 and 5. At 48 WAP, which was the harvest time, ACMD severity score was recorded to assess the degree of recovery from the disease among the accessions. The mean scores reduced to 1.7 and 1.9 for 2007 and 2008, respectively.

However, in both years, 23 had severity score of 1, 12 were scored 2, five had a score of 3 while three of them had a score of 4. None of the accessions was scored the most severity score of 5.

The overall mean CMD severity responses recorded for all the 38 cassava accessions at different sampling dates and time revealed varying levels of resistance or susceptibility (Figure 1). The accessions were thus grouped into the five disease severity classes. Three genotypes were classified as highly resistant (HR) with a mean score of 1, nine as resistant (R) with a mean score of 2, 12 as susceptible (S) with a mean score of 3 and 14 as highly susceptible (HS) with mean scores of 4 and 5.

Whitefly population

At six weeks after planting (WAP), the overall mean adult whitefly population was 9.7 whiteflies plant^{-1}, with a range of 1.8 to 28.4 whiteflies plant^{-1} in 2007 (Table 4). More

Table 3. Severity of cassava mosaic disease (CMD) infections on 38 cassava accessions during 2007 and 2008 cropping seasons.

Cassava accession	2007				2008			
	WAP				WAP			
	6	12	20	48	6	12	20	48
OFF 146	3.7	4.2	3.1	2.1	5.0	4.2	3.7	3.1
AFS 136	3.0	2.7	2.2	1.9	4.7	3.7	3.1	1.4
ADW 063	4.0	3.1	3.2	1.2	5.0	4.4	2.8	1.2
DMA 002	4.7	5.0	4.0	1.0	5.0	4.1	4.1	1.3
AFS 001	4.0	3.8	2.7	2.8	4.2	4.0	5.0	4.3
AFS 027	3.1	4.1	3.0	2.2	3.1	4.4	4.3	1.0
OFF 058	4.3	2.7	2.8	2.1	4.0	3.2	4.1	3.1
DMA 066	3.1	4.1	3.1	1.3	4.1	3.0	3.3	1.0
ADW 004	4.6	5.0	4.1	4.0	5.0	4.1	4.4	1.2
AFS 131	4.4	4.3	3.2	3.2	5.0	4.0	3.6	1.1
KW 148	2.1	3.1	2.0	1.0	3.8	3.2	2.8	1.2
KW 181	3.3	3.3	3.0	2.1	4.7	4.8	4.2	1.4
ADW 051	3.1	3.1	2.1	1.0	3.3	2.4	3.1	2.4
KW 001	1.5	2.8	1.8	1.0	4.0	2.1	3.3	1.1
KW 085	1.0	1.0	1.0	1.0	1.0	1.0	2.2	1.0
OFF 029	4.6	4.8	3.5	1.8	5.0	5.0	5.0	3.2
ADW 053	2.9	3.1	1.6	1.0	3.1	2.3	4.3	1.3
OFF 086	3.1	2.6	3.0	2.0	4.7	3.4	3.1	2.2
OFF 145	2.2	3.3	4.0	3.7	4.0	4.0	5.0	4.1
KW 161	3.1	2.4	3.1	1.0	4.2	3.1	4.1	2.0
OFF 025	1.8	3.9	4.3	2.0	5.0	4.3	4.8	3.2
OFF 023	2.8	2.6	2.0	2.0	3.1	4.0	1.7	1.3
OFF 063	1.0	1.7	2.1	1.0	2.3	2.0	1.8	1.2
AFS 048	2.1	2.1	1.7	1.0	2.1	2.3	2.1	2.3
KW 070	3.8	3.0	5.0	1.0	4.3	4.6	4.8	1.0
AFS 041	2.0	2.1	2.0	2.7	1.5	2.4	2.2	2.1
OFF 093	3.0	3.0	2.8	1.0	3.2	4.3	4.2	1.0
OFF 019	2.3	2.0	2.1	1.0	3.3	2.8	2.4	2.0
AFS 126	4.1	3.7	3.9	1.0	5.0	4.1	5.0	4.4
NKABOM[a]	2.4	2.3	1.5	1.1	1.9	1.9	2.0	2.0
OFF 136	2.1	3.0	2.1	1.7	2.2	2.1	2.1	2.0
UCC 517	2.7	2.1	2.0	1.8	3.4	4.1	3.2	2.0
UCC506	2.2	2.0	1.3	1.1	1.6	3.2	4.1	1.3
B. BOTAN[a]	1.0	1.0	2.0	1.0	1.6	3.5	2.6	2.4
CAPEVARS[a]	1.0	1.0	1.1	1.0	1.0	1.0	1.0	1.0
ADEHYE	1.0	1.0	1.2	1.0	1.0	1.0	1.0	1.0
UCC 470	2.0	2.2	2.0	1.2	2.2	1.4	2.0	2.0
UCC 153	1.8	2.0	3.0	1.3	3.0	2.2	2.3	2.0
Mean	2.8	2.9	2.6	1.6	3.4	3.2	3.3	1.9
Range	1 - 5	1 - 5	1 - 5	1- 4	1 - 5	1 - 5	1 - 5	1 - 4
%CV	39.3	35.9	38.5	52.9	37.1	33.3	35.3	52.6

WAP = Weeks after planting.

than 50% of the cassava accessions had values below the overall mean value for 2007. However, in 2008 at 6 WAP, the overall mean was 93.2 whiteflies plant^{-1} with a range of 25.4 to 209.9. The mean in 2008 was almost 10 times higher than that for 2007. Capevars had the highest mean number of whiteflies plant^{-1}, being 28.4 and 209.9

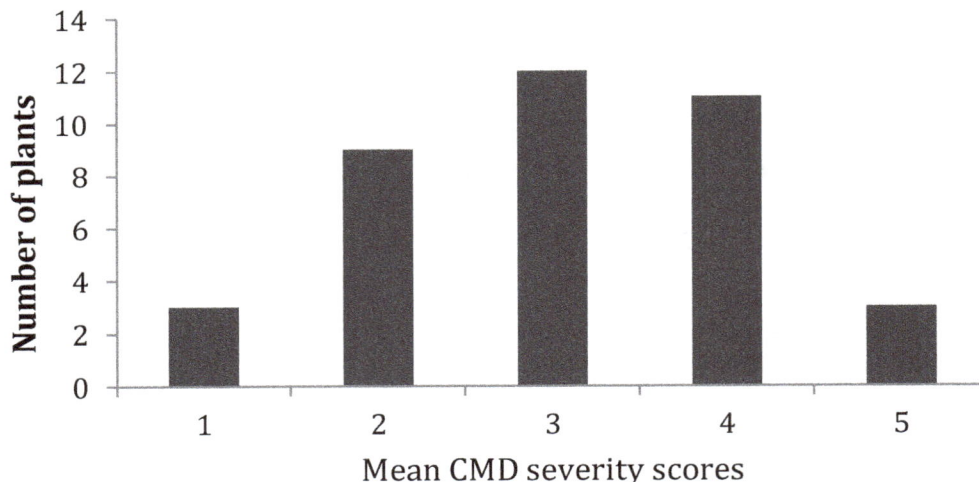

Figure 1. Distribution of 38 cassava accessions in CMD severity classes of 1 to 5. A score of 1 denotes no symptom while 5 indicates a display of severe mosaic symptoms, based on the mean CMD severity responses.

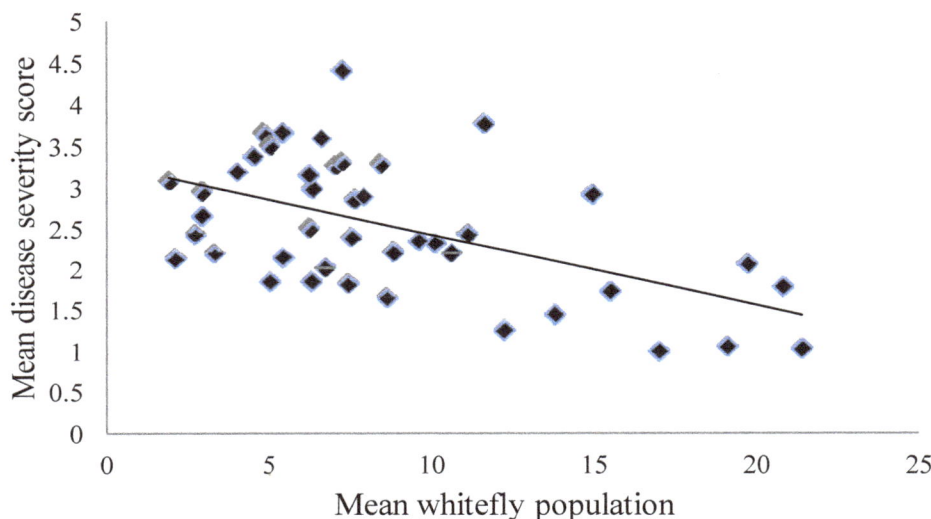

Figure 2. Relationship between mean whitefly population and mean score of cassava mosaic disease (CMD) during 2007 crop season (r = -0.543; $P < 0.05$).

for 2007 and 2008, respectively (Table 4). The lowest count was recorded on OFF 086 with a mean value of 1.8 and 25.4 for 2007 and 2008, respectively.

The whitefly population for most accessions reduced at 8WAP for both years. The mean values were 8.7 for 2007 and 33.9 for 2008. Adehye (24.4) and AFS 001 (52.7) had the highest mean counts for 2007 and 2008, respectively.

The whitefly population reduced further for most of the genotypes at 10 WAP. The mean counts ranged from 1.1 to 18.6 and 5.5 to 35.3 for 2007 and 2008, respectively (Table 4). The most infested genotypes were KW 148 for

2007 and AFS 001 for 2008. Overall, AFS 027 was the least infested by whiteflies and Capevars was the most infested in 2007. However, in 2008, genotype AFS136 was the least infested and Capevars cultivar was again the most infested. The infestation in 2008 also was clearly higher than in 2007.

Relationships between whitefly population and disease severity score

Interestingly, in both 2007 and 2008 crop seasons (Figures 2 and 3), the mean whitefly populations

Table 4. Mean number of adult whiteflies on 38 genotypes of cassava during 2007 and 2008 crop seasons.

Cassava accession	2007 WAP				2008 WAP			
	6	8	10	Mean	6	8	10	Mean
OFF 146	7.6	9.0	4.5	7.0	53.9	31.4	26.1	37.1
AFS 136	18.0	11.4	3.8	11.1	28.0	29.0	5.6	20.8
ADW 063	12.6	5.6	4.7	7.6	55.0	25.1	15.8	32.0
DMA 002	3.0	5.6	5.7	4.8	43.0	46.1	12.5	33.9
AFS 001	10.6	7.6	3.4	7.2	31.4	52.7	35.3	39.8
AFS 027	2.2	2.4	1.1	1.9	49.6	33.0	9.3	30.6
OFF 058	3.4	3.0	2.4	2.9	56.6	23.2	15.4	31.7
DMA 066	11.6	8.4	3.8	7.9	79.9	46.9	27.0	51.2
ADW 004	8.8	8.0	4.8	7.2	49.6	31.9	27.5	36.3
AFS 131	16.2	13.3	5.2	11.6	39.2	27.4	11.8	26.1
KW 148	25.2	15.3	18.6	19.7	98.7	38.5	9.7	48.9
KW 181	26.4	10.8	7.5	14.9	51.7	42.2	25.8	39.9
ADW 051	13.6	8.2	8.6	10.1	70.3	33.6	13.6	39.2
KW 001	28.4	17.2	16.9	20.8	90.3	44.5	25.7	53.5
KW 085	5.2	6.0	4.9	5.4	90.0	28.4	8.6	42.3
OFF 029	6.2	5.6	4.5	5.4	97.6	33.3	11.8	47.6
ADW 053	20.4	16.6	14.0	17.0	71.7	42.9	12.3	42.3
OFF 086	1.8	3.6	3.4	2.9	25.4	30.0	27.2	27.5
OFF 145	10.0	8.8	6.3	8.4	81.8	25.4	11.7	39.6
KW 161	6.5	8.6	7.5	7.5	106.8	31.7	24.4	54.3
OFF 025	5.8	8.2	4.9	6.3	47.0	27.1	12.3	28.8
OFF 023	10.3	10.8	7.8	9.6	127.5	26.5	12.4	55.4
OFF 063	14.0	13.4	14.0	13.8	89.7	45.5	5.5	46.9
AFS 048	14.2	16.6	15.6	15.5	163.8	35.9	15.0	71.6
KW 070	3.4	4.6	3.9	4.0	55.6	21.3	15.0	30.6
AFS 041	11.2	7.6	12.9	10.6	162.8	33.5	10.0	68.8
OFF 093	2.6	2.2	3.3	2.7	134.9	31.6	24.1	63.5
OFF 019	6.6	7.6	4.7	6.3	143.7	34.3	10.1	62.7
AFS 126	4.6	8.6	5.5	6.2	105.6	41.5	14.6	53.9
NKABOM[a]	8.0	8.8	9.6	8.8	165.7	32.6	11.9	70.1
OFF 136	5.8	8.2	8.2	7.4	175.9	32.7	12.1	73.5
UCC 517	3.0	1.0	2.2	2.1	128.2	35.1	9.1	57.5
UCC506	10.2	6.0	9.7	8.6	161.5	22.4	14.7	66.2
B. BOTAN[a]	11.6	16.2	8.9	12.2	162.5	26.5	17.0	68.6
CAPEVARS[a]	16.8	24.4	16.1	19.1	140.3	38.4	19.6	66.1
ADEHYE	26.6	21.6	16.0	21.4	209.9	33.9	26.7	90.2
UCC 470	7.4	1.4	6.2	5.0	93.6	43.9	27.0	54.8
UCC 153	2.2	7.0	10.8	6.7	97.9	30.1	26.2	51.4
Mean	9.7	8.7	7.6	8.7	93.2	33.9	16.4	47.8
Range	1.8 -28.4	1.0 - 24.4	1.1 - 18.6	1.9 - 21.4	25.4-209.9	21.3- 52.7	5.5 - 35.3	20.8-90.2
% CV	75.3	60.9	59.2	59.8	49.5	33.9	44.5	32.4

[a] Released varieties; WAP=weeks after planting.

significantly (P <0.05) negatively correlated with mean CMD severity scores. That is, on the average, higher populations of whitefly were found on the resistant cultivars than on the susceptible cultivars.

Detection by PCR of 4 variants of ACMV

All four ACMV-specific primer pairs (associated with the four variants of ACMV), produced allelic bands in the

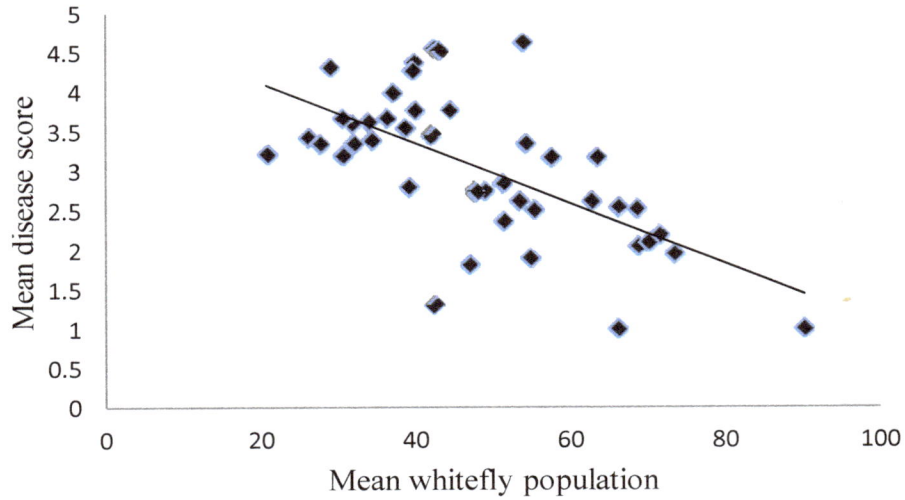

Figure 3. Relationship between mean whitefly population and mean score of cassava mosaic disease (CMD) during 2008 crop season. (r = -0.634; *P* < 0.05).

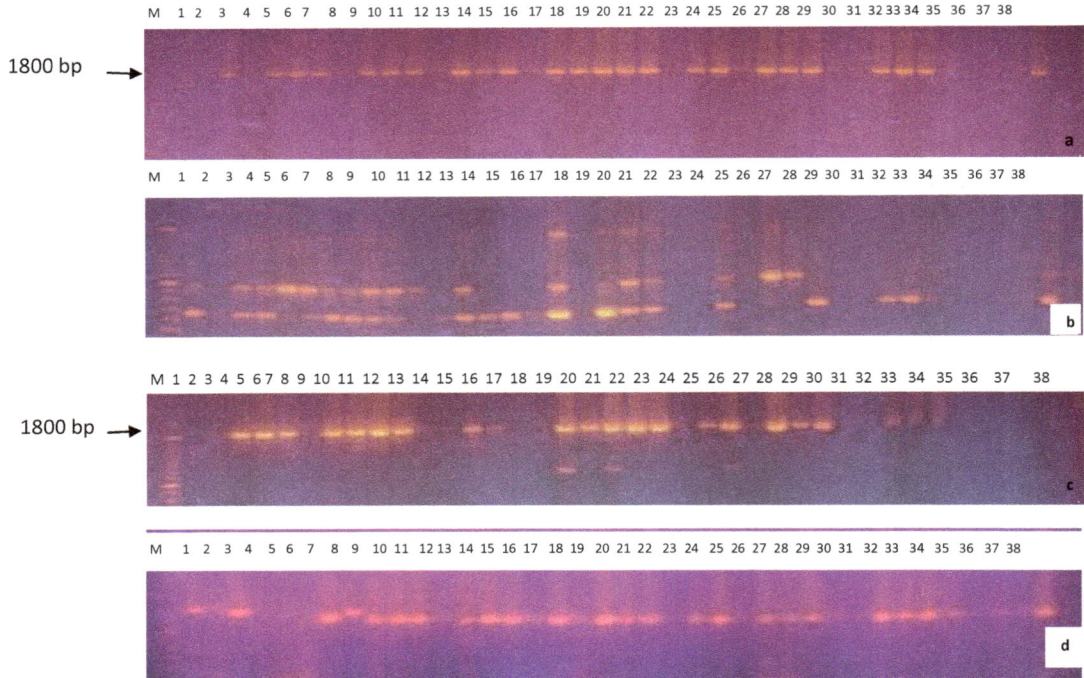

Figure 4. PCR amplification products for ACMV-specific primers: ACMV-F1/ACMV-R1 (a), ACMV-F2/ACMV-R2 (b), ACMV-AL1/F/ACMV-ARO/R(c) and ACMV-1/ACMV-2 (d) - resolved by PAGE and stained with ethidium bromide. M = 1kb+ ladder; 1-38 represent the various cassava accessions. Arrow indicates specific band for ACMV resistance.

accessions. The ACMV-specific primer pair that was most efficient in detecting the virus was ACMVF1/ACMV-R1, which detected the virus in 34 (89.5%) out of the 38 cassava accessions, whilst the primers ACMV-1/ ACMV-2, ACMV-F2/ACMV-R2, and ACMV-AL1/F/ACMV-ARO/R detected the virus in 26(68.4%), 24(63.2%) and 22(57.9%) accessions, respectively (Figure 4). With the exception of genotype Capevars, all the samples were

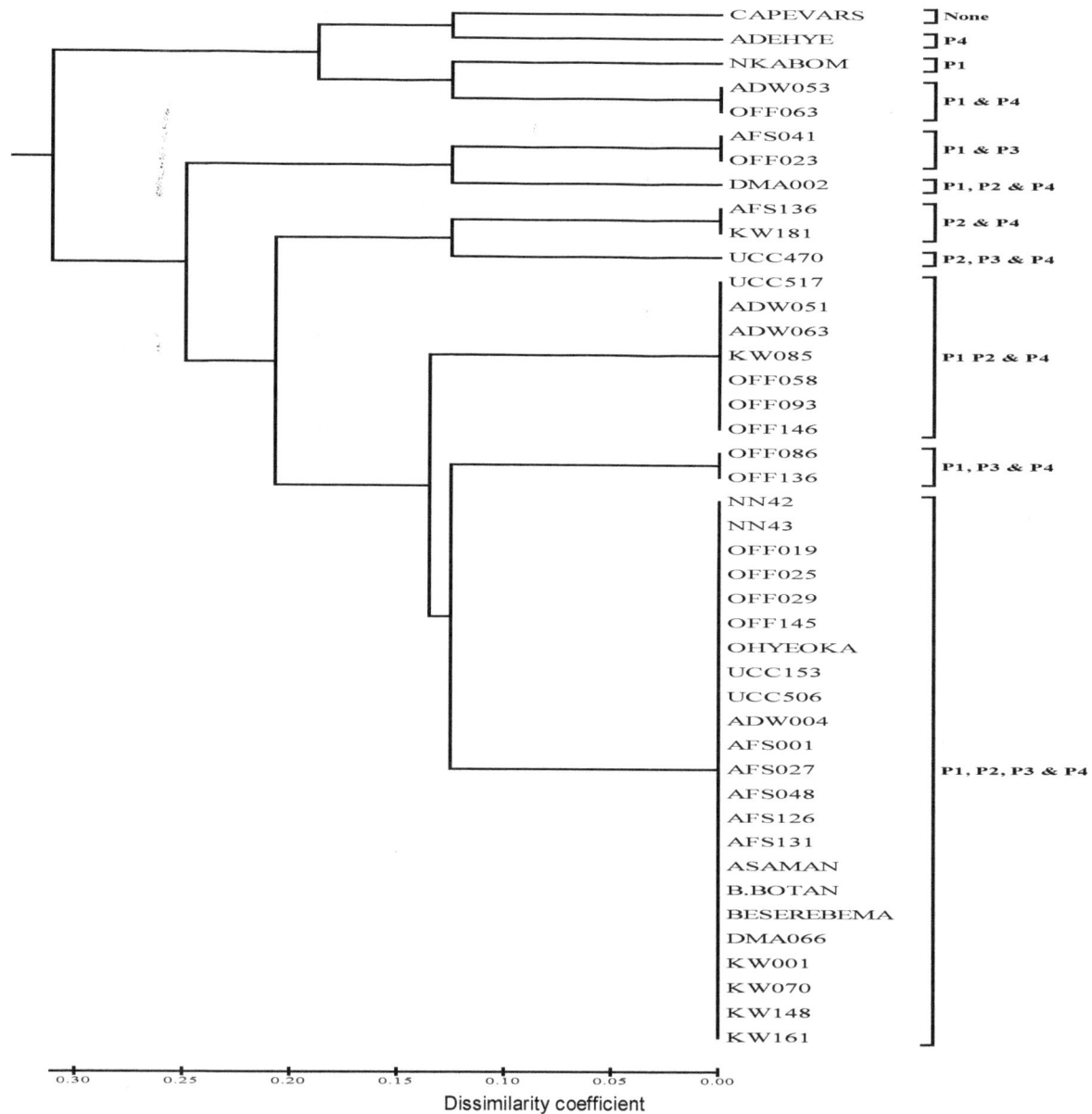

Figure 5. Genetic differences among the 38 cassava accessions based on PCR products of four ACMV primer pairs using the unweighted pair group method with arithmetic averages. P1, P2, P3 and P4 represent ACMV variants ACMV1, ACMV2, ACMV-AL, and AMCV3, respectively.

infected with one or more of the ACMV strains. The cassava genotypes were infected with two or more of the ACMV variants, with the exception of Adehye and Nkabom, which were infected with only one ACMV variant (ACMV1 and ACMV3, respectively).

The cassava genotypes were clustered into 11 groups at a similarity coefficient of 0.13 based on the PCR amplification products, indicating that the cassava genotypes were genetically diverse (Figure 5). The cluster size ranged from 1 to 23 cassava accessions. Cluster 11 had the highest number of accessions (Figure 5).

Detection of *CMD2* resistance gene

From the results obtained from PCR reactions with ACMV-specific primers and field screening for CMD resistance, four genotypes were selected for further screening with markers associated with the *CMD2* gene that confers resistance to CMD to ascertain their source of resistance. All the four accessions selected had bands of alleles of all the four markers associated with the *CMD2* gene (Figure 6). However, the bands present were more intense in two markers (NS169 and RME1), which

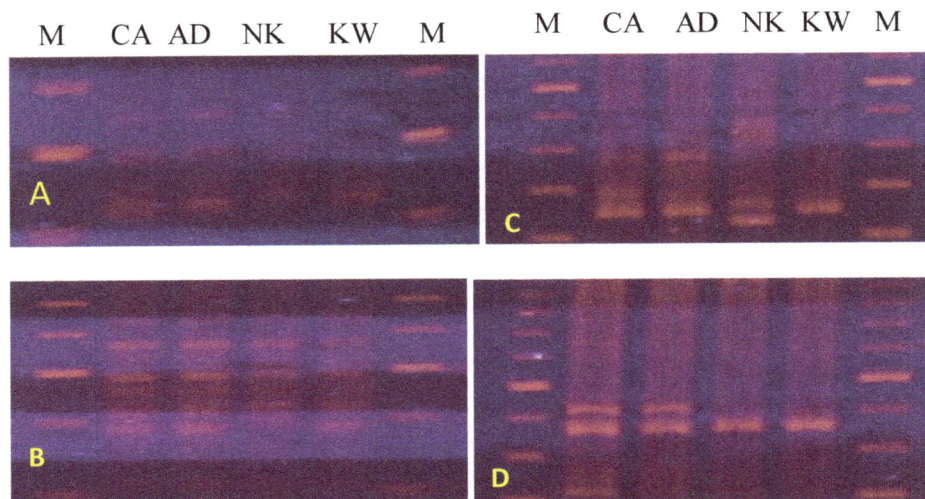

Figure 6. PCR amplification products of four markers associated with *CMD2* resistance gene (SSRY28 (A), NS158 (B), NS169 (C) and RME1 (D) resolved by PAGE stained with ethidium bromide among 4 cassava accessions - Capevars (CA), Adehye (AD), Nkabom (NK) and KW085 (KW). M is the standard marker.

are closer to the gene than the SSRY28 and NS158 markers, indicating that they were more efficient in detecting the *CMD2* gene than the latter two.

DISCUSSION

Morphological screening of the 38 cassava genotypes for CMD resistance based on the 1-5 disease rating (IITA, 1990; Ariyo et al., 2005) and classification according to Lokko et al. (2005) revealed one highly resistant geno-type (Capevars) and three moderately resistant geno-types (Adehye, Nkabom and KW 085) (Table 3). How-ever, the subsequent resistance screening using PCR with CMG strain-specific primers showed that only one genotype, Capevars, was resistant whilst the others were infected with ACMV (Figures 4 and 5). This suggests that the three genotypes (Adehye, Nkabom and KW 085) are tolerant to ACMV infection whereas Capevars was a resistant genotype. Thus, field selection of resistance should be complemented with virus detection methods such as PCR. The reason could be that the field resis-tance, as shown by lack of symptoms, is not necessarily an indication of resistance to virus infection as has been reported by Ogbe (2001). Therefore, the mean symptom severity scores calculated for breeding lines has a limitation, in that, the virus incidence and symptom severity are not clearly distinguished; and symptomless plants plants could be CMD-free 'escapes', or they could be extremely tolerant (Thresh and Cooter, 2005). More-over, a low average score for a progeny or selection could mean that a few plants are infected and show severe symptoms, or that many succumb but are only slightly affected.

The ACMV-specific primer ACMVF1/ACMV-R1 was more efficient in detecting the virus in the cassava genotypes, since it detected the virus in more samples than the primers ACMV-1/ ACMV-2, ACMV-F2/ACMV-R2, and ACMV-AL1/F/ACMV-ARO/R. Whilst primer ACMVF1/ACMV-R1 detected the virus in 34 (89.5%) out of the 38 cassava accessions, the primers ACMV-1/ACMV-2, ACMV-F2/ACMV-R2, and ACMV-AL1/F/ACMV-ARO/R detected the virus in 26 (68.4%), 24 (63.2%) and 22 (57.9%) accessions respectively. In screening F_1 progeny of cassava against CMD infection, Lokko et al. (2005) also observed that the ACMV primer ACMV-F1/ACMV-R1 detected the virus in more samples than the primer ACMV-AL F/ACMV-AROR. This suggests that the ACMV1 strain detected by the primer ACMVF1/ACMV-R1 as reported by Zhou et al. (1997) is the most dominant virus among the ACMV variants detected in the study.

The detection of the resistance gene (*CMD2*) using linked SSR markers, in the four field-resistant cassava genotypes (Capevar, Adehye, KW058 and Nkabom) suggests that the *CMD2* gene is, at least, partly responsible for both CMD resistance and field tolerance. In this case Capevars can be said to be a highly resistant genotype, whereas Adehye, KW058 and Nkabom, which showed mild field symptoms are tolerant genotypes. The dominant nature of *CMD2* and its effectiveness against a wide spectrum of viral strains makes its deployment very appealing in protecting cassava against the actual or potential ravages of CMD in Africa (Boateng, 2010). Knowledge of the markers associated with this resistance gene will also facilitate the use of marker- assisted selection in a cassava breeding programmes for the development of resistant lines. It was observed in this study, that markers RMEI and NS158 were more reliable for the detection of the CMD2 resistance gene than markers SSRY28 and NS158, as the former gave more intense bands in the gel than the latter two.

Capevars, the CMD-resistant cassava cultivar has since been released (Tetteh et al., 2005). Currently, the Government of Ghana, through the Ministry of Food and Agriculture, is multiplying the Capevars cultivar to be distributed to farmers, especially, those from the Western Region (J.P. Tetteh, pers. comm.).

The highest mean severity score for 2007 was recorded at 12 WAP. This finding agrees with Leuschner (1978) and Ogbe et al. (1996) that high incidence of CMD is achieved at 12 WAP. However, in 2008 the highest mean severity was recorded at 6 WAP. It might be due to the fact that the cuttings used were obtained from the previous crop, and these might have been already infected. This confirms the reports of Fargette et al. (1988) that plants are generally more susceptible to secondary infection.

Most (35 out of 38) of the cassava genotypes showed mixed infection with the four different ACMV variants, and this can have serious consequences for the management of CMD. It has been reported that mixed infections provide the precondition for recombination, which may contribute to the appearance of more severe viral strains (Ribeiro *et al.*, 2003). Zhou et al. (1997) has shown that EACMV-Ug, associated with the severe cassava mosaic disease in Uganda, has arisen by interspecific recombination of EACMV and ACMV. Mixed genotypes infections have been reported in many host-pathogen interactions (Read and Taylor, 2001; Hodgson et al., 2004; Schurch and Roy, 2004).

The whitefly, *Bemisia tabaci*, is one of the most important insect pests in world agriculture, because of its direct feeding, contamination from honeydew, and ability to transmit plant viruses (Perrings, 2001). Additional evidence of differences in whitefly infestation among a range of cassava accessions at different locations in Ghana were also found in the present study. The adult whitefly population was high at six WAP in both years. A higher number of whiteflies were found on resistant genotypes in this study, which agrees with Otim Nape et al. (2005), who recorded higher populations of *B. tabaci* on the cassava mosaic disease-resistant genotypes than in susceptible ones. Similar observations have been made by Legg et al. (2003), and are attributed to the whitefly preference for the resistant varieties of cassava. The leaves of resistant plants were broader and softer than the susceptible ones, whose leaves were mis-shapen, highly reduced and showed severe mosaic symptoms. According to Sserubombwe et al. (2001), Omongo (2003) and Ariyo et al. (2005), such leaves are usually avoided by the whitefly and this might account for the whitefly preference for the resistant plants in this study. Otim-Nape et al. (1994) has also reported the lack of any significant correlation between whitefly numbers

and mosaic severity when they studied the effects of African cassava mosaic geminivirus on the main cassava varieties grown in three districts of western Uganda. On the contrary, we observed a significant negative correlation between the whitefly population and the CMD severity scores. This further supports the findings earlier made by Sserubombwe et al. (2001), Omongo (2003) and Ariyo et al. (2005).

Conclusion

Out of 38 cassava genotypes screened against CMG infection, three tolerant cassava genotypes (Adehye, KW058 and Nkabom) and a highly resistant genotype, (Capevars) were identified. Apart from Capevars, between 1 and 4 variants of ACMV (ACMV1, ACMV2, ACMV-AL, and ACMV3) were detected in the cassava genotypes including the tolerant ones. This suggests that field selection of resistance should be complemented with virus detection methods such as PCR test. Most (35 out of 38) of the cassava genotypes showed mixed infections with two or more ACMV variants, which could have serious consequences for the management of the CMD in Ghana. A higher number of whiteflies were found on resistant genotypes than the susceptible genotypes in this study, which confirms that the presence of whiteflies per se may not be an indication of possible infection with the ACMV.

Conflict of Interests

The author(s) have not declared any conflict of interests.

REFERENCES

Abdullahi I, Atiri GI, Dixon AGO, Winter S, Thottapily G (2003). Effects of Cassava Genotype, Climate and the *Bemisia tabaci* vector population on the development of African Cassava Mosaic Geminvirus (ACMV). Acta Agron. Hung. 51(1):37-46.

Akano AO, Dixon AGO, Mba C, Barrera E, Fregene M (2002). Genetic mapping of a dominant gene conferring resistance to the cassavamosaic disease (CMD). Theor. Appl. Genet. 105:521-525.

Akinlosotu TA (1985). Studies on the control of the cassava mealybug (*Phenacoccus manihoti*) and green spider mite (*Mononychellus tanajoa*) in south-western Nigeria. J. Root Crops 9:33-43.

Ariyo OA, Dixon AGO, Atiri GI (2005). Whitefly *Bemisia tabaci* (*Homoptera*: Aleyrodidae) infestation on cassava genotypes grown at different ecozones in Nigeria. J. Econ. Entomol. 98(2):611-617.

Asamoa GK (1973). Soils of the proposed farm site of the University of Cape Coast. Soil Research Institute (C.S.I.R, Ghana) Technical Report No. 88

Bi H, Aileni M, Zhang P (2010) .Evaluation of cassava varieties for cassava mosaic disease resistance jointly by agro-inoculation screening and molecular markers. Afr. J. Plant Sci. 4(9):330-338.

Boateng PA (2010). Using marker-assisted selection technique as a tool to identify cassava mosaic disease resistant cultivars in first-backcross populations. MSc. Thesis, Department of Crop and Soil Sciences, Kwame Nkrumah University of Science and Technology, Kumasi, Ghana.

Braima J, Yaninek J, Neuenschwander P, Cudjoe A, Modder W, Echendu N, Muaka T (2000). Pest Control in Cassava Farms: IPM Field guide for extension agent. Lagos: Wordsmithes Printers.

Clerk GC (1974). Crops and their diseases in Ghana. Tema, Ghana Publishing Corporation.

Dellaporta SL, Woods J, Hicks JB (1983). A plant DNA mini-preparation: version II. Plant Mol. Biol. Rep. 1:19-21.

Fargette D, Fauquet C, Thouvenel JC (1985). Field studies on the spread of African cassava mosaic. Ann. Appl. Biol. 106:285-294.

Fargette D, Fauquet C, Thouvenel JC (1988) Yield losses induced by African cassava mosaic virus in relation to the mode and the date of infection. Trop. Pest Manag. 34:89-91.

Fargette D, Colon LT, Bouveau R, Fauquet C (1996). Components of resistance of cassava to African cassava mosaic virus. Eur. J. Plant Pathol. 102:645-654.

Fauquet C, Fargette D (1990). African Cassava Mosaic Virus: Etiology, Epidemiology, and Control. Laboratoire de Phytovirologie, ORSTOM, Abidjan, Ivory Coast. Plant Dis. 74:404-411.

Fauquet C, Mayo M, Maniloff J, Desselberger U, Ball L eds (2005). *Virus Taxonomy VIIIth Report* of the International Committee on Taxonomy of Viruses. London: Elsevier/Academic.

Fauquet C, Stanley J (2003). Geminivirus classification and nomenclature progress and problems. Ann. Appl. Biol. 142:165-189.

Fondong VN, Pita JS, Rey C, Beachy RN, Fauquet, C. M. (1998). First report of the presence of East African cassava mosaic virus in Cameroon. Plant Dis. 82:1172.

Fondong VN, Pita JS, Rey MEC, de Kochko A, Beachy R N, Fauquet CM (2000). Evidence of synergism between African cassava mosaic virus and a new double-recombinant geminivirus infecting cassava in Cameroon. J. Gen. Virol. 81:287-97.

Fontes EPB, Luckow VA, Hanley-Bowdoin L (1992). A geminivirus replication protein is a sequence specific DNA binding protein. Plant Cell 4:597-608.

Food and Agriculture Organisation (FAO) (2014). Empowering cassava value chain actors to contribute to increased food security. Rome, Italy: FAO. Retrieved from Food and Agriculture Organisation website:http://www.fao.org/archive/from-the-field/detail/en/c/214024/

Food and Agricultural Organisation of United Nations (FAO) (2008). Facts and figures. Rome, Italy http// www.faostat.org.

Fregene M, Okogbenin E, Mba C, Angel F, Suarez MC, Guitierez J, Chavarriaga P, Roca W, Bonierbale M, Tohme J (2001). Genome mapping in cassava improvement: Challenges, achievements and opportunities. Euphytica 120:159-165.

Hahn SK, Terry ER, Leuschner K (1980). Breeding cassava for resistance to cassava mosaic virus disease. Euphytica 29:673-683.

Hahn SK, John C, Isoba G, Ikoun T (1989). Resistance breeding in root and tuber crops at the International Institute for Tropical Agriculture (IITA), Ibadan Nigeria. Crop Prot. 8:147-168.

Harrison BD, Zhou X, Otim-Nape G W, Liu Y, Robinson DJ (1997). Role of a novel type of double infection in the geminivirus-induced epidemic of severe cassava mosaic in Uganda. Ann. Appl. Biol. 131(3):437-448.

Hill BG (1968). Occurrence of Bermisia tabaci in the field and its relation to the leaf curl disease of tobacco. South Afr. J. Agric. 11:583-594.

Hodgson DJ, Hitchman RB, Vanbergen AJ, Hails RS, Possee RD, Cory JS (2004). Host ecology determines the relative fitness of virus genotypes in mixed-genotype nucleopolyhedrovirus infection. J. Evol. Biol. 17:1018-1025.

IITA (International Institute of Tropical Agriculture), (1990). Annual Reports, IITA, Ibadan, Nigeria.

Lamptey JNL, Okoli OO, Rossel HW, Frimpong-Manso PP (2000). A method for determining tolerance of cassava genotypes to African cassava mosaic disease in the screenhouse. Ghana J. Agric. Sci. 33:29-32.

Lamptey JNL, Okoli OO, Frimpong-Manso PP (1998). Incidence and severity of African cassava mosaic disease (ACMD) and cassava bacterial blight (CBB) on some local and exotic cassava varieties in different ecological zones of Ghana. Ghana J. Agric. Sci. 31 (1):35-43.

Leather RJ (1959). *Disease of economic plants in Ghana other than cocoa*. Ghana Ministry of Food and Agriculture, Division Bulletin No 1.

Legg JP, Mallowa S, Sseruwagi P (2003). First report of physical damage to cassava caused by the whitefly, *Bermisia tabaci* (Gennadius) (Gehiptera: Sternorrhyncha: Aleyrodidae). Book of Abstracts. 3rd International Bermisia Workshop. Barcelona 17-20 March, 2003.

Legg J, Fauquet C (2004). Cassava mosaic geminiviruses in Africa. Plant Mol. Biol. 56:585-599.

Leuschner K (1978). Whiteflies: biology and transmission of African cassava mosaic disease. In: T. Brekelbaum, A. Bellotti, T. C. Lozano (Eds), Proceedings of the Cassava Protection Workshop (pp 51-58).Columbia: CIAT.

Liu K, Muse SV (2005). PowerMarker: Integrated analysis environment for genetic marker data. Bioinformatics 21(9):2128-2129.

Lokko Y, Danquah EY, Offei, SK, Dixon AGO, Gedil MA (2005). Molecular markers associated with a new source of resistance to the cassava mosaic disease. Afr. J. Biotech. 4 (9):873-881.

Maruthi M N, Seal S, Colvin J, Briddon R W, Bull S E (2004). *East African cassava mosaic Zanzibar virus* - a recombinant begomovirus species with a mild phenotype. Archives Virol. 149:2365-2377.

Matthew AV, Muniyappa,V (1992). Purification and characterization of *Indian cassava mosaic virus*. Phytopathology 135:99-308

Moses E, Asafu-Agyei JN, Adubofour K, Adusei A (2007). *Guide to identification and control of cassava diseases*. Kumasi, Ghana: CSIR-Crops Research Institute.

Ministry of Food and Agriculture (MoFA) (2011). Agriculture in Ghana: Factys and Figures. *Annual Report*. Accra, Ghana: Statistcis, Research and Information Directorate, MoFA.

Nei M (1983). Estimation of genetic distances and phylogenetic trees from DNA analysis. Proc. 5th World Cong. Genet Appl. Livestock Prod. 21:405-412.

Nicholas RFW (1947). Breeding cassava for virus resistance. East Afr. Agric. J. 15:154-160.

Offei SK, Owuna-Kwakye M, Thottappilly G (1999). First Report of East African Cassava Mosaic Begomovirus in Ghana. Plant Dis. 83 (9):877.

Ogbe FO, Nnodu E C, Odurukwe SO (1996). Control of African cassava mosaic disease incidence and severity. Tropical Sci. 36:174-181.

Ogbe FO, Atiri GI, Robinson D, Winter S, Dixon AO, Quin FM, Thottappilly G (1999). First report of East African cassava mosaic begomovirus in Nigeria. Plant Dis. 83:398-402.

Ogbe FO (2001). Survey of cassava begmoviruses in Nigeria and the response of resistant cassava genotypes to African cassava mosaic begmovirus infection. (PhD thesis, University of Ibadan, Ibadan, Nigeria).

Omongo CA (2003). *Cassava whitefly, Bemisia tabaci, behaviour and ecology in relation to the spread of the cassava mosaic pandemic in Uganda*. (Ph.D. Thesis, University of Greenwich, UK).

Otim. M, Legg J, Kyamanywa S, Polaszek A, Gerling D (2005). Population dynamics of *Bemisia tabaci* (Homoptera: Aleyrodidae) parasitoids on cassava mosaic disease-resistant and susceptible varieties. Biocontrol Science and Tech. 16(2):205-214.

Otim-Nape GW, Shaw MW, Thresh JM (1994). The effects of African cassava mosaic gemini virus on the growth and yield of cassava in Uganda. Trop. Sci. 34:43-54.

Payne RW, Murray DA, Harding SA, Baird DB, Soutar DM (2009). Genstat for Windows: *Introduction*. In: Hemel Hempstead, Hemel Hempstead, UK: VSN International.

Perring TM (2001). The *Bemisia tabaci* species complex. Crop Protect. 20:725-737.

Read A, Taylor LH (2001). The ecology of genetically diverse infections. Science 292:1099-1102.

Ribeiro SG, Ambrozevicius LP, Avila AC, Bezerra IC, Calegario RF, Fernandes JJ, Lima MF, de Mello RN, Roche H, Zerbini FM (2003). Distribution and genetic diversity of tomato infecting begomoviruses in Brazil. Archives of Virol. 148:281-295.

Schurch S, Roy BA (2004). Comparing single- vs. mixed-genotype infections of *Mycophaerella graminicola* on wheat: effects on pathogen virulence and host tolerance. Evolutionary Ecol. 18:1-14.

Sneath PHA, Sokal RR (1973). Numerical taxonomy- the principles and practice of numerical classification. San Francisco: Freeman.

Sserubombwe WS, Thresh JM, Otim-Nape GW, Osiru DSO (2001). Progress of cassava mosaic virus disease and whitefly vector population in single and mixed stands of four cassava varieties grown under epidemic conditions in Uganda. Ann. Appl. Biol. 135:161-170.

Terry ER, Hahn SK (1990). The effect of cassava mosaic disease on growth and yield of a local and an improved variety of cassava.

Tropical Pest Manag. 26:34-37.

Tetteh JP, Carson AG, Taah KJ, Addo-Quaye AA, Opoku-Asiama Y, Fiscian P, Buah JN, Asare-Bediako E, Tetteh EO, Okai D, Teiko O (2005). Evaluation report on cassava elite lines proposed for release, presented to National Varietal Release Committee. Department of Crop Science, School of Agriculture, University of Cape Coast. p. 16.

Thresh JM, Cooter TJ (2005). Strategies for controlling Cassava Mosaic Disease in Africa. Plant Pathol. 54:587-614.

Thresh JM, Otim-Nape GW, Nichols RFW (1994). Strategies for controlling African cassava mosaic geminivirus. Adv. Dis.Vector. Res. 10:215-236.

Thresh JM, Otim-Nape GW, Legg JP, Fargette D (1997). *African cassava mosaic virus* disease: The magnitude of the problem. Afr. J. Root Tuber Crops 2:13-18.

Webster GL (1994) Synopsis of the genera and suprageneric taxa of Euphorbiaceae. Ann. Missouri Bot. Garden 81:33-144.

Zhou X, Liu Y, Calvert L, Munoz C, Otim-Nape GW, Robinson DJ, Harrison BD (1997). Evidence that DNA-A of a geminivirus associated with severe cassava mosaic disease in Uganda has arisen by interspecific recombination. J. Gen. Virol. 78 (8):2101-11.

Zhou X, Robinson DJ, Harrison BD (1998). Types of variation in DNA A among isolates of *East African cassava mosaic virus* from Kenya, Malawi and Tanzania. J. Gen. Virol. 79:2835-2840.

25

Molecular characterization of influenza A (H7N9) virus from the first imported H7N9 infection case in Malaysia

Jeyanthi S.[1], Tengku Rogayah T. A. R.[1], Thayan R.[1], Az-UlHusna A.[1], Aruna A.[1], Khebir B. V.[2], Thevarajah B.[3], Maria S.[4] and Zainah S.[1]

[1]Virology Unit, Institute for Medical Research, Jalan Pahang, 50588 Kuala Lumpur, Malaysia.
[2]Disease Control Division, Ministry of Health Malaysia, Putrajaya. Malaysia.
[3]Kumpulan Perubatan Johor Hospital,Kota Kinabalu, Sabah, Malaysia.
[4]Sabah Health Department, Kota Kinabalu, Sabah Malaysia.

H7N9 is an avian strain of the species Influenza virus that circulates among avian populations. Occasionally, some variants of this strain were known to infect humans. On March 30, 2013, a novel avian influenza A H7N9 virus that infects human beings was identified in China. In February 2014, the first case of H7N9 infection outside China was reported in Malaysia involving a Chinese tourist. This study was aimed to characterize the first case of H7N9 in Malaysia by means of molecular identification, sequencing of hemagglutinin (HA) and neuraminidase (NA) genes, and phylogenetic analysis. The patient was confirmed positive for H7N9 virus by real-time RT-PCR (rRT-PCR). Subsequently, the samples were sequenced and mutation analysis identified R65K, E122K, L186I and N285D mutations in *HA* gene and M26I, R78K and V345I mutations in *NA* gene. We reported the emergence of a new mutation L186I, not found in the current database of any H7N9 sequences. Mutations associated with drug resistance were not found in this patient. Phylogenetic analysis revealed that the *HA* gene is closely related to the group of strains from Guangzhou, whereas *NA* gene is closely related to the group of strains from Guangdong. The present study provides crucial information on the first case of H7N9 outside China and the diversity of this strain from other reported H7N9 strains by molecular analysis.

Key words: H7N9, avian strain, China, molecular analysis, influenza virus.

INTRODUCTION

In March 2013, cases of novel Influenza A (H7N9) were first identified in China involving three urban residents of Shanghai and Anhui (Shuihua et al., 2013). The novel avian-origin reassortant influenza A (H7N9) virus was identified in patients who were hospitalized due to severe lower respiratory tract disease of unknown cause (Gao et al., 2013). As at 3 March 2014, the fatality rate was 379 cases (WHO, 2013). Most of these infections are believed to have occurred due to exposure to infected poultry or contaminated environments. To date, no evidence of sustained person-to-person transmission of H7N9 has been reported (WPRO, 2014). However, there were several family-clustered cases with unsustained person-to-person transmission of H7N9 reported. As of 1st December 2013, four family-clusters had been identified in three areas in China (Li et al., 2014).

On February 12, 2014, Malaysia confirmed and reported the first case of influenza A (H7N9) outside China (Centre for Disease Control, 2014). The import case involved a 67-year-old female Chinese tourist, who had travelled from Guangdong, China, to Kuala Lumpur on February 4, then to Sandakan, Sabah the next day before going to Kota Kinabalu, Sabah on February 6. The patient was previously treated by a general practitioner (GP) for symptoms of fever, cough, fatigue and joint pain in China on January 30, 2014, four days before travelling to Malaysia. On February 5, she sought treatment at a GP in Sandakan for similar complaints and was given symptomatic treatment. On February 7, as her condition worsened, she was brought to a district hospital in Sabah and upon family request; she was referred to a private hospital in Sabah and admitted to intensive care unit (ICU). On February 9, the first specimen was tested for suspected avian influenza A (H7N9) and on February 11, our team at Institute for Medical Research (IMR) in Kuala Lumpur tested her second specimens to be positive for H7N9 avian influenza virus nucleic acid. Currently, patient has recovered and was discharged from the hospital. In this study, we report the identification of the first case of influenza A (H7N9) Malaysia, which is also the first case outside China and molecular characterization of the virus by direct sequencing of the hemagglutinin (*HA*) and neuraminidase (*NA*) genes.

MATERIALS AND METHODS

Clinical specimen

Two types of clinical specimens from the patient which include throat swab (TS) and tracheal aspirate (Tasp) were obtained from the private hospital, Kota Kinabalu, Sabah where the patient was admitted to ICU. The first batch of specimens consisting of 2TS (TS1 and TS2) were received at ambient temperature on February 9, 2014. The second batch consisting of 2 Tasp and 1 TS (Tasp1, Tasp2 and TS3) received on February 11, was sent in ice.

Isolation of viral nucleic acid

Viral RNA extraction was performed using the QIAamp Viral RNA Mini Kit (Qiagen), according to the manufacturer's instructions. The isolation procedure was based on spin-column method. A final elution volume of 50 µL containing viral RNA from each specimen was used as template in the one step Real-time RT- PCR amplification.

Real-time reverse transcriptase PCR (rRT-PCR)

The clinical specimens received on February 9, were tested for flu A, Flu B, H1, H3, H7 CNIC, N9 CNIC, pdmA and pdmH1 by real-time RT-PCR assay using sets of specific primers and probes obtained from Centre of Disease Control (CDC), Atlanta and Chinese National Influenza Center (CNIC). Clinical specimens received on February 11 were tested together with the previous specimens using Flu A, H7 CDC, H9 CDC, H7 CNIC and H9 CNIC primers and probes. Positive controls, extraction controls and reagent controls were included in each run. All amplification reactions were performed using the SuperScript III one-step RT-PCR kit (Invitrogen, USA) in a 96-well real-time PCR thermal cycler (Bio Rad, USA). The assay was undertaken at 50°C for 30 min, 95°C for 2 min and 45 cycles of 95°C for 15 s and 55°C for 30 s. Reaction setup was composed of 12.5 µL of 2x RT-PCR Mix, 0.5 µL of each primers (40 µM), 0.5 µL of respective probes (20 µM), 0.5 µL of RT enzyme, 5.5 µL of sterile distilled water and 5 µL of extracted RNA.

cDNA synthesis

The isolated RNA of Influenza A (H7N9) virus was subjected to cDNA synthesis using Super Script III First Strand Synthesis kit (Invitrogen, USA) according to manufacturer's instruction. An influenza specific universal reverse transcriptase oligonucleotide, uni12 (5'AGC AAA AGC AGG 3') was used in this assay (Hoffmann et al., 2001). The cDNA was then used as template in the amplification of HA and *NA* gene by conventional PCR.

Conventional PCR amplification of HA and NA

Amplification of the HA and NA using the WHO Collaborating Centre (WHOCC), Melbourne primers was performed in a thermal cycler (BioRad, USA) with the following condition: 95°C for 2 min; 40 cycles at 95°C for 30 s, 57°C for 30 s and 72°C for 1 min, and a final heating at 72°C for 10 min. Annealing temperature for amplification of NA was optimized to 60°C. Assay reactions were carried out in a final volume of 25 µL containing 5 µL of 1x Buffer (Promega, USA), 4 µL of MgCl₂ (Promega, USA), 0.5 µL dNTP (Promega, USA), 1.0 µL of each primers (10 µM), 0.5 µL of Taq Polymerase (Promega, USA), 8 µL of distilled water and 5 µL of cDNA. Amplification of the HA and NA using the primer sequences obtained from CDC was performed with the following condition: 94°C for 3 min; 35 cycles at 94°C for 20 s, 60°C for 30 s and 72°C for 30 s, and a final heating at 72°C for 1 min. Assay reaction was composed of 12.5 µL of MiFi Mix (Bioline, UK), 1.2 µL of each primers (10 µM), 1.0 µL of MgCl₂ (Promega, USA), 6.1 µL of distilled water and 3 µL of cDNA.

Agarose gel electrophoresis

The 25 µL of each amplified PCR products were analyzed using 2.0% agarose gel (Promega, USA) pre-stained with Red Safe dye (Intron Biotech, Korea). Gel electrophoresis was performed in 1x TBE buffer at 90 V for 40 min and visualized under UV illumination. The expected amplicons were extracted from the agarose gel by Gel Extraction Kit (Qiagen, USA) according to manufacturer's instruction. Final elution contained 15 µL of purified PCR amplicons from which 2 µL was reanalyzed on 2% agarose gel to confirm that the purification step was performed precisely.

Sequencing

Prior to sequencing, the purified amplicons were subjected to cycle sequencing under the following condition: 40 cycles of 96°C for 2 min, 50°C for 5 s and 60°C for 4 min. The assay setup composed a Final volume of 10 µL consisting of 2 µL of Big Dye Terminator

(Applied Biosystem, USA), 2 µL of Buffer (Applied Biosystem, USA), 1 µL of either sense or antisense primers (4 µM), 4 µL of purified PCR amplicon and 1 µL of distilled water. The PCR amplicons which were amplified by WHOCC, Melbourne primers were cycle sequenced using universal primers (M13F: 5'TGTAAAACGACGGCCAGT3' and M13R: 5'CAGGAAACAGCTATGACC3'), whereas the amplicons which were amplified by CDC primers were cycle sequenced using the same primers in the PCR amplification step. All reactions were purified by dye-ex purification kit (Qiagen, USA) according to the manufacturer's instruction after which was concentrated by vacuum spin. A 10 uL of HIDI formamide (Applied Biosystem, USA) was added to each concentrated reaction tubes. Subsequently, these reaction mix were transferred to a 96 well plate, sealed and denatured at 95°C for 2 min and finally subjected to sequencing in genetic analyzer ABI 3730 (Applied Biosystem, USA).

Data analysis

All sequencing raw data were first processed and analyzed by Cromas Lite 2.1.1 software. Sense and antisense sequences were then aligned to produce full length of HA and NA gene sequences using CLUSTAL Omega software (http://www.ebi.ac.uk/Tools/msa/clustalo/). The reference sequences used in the alignment were influenza A virus (A/Guangdong/1/2013(H7N9)) segment 4 hemagglutinin (HA) gene (GenBank Accession number: KF662943.1) and influenza A virus (A/Guangdong/1/2013(H7N9)) segment 6 neuraminidase (NA) gene (GenBank Accession number: KF662949.1). A BLAST search was performed for the aligned sequences in NCBI database to indicate the closest match. The assembled sequences were also analyzed in FluSurver database (http://flusurver.bii.a-star.edu.sg/) for the presence of mutation.

Phylogenetic tree

Phylogenetic trees were constructed using neighbor joining method (bootstrap replication 1000x) to display the relationship and genetic variation of the HA and NA genes among the various influenza A (H7N9) isolates available in GenBank database. This was performed using MEGA 6.06 software.

RESULTS

Real-time RT-PCR data

The amplification is regarded positive when CT value is ≤38. The first batch of patient specimens (TS 1 and TS 2) tested on February 9 were positive for flu A and H7 but was negative for N9. Thus, a second batch of specimens was requested from the clinician considering the possibility for degradation of specimens due to broken cold-chain during transportation of the first batch.

The second batch of specimens which was tested on February 11 clearly indicated a positive result for H7N9 with a strong CT value for tracheal aspirate (Tasp). The CT values and the amplification curves of the second test are shown in Figure 1. In all run, positive controls were successfully amplified and negative controls showed no amplification.

HA and NA genes amplified by conventional PCR

Amplification of HA gene using WHOCC Melbourne primers were successful for both HA1 (~860 bp) and HA2 (~890 bp) segments using cDNA synthesized from Tasp 2. The CDC primers failed to amplify the HA region. The NA1 segment (~800 bp) was successfully amplified by WHOCC Melbourne primers using cDNA of Tasp2 but failed to amplify the NA2 segment. Therefore, CDC primers were used as an alternative targeting four segments and all yielded the expected amplicons: NA1 (~290 bp), NA2 (~550 bp), NA3 (~520 bp) and NA4 (~270 bp).

Sequencing data

Sequencing and alignment of the HA and NA genes produced a length of 1664 and 1321 bp respectively. The sequences were deposited in GISAID (Accession numbers: EPI 509111 for influenza A virus (A/Malaysia/228/2014(H7N9)) segment 6 neuraminidase (NA) gene and EPI 509205 for influenza A virus (A/Malaysia/228/2014(H7N9)) segment 4 hemagglutinin (HA) gene. The BLAST search of these sequences revealed the closest match with influenza A virus (A/environment/Guangzhou/1/2014(H7N9)) segment 4 HA gene for HA and Influenza A virus (A/Guangdong/05/2013(H7N9)) segment 6 NA gene for NA.

Phylogenetic analysis

The phylogram as shown in Figure 2 clustered the Influenza A Virus (A/Malaysia/228/2014(H7N9)) segment 4 hemagglutinin (HA) gene into the group of Guangzhou strains whereas the Influenza A Virus (A/Malaysia/228/2014(H7N9)) segment 6 neuraminidase (NA) gene was clustered into the group of Guangdong strains (Figure 3). In both phylogram, these strains were observed to be highly divergent from the A/Shanghai/1/2013 which is the isolate from the first case of H7N9 in China during the 2013 outbreak.

Mutation analysis

The FluSurver computed all mutations detected in our sequences. Results displayed include details such as position, involvement of mutation, frequency of appearance and previous literature reviews of the particular mutation. It was found that the A/Malaysia/228/2014(H7N9) segment 4 HA gene contained mutations R65K, E122K, L186I and N285D whereas A/Malaysia/228/2014(H7N9) segment 6 neuraminidase (NA) gene contained mutations M26I, R78K and V345I. The HA mutations found in this study were mostly involved in viral oligomerization and NA

Amplification of FLU A

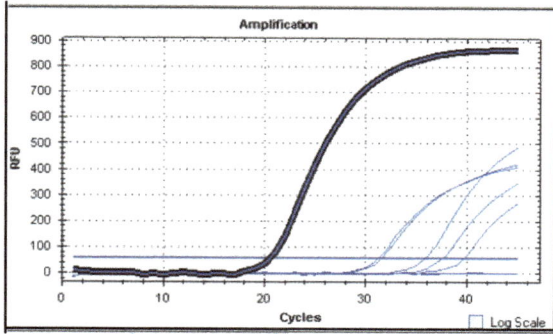

Fluor	Content	Sample	C(t)
FAM	Neg Ctrl	NC Ext	N/A
FAM	Neg Ctrl	NC Rgt	N/A
FAM	Pos Ctrl	PC	20.18
FAM	Unkn	RM039 Swab 1	39.30
FAM	Unkn	RM039 Swab 2	37.05
FAM	Unkn	RM039 Swab 3	35.40
FAM	Unkn	RM039 T.Asp 1	30.80
FAM	Unkn	RM039 T.Asp 2	31.20

Amplification of H7 with CDC

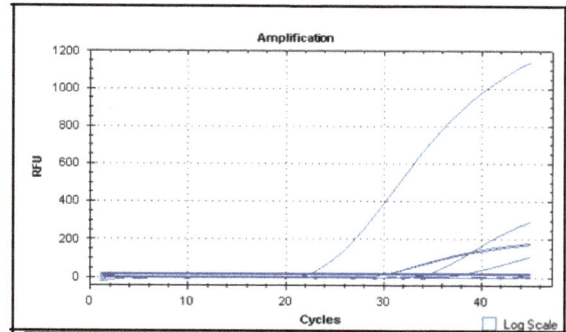

Fluor	Content	Sample	C(t)
FAM	Neg Ctrl	NC Ext	N/A
FAM	Neg Ctrl	NC Rgt	N/A
FAM	Pos Ctrl	PC	23.43
FAM	Unkn	RM039 Swab 1	N/A
FAM	Unkn	RM039 Swab 2	35.83
FAM	Unkn	RM039 Swab 3	40.53
FAM	Unkn	RM039 T.Asp 1	32.40
FAM	Unkn	RM039 T.Asp 2	32.31

Amplification of H7 with CNIC oligonucleotides

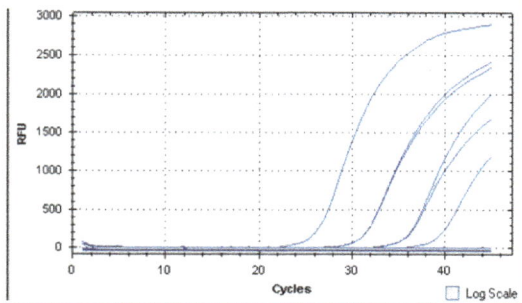

Fluor	Content	Sample	C(t)
FAM	Neg Ctrl	NC Ext	N/A
FAM	Neg Ctrl	NC Rgt	N/A
FAM	Pos Ctrl	PC	25.76
FAM	Unkn	RM039 Swab 1	39.43
FAM	Unkn	RM039 Swab 2	36.07
FAM	Unkn	RM039 Swab 3	36.00
FAM	Unkn	RM039 T.Asp 1	31.21
FAM	Unkn	RM039 T.Asp 2	31.24

Amplification of N9 with CDC

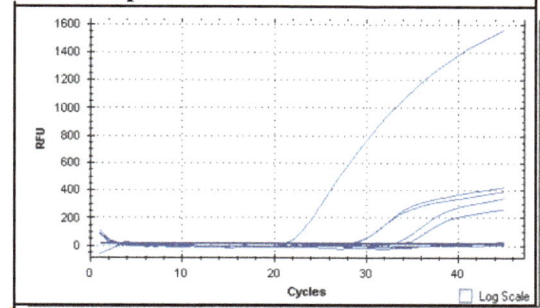

Fluor	Content	Sample	C(t)
FAM	Neg Ctrl	NC Ext	N/A
FAM	Neg Ctrl	NC Rgt	N/A
FAM	Pos Ctrl	PC	22.52
FAM	Unkn	RM039 Swab 1	N/A
FAM	Unkn	RM039 Swab 2	35.94
FAM	Unkn	RM039 Swab 3	34.75
FAM	Unkn	RM039 T.Asp 1	30.46
FAM	Unkn	RM039 T.Asp 2	30.45

Amplification of N9 with CNIC

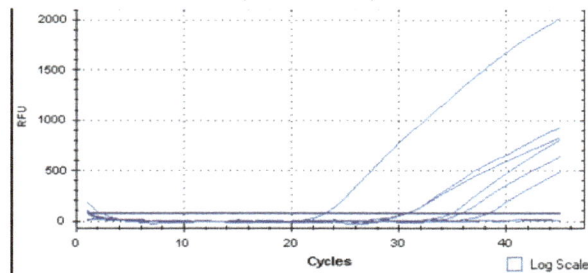

Fluor	Content	Sample	C(t)
FAM	Neg Ctrl	NC Ext	N/A
FAM	Neg Ctrl	NC Rgt	N/A
FAM	Pos Ctrl	PC	23.58
FAM	Unkn	RM039 Swab 1	38.87
FAM	Unkn	RM039 Swab 2	36.34
FAM	Unkn	RM039 Swab 3	35.34
FAM	Unkn	RM039 T.Asp 1	31.43
FAM	Unkn	RM039 T.Asp 2	31.34

Figure 1. Real-time RT-PCR results for amplification of H7N9 virus (all amplification were performed in a single run).

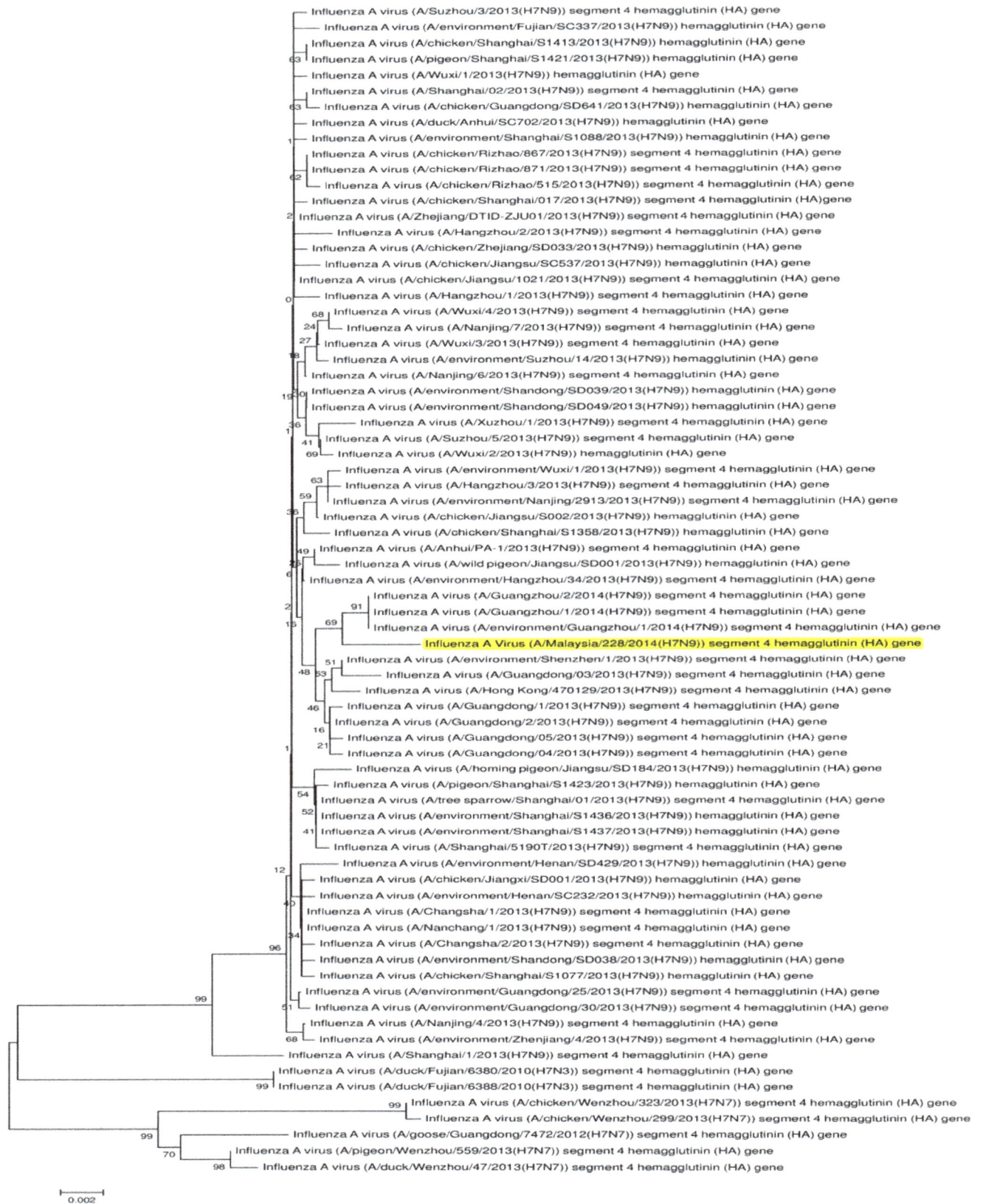

Figure 2. Phylogram showing the divergence of influenza A virus (A/Malaysia/228/2014(H7N9)) segment 4 hemagglutinin (HA) gene from other strain.

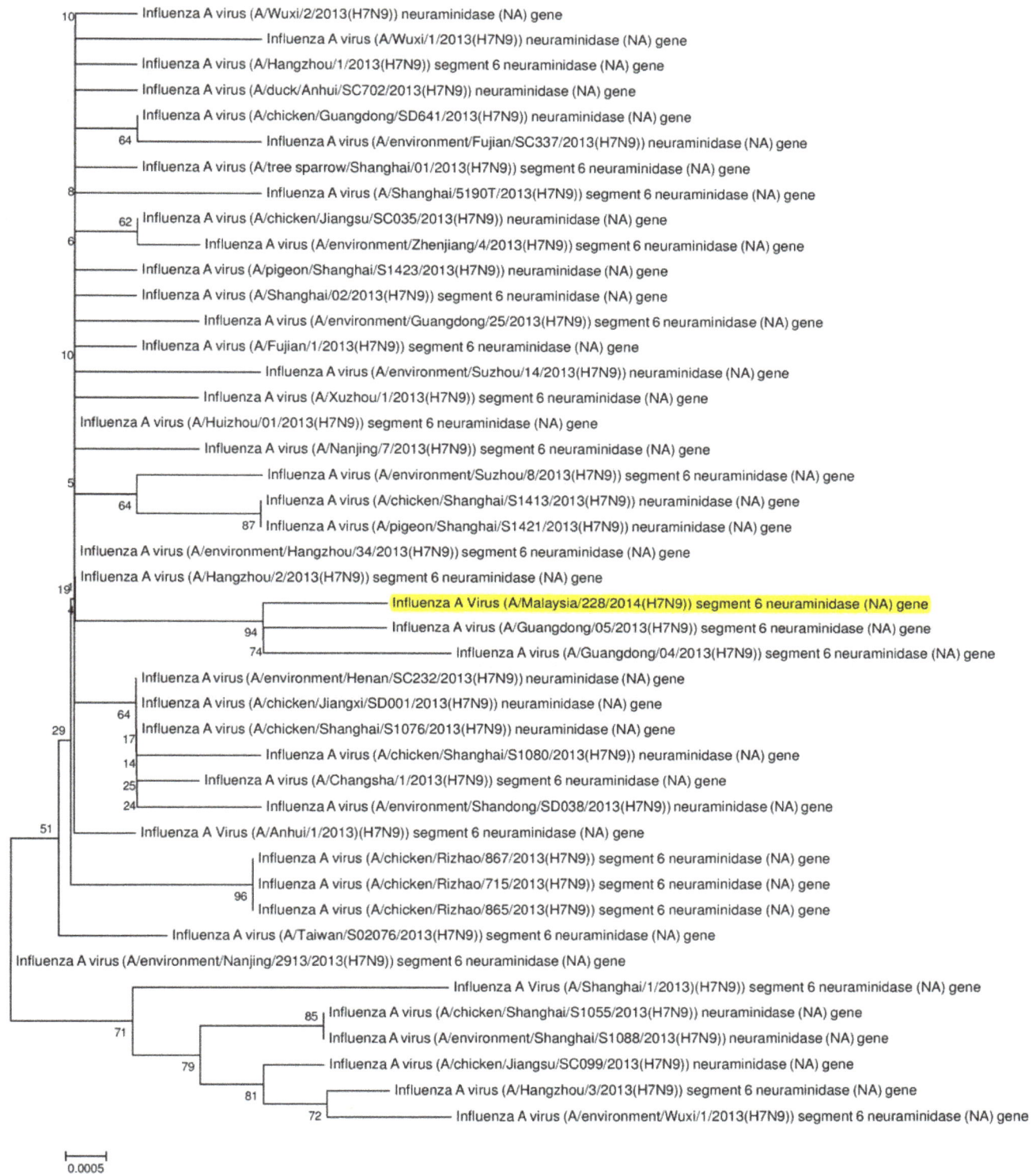

Figure 3. Phylogram showing the divergence of influenza A virus (A/Malaysia/228/2014(H7N9)) segment 6 neuraminidase (NA) gene from other strain.

mutations were mainly involved in small ligand binding. None of the neuraminidase inhibitor resistant mutations were found in these strains. The details of the mutations are summarized in Table 1.

Table 1. Mutations found in this study.

HA mutation	Frequency of mutation found globally	Involvement of mutation
E122K	Frequency of 1.12% of all samples with HA sequence. Occurred one time in one country in the strainA/Chicken/Jiangxi/SD001/2013(H7N9).	Involved in viral oligomerization interfaces and in a T- cell epitope presented by MHC molecules
L186I	New mutation not found in any of the H7N9 sequences in current GenBank database.	Unknown
N285D	Frequency of 2.25% of all samples with HA sequence. Occurred two times in one country in strains A/Huizhou/01/2013(H7N9) and A/Guangdong/1/2013/(H7N9).	Involved in viral oligomerization interfaces, binding small ligands and antibody recognition sites
R65K	Frequency of 17.20% of all samples with HA sequence. Occurred 16 times in one country. The first strain with this mutation was A/chicken/Zhiejiang/DTID-ZJUO1/2013/(H7N9) collected in April 2013 and the most recently occurred in strain A/Guangzhou/2/2014/(H7N9) collected in Jan 2014.	Involved in viral oligomerization interfaces and binding small ligands
NA mutations	**Frequency of mutation found globally**	**Significance of mutation**
M26I	Frequency of 95.40% of all samples with NA sequence. Occurred 83 times in three countries. The first strain with this mutation was A/Changsa/1/2013/(H7N9) collected in March 2013 and recent occurrence in A/Guangdong/05/2013/(H7N9).	Involved in binding small ligand
R78K	Frequency of 3.45% of all samples with NA sequence. Occurred three times in one country. First occurrence is in the strain A/Guangdong/02/2013/ (H7N9) and most recent presence was in A/Guangdong/05/2013/(H7N9).	Unknown
V345I	Frequency of 3.45% of all samples with NA sequence. Occurred three times in one country. First occurrence is in the strain A/chicken/ Shanghai/S1055/2013/(H7N9) and most recently found in A/chicken/Shanghai/S1053/2013/(H7N9).	Involved in small ligand binding

Information on the distribution of the mutations and the involvement in biological process were extracted from FluSurver database (http://flusurver.bii.a-star.edu.sg).

DISCUSSION

In this study, we reported an imported case of Influenza A(H7N9) in Malaysia. Molecular characterization of the H7N9 virus extracted from clinical specimen of the patient were carried out by real-time RT-PCR detection, amplification by conventional PCR, sequencing of the HA and NA genes and phylogenetic analysis. The real-time RT-PCR data showed a substantially strong CT value for tracheal aspirate specimens (Tasp1 and Tasp2) as compared to throat swabs (TS1, TS2 and TS3). This indicated that tracheal aspirate has higher viral RNA yield, a reason why it is generally regarded as the specimen of choice for detection of lower respiratory infection (Drosten et al., 2013). Lower respiratory tract specimens such as tracheal aspirate can produce high viral load because influenza virus shedding is no longer in the upper respiratory tract as the duration of infectiousness prolongs. Therefore, a negative viral yield on upper respiratory tract specimens does not necessarily conclude absence of the virus. To increase the likelihood of detecting the virus, multiple samples from

multiple sites should be collected over the course of the illness. Moreover, it is noted that the first batch of specimens was not received in an optimum condition, whereby the cold-chain was not maintained. This could have triggered the false negative result for N9 during the first real-time RT-PCR amplification. The requested second batch of samples was properly shipped and real-time RT-PCR clearly indicated a positive result for H7N9.

Due to unavailability of culture isolate in our study, amplification of the HA and NA genes by means of conventional PCR from direct specimen was laborious and time consuming as it required optimization from many aspects. The primers used in this step showed variability in amplifying the H7N9 virus from the clinical specimens. For instance, the WHOCC Melbourne primers for NA amplification could not amplify the N2 segment whereas the CDC primers failed to amplify the HA segments. This could be mainly due to some variation that had occurred in the new strain that had prevented the primer to bind to the sequences. It has been reported that the novel H7N9 strains could be mutating up to eight times faster than an average flu virus

(Cheepsattayakorn and Cheepsattayakorn, 2013). Apart from that, some of the primers have been validated by amplification with culture but have not been tested with clinical specimens. Therefore, the efficiency in amplifying viral genetic material from clinical specimen is questionable.

The BLAST search and phylogenetic analysis suggested that the *NA* and *HA* genes of the new strain of A/Malaysia/228/2014/(H7N9) clustered to the Guangdong and Guangzhou group of strains respectively. This finding is consistent with the fact that the patient originated from the Guangdong province and Guangzhou being the capital of this province. Initially, the *HA* gene sequence of A/Malaysia/228/2014 were found to have clustered into the Guangdong group of isolates, however, with the recent addition of the Guangzhou strains to the GenBank database, a reconstructed phylogram showed that it was more closely related to the Guangzhou strains.

The mutation analysis in the HA and NA sequences of A/Malaysia/228/2014/(H7N9) did not discover drug resistant associated mutations. The patient was initially given oseltamivir treatment, and recently switched to zanamivir, gradually recovering and reported to be in stable condition. This unlikely have produced drug resistant mutations within a short period of time. The R292K (R294K in N9 numbering) mutation is one of the most commonly identified mutations among seasonal H3N2 isolates with dramatically reduced sensitivity to oseltamivir, intermediate resistance to peramivir, and slightly reduced sensitivity to zanamivir (Gubareva, 2004). This mutation was also discovered in A/Shanghai/1/2013/(H7N9), the first case of H7N9 infection in China. However, surveillance study suggested that the emergence of NA mutations conferring resistance to NA inhibitors has reportedly been low, with the exception of the naturally emergent H274Y NA mutation in H1N1 seasonal influenza (Whitley et al., 2013).

Some other mutations were found in the HA and NA sequences of the A/Malaysia/228/2014 as shown in Table 1. All mutations except one have been discovered in other strains of H7N9 at least once. Among the HA mutations discovered in this strain, the R65K was found to have occurred more commonly in other strains of H7N9 reported thus far, whereas M261 had higher prevalence of occurrence in *NA* gene. The significance and function of these reported mutations were not well understood, however most are thought to be involved in viral oligomerization and ligand binding. Previous study had demonstrated that evolutionary variation involved in an oligomerization interface of the influenza A virus neuraminidase were essential for viral survival (Mok et al., 2013). Involvement in ligand binding mechanism is crucial for the virus to substantially interact with host receptor, sialic acid (Taylor and von Itzstein, 1994).

The novel mutation found in the *HA* gene of our new strain was L186I, a substitution of leucine to isoleucine

(CTA→ATA). The occurrence of this mutation globally was not documented, however, our alignment results with other influenza strains, revealed that L186I has previously occurred in Influenza A virus (A/chicken/Wenzhou/323/2013(H7N7)) segment 4 *HA* gene and Influenza A virus (A/chicken/Wenzhou/299/2013(H7N7)) segment 4 *HA* gene. This may suggest that the H7N9 and H7N7 Wenzhou viruses have similar, but independent evolutionary origins. Surveillance study showed that the hemagglutinin genes from these two lineages originated from H7 viruses that have been introduced to and established among the domestic ducks in China since 2010 (Lam et al., 2013).

In conclusion, the present study provides crucial information on the first case of H7N9 outside China and the diversity of this strain from other reported H7N9 strains by molecular analysis.

ACKNOWLEDGEMENT

The authors would like to thank the Director General of Health Malaysia for his permission to publish this paper. We would also like to extend our gratitude to Dr. Shahnaz Murad, the Director of Institute for Medical Research Malaysia for her support. We acknowledge the group of experts, Dr. Aeron Hurt, Dr. Deng Yi-Mo and Dr. Chantal Bass from the WHO Collaborating Centre for Reference and Research on Influenza, VIDRL, Melbourne and Dr. Frank Konings from WHO, Manila for their technical support and exchange of knowledge.

Conflict of Interest(s)

Authors have no financial interests related to the material in the manuscript.

REFERENCES

Centre for Disease Control (CDC) (2014). H7N9 Case Detected in Malaysia. 12 February 2014.

Cheepsattayakorn A, Cheepsattayakorn R (2013). Novel Avian Flu A (H7N9): Clinical and Epidemiological Aspects, and Management. Virol Mycol. 2(121).

Drosten C, Seilmaier M, Corman VM, Hartmann W, Scheible G, Sack S, et al. (2013). Clinical features and virological analysis of a case of Middle East respiratory syndrome coronavirus infection. Lancet Infect Dis. 13(9):745-751.

Gao R, Cao B, Hu Y, Feng Z, Wang D, Hu W, Chen J et al. (2013). Human Infection with a Novel Avian-Origin Influenza A (H7N9) Virus. N Engl J Med. 368(20):1888-1897.

Gubareva LV (2004). Molecular mechanisms of influenza virus resistance to neuraminidase inhibitors. Virus Res. 103: 199 -203.

Hoffmann E, Stech J, Guan Y, Webster RG, Perez DR (2001). Universal primer set for the full-length amplification of all influenza A viruses. Arch Virol. 146(12):2275-89.

Lam TT, Wang J, Shen Y, Zhou B, Duan L, Cheung CL, et al (2013). The genesis and source of the H7N9 influenza viruses causing human infections in China Nature. 502:241-246.

Li Q, Zhou L, Zhou M, Chen Z, Li F, Wu H, Xiang N, et al. (2014). Epidemiology of Human Infections with Avian Influenza A(H7N9) Virus in China. N Engl J Med. 370:520-532.

Mok CK, Chen GW, Shih KC, Gong TN, Lin SJ, Horng JT (2013). Evolutionary conserved residues at an oligomerization of the influenza A virus neuraminidase are essential for viral survival. Virol. 447(1-2):32-44.

Shuihua L, Yufang Z, Tao L, Yunwen H, Xinian L, Xiuhong X, Qingguo C, et al. (2013). Clinical Findings for Early Human Cases of Influenza A (H7N9) Virus Infection, Shanghai, China. Emerging Inf Dis. 19(7):1142-1146.

Taylor NR, von Itzstein M (1994). Molecular modeling studies on ligand binding to sialidase from influenza virus and the mechanism of catalysis. J Med Chem. 37(5):616-624.

Whitley RJ, Boucher CA, Lina B, Nguyen-Van-Tam JS, Osterhaus A, Schutten M, Monto AS. 2013. Global assessment of resistance to neuraminidase inhibitors, 2008-2011: the influenza resistance information study (IRIS). Clin. Infect Dis. 56:1197-1205.

WHO. Global Alert and Response: Human infection with avian influenza A(H7N9) virus-update (2013). Accessed: 3rd March 2014.

WPRO. Human Infection with Avian Influenza A(H7N9)- update (2014). Accessed: 16 May 2014.

26

Assessment of the current status of HIV virus and predisposing factors among students at Dilla University and Dilla Referral Hospital, Ethiopia

Fekadu Alemu

Department of Biology, College of Natural and Computational Sciences, Dilla University, P.O.Box. 419, Dilla, Ethiopia.

Acquired immunodeficiency syndrome (AIDS) cannot be transmitted by causal contact, air, food and water. People with acquired immunodeficiency syndrome have human immunodeficiency virus in their blood and body fluids that can enter the blood stream of an uninfected person upon contact with infected body fluids or sexual contact with an infected person. Primary data was collected by using structured questionnaire, interviewing and reviewing secondary data from Dilla Referral Hospita. Majority of students 137 (67.15%) had an idea on HIV virus that it is not curable and a deadly killer disease while 36 (17.64%) had an idea on HIV that it is curable. Majority of students (158, 77.46%) had discussed about HIV/AIDS with other people. Accordingly to the information in the questionnaire by the Dilla University students, most of the students (96.00%) had heard about HIV/AIDS. Majority of students heard and got information about HIV from mass media followed by health center, books, and combination of all lists, parents and friends: 46.00, 18.00, 12.00, 10.00, 6.00 and 2.00%, respectively. According to the secondary date obtained from Dilla Referral Hospital, the majority of the people that live with HIV virus were of the productive age (15-49 age) group in both male and female. Among these the number of female infected with HIV were higher than for male: 42.63, 44.35, 37.76, 44.37, 42.70, 37.62 and 40.40%, and 49.38, 52.17, 46.90, 44.70, 44.32, 52.47 and 39.40%, respectively for each year from 2008-2014. Therefore, HIV virus prevalence among Dilla Referral Hospital patients was on the decrease, and the students of Dilla University were very aware of HIV from well gathered information from different sources and also discussed with their partners.

Key words: Dilla University, human immunodeficiency virus (HIV), prevalence, sexual, students.

INTRODUCTION

Immunodeficiency syndrome (AIDS) caused by the human immunodeficiency virus (HIV) is one of the greatest public health and social problems threatening the human race. Globally, AIDS is now the fourth leading cause of mortality; 3.1 million deaths have been attributable to AIDS in 2002 alone, of which 1.2 million occurred in women (Nancy et al., 2005). According to the Joint UN Committee on HIV/AIDS (UNAIDS, 2004), an estimated 38 million people worldwide were living with HIV in 2003, of which 5 million were newly infected. More than 95% of HIV-infected people live in the developing world, most in Sub-Saharan Africa (Nancy et al., 2005). Worldwide, women now represent 50% of all adults living with HIV and AIDS and this proportion had increased over time (UNAIDS, 2002). Improved data have revealed that the prevalence rates in southern Africa are staggering: 20-26% of adults (aged 15-49 years) are infected; in some regions 20-50% of pregnant women were infected and are likely to transmit infection to one third of their offspring (Nancy et al., 2005).

Epidemiologic studies have demonstrated that HIV is transmitted by three primary routes: sexual, parenteral (blood-borne), and perinatal (Nancy et al., 2005). Factors that increase the risk of exposure to blood, such as genital ulcer disease (Cameron et al., 1989; Plummer et al., 1991), trauma during sexual contact (Marmor et al., 1986), and menstruation of an HIV-infected woman during sexual contact (European Study Group, 1992; Nair et al., 1993; St Louis et al., 1993) may all increase the risk of transmission. Sexual transmission of HIV from an infected partner to an non-infected partner can occur through male-to-female, female-to-male, male-to-male, and female-to-female sexual contact. Worldwide, sexual transmission of HIV is the predominant mode of transmission (Quinn, 1996).

The first step in infection is HIV binding to target cells, followed by its transportation to regional lymphnodes, where it replicates and establishes a productive and permanent infection. In the last few years it has been demonstrated that in the early phases of infection, HIV preferentially targets $CCR5^+ CD4^+$ memory T lymphocytes in the gastrointestinal tract (Brenchley et al., 2004). This results in a rapid, massive and possibly permanent destruction of CD4 cells, rupture of the intestinal mucosa and penetration of microbial translocation products in the systemic circulation.

HIV disrupts the proper functioning of the immune system. A weakened immune system allows the development of a number of different infections and cancers which cause illness and death in people with AIDS. HIV also infects and causes direct damage to other types of cells: for example, damage to the lining of the intestine can contribute to wasting (severe weight loss) and, damage to nerve cells can cause neurological problems (Nelson, 1988; Pomerantz, 1987; Elder and Sever, 1988).

The first documented report of HIV/AIDS case in Ethiopia was recorded in 1986 (Hladik, 2005). The HIV/AIDS epidemics have since evolved into a generalized epidemic with AIDS as the leading cause of morbidity and mortality among adults. Ethiopia has just over 1% of the world's population but contributes to 7% of the world's HIV/AIDS cases (World Bank, 2004). In terms of the number of people infected with HIV, Ethiopia is the fifth rank after South Africa, Nigeria, Kenya and Zimba-

bwe, and it is the second (after Nigeria) in terms of the number of children orphaned by the AIDS epidemic. More than 90% of the infections in Ethiopia takes place among aged ranges between 15 to 49; the most economically productive segment of the population. The prevalence rate of HIV for 2004 was estimated 13.2% for urban areas and 2.3% for rural Ethiopia to give a national average of 3.7%. Therefore, the aim of the current study was necessary to know the current status of HIV infected individual in the study area and exploring knowledge attitude and practice among ths study participants towards HIVAIDs.

MATERIALS AND METHODS

Study design, area, time and population

This study was conducted in Dilla University and Dilla Referral Hospital found in Gedeo zone in the south west of the country. Dilla is located on the main road from Addis Ababa to Nairobi. The Dilla town is 361 and 90 Km away from Addis Ababa and Hawassa town, respectively. The study was conducted with the purpose of exploring knowledge, attitude and practice of HIV among the main campus students of Dilla University and its prevalence at Dilla referral Hospital from November to July 2014.

The study employed mixed methods of quantitative and qualitative approaches for gathering information from randomly selected students from different colleges and departments at Dilla University. Around 204 students were randomly and systematically selected for interview and 50 questionnaires were distributed as well as filled by these students at Dilla University.

Method of data collection

The study utilized both published and unpublished materials. Primary data were collected through structured questionnaire and interviewing the volunteer students. Around 204 volunteer students were interviewed and 50 questionnaires were distributed and recollected after being filled by volunteer students at Dilla University. Secondary data were collected from Dilla Referral Hospital from registration documents for voluntary counseling test (VCT) of HIV Virus. The date ranged from year 2008-2014 and about 14,668 individuals were registered for voluntary counseling test of HIV Virus at Dilla Referral Hospital.

Ethical clearance

The study protocol was reviewed and ethically approved by Dilla University ethical and clearance committee. Before data collection, an informed consent was obtained from respondents. The confidentiality of the respondents was maintained.

Data analysis

Data entry and analysis was performed using the statistical package for Social Sciences for Windows SPSS (version 16.0). For analysis of the percentage and total HIV infected individuals tested at Dilla Referral Hospital, the results were expressed graphically and as tables.

Table 1. Students' response to the following interviews at Dilla University, Ethiopia at 2014.

Type of question to the student (interview question)	Way of answering the interview question	Sex	
		Male (n,%)	Female (n,%)
What do you know about HIV/AIDS?	Not curable	73 (35.78)	64 (31.37)
	Curable	16 (7.84)	20 (9.80)
	From super natural power	18 (8.82)	13 (6.37)
Do you think that using condom can prevent the transmission of HIV/AIDS?	Yes	92 (55.09)	73 (35.78)
	No	15 (7.35)	24 (11.76)
Do you agree that a student member, who has HIV virus in his/her blood should be isolated from others?	Agree	7 (3.43)	6 (2.94)
	Disagree	100 (49.02)	91 (44.61)
Do you agree that a student member, who has HIV in his/her blood should keep his/her status secret to other student?	Agree	15 (7.35)	9 (4.41)
	Disagree	92 (45.10)	88 (43.14)
Did you discuss about HIV/AIDS with other people?	Yes	83 (40.69)	75 (36.77)
	No	24 (11.76)	22 (10.78)
If your answer to above question is yes, with whom did you discuss?	Friends	50 (31.65)	35 (22.15)
	Anti-HIV/AIDS	12 (7.59)	16 (10.13)
	Parents	11 (6.97)	10 (6.33)
	Health profession	5 (3.16)	7 (4.43)
	Teachers	5 (3.16)	7 (4.43)
Have you ever tested for HIV?	Yes	82 (40.20)	70 (34.31)
	No	25 (12.25)	27 (13.24)

N, number of respondents;% percentage.

RESULT

Socio-demographic data

This study was conducted between January and June 2014. Two hundred and four respondents participated. All of them responded to the structured questionnaire on the knowledge, attitude and behavior related variables of the assessment. The number of voluntary counselling and test (VCT) individuals for HIV Virus in Dilla Referral Hospital in 2008, 2009, 2010, 2011, 2012, 2013, 2014 was 2698, 2602, 2594, 2045, 1761, 1638, 1330, respectively. The number of HIV positive individual in 2008, 2009, 2010, 2011, 2012, 2013 and 2014 were 401, 345, 339, 302, 185, 202 and, 99 accordingly.

Interview and questionnaires results

Table 1 shows that the majority of students (137, 67.15%) were of the idea that HIV virus was not curable or is deadly disease while (36, 17.64%) had an idea that, HIV is curable. Similarly some students were of the idea that, HIV patients were cured by faith (the power of God)

(31, 15.19%) if they believed in a super natural power as indicated in Table 1. As indicated in Table 1 the majority of students (100, 49.02%) male and (91, 44.61%) female did not believe in isolating HIV positive individuals while a small minority (7, 3.43%) male and (6, 2.94%) female believed in isolating those HIV positive individuals.

As shown in the Table 1, the majority of students (158, 77.46%) responded that, they discussed about HIV/AIDS with other people. They discussed it with their friends, parents, at anti HIV/AIDS club, health professionals and teachers with 53.80, 17.72, 13.30, 7.59 and 7.59%, respectively. As shown in the Table 1, 40.20% male and 34.31% female students had been tested for HIV virus counseling while 12.25% male and 13.24% female students did not attempt the test.

Distributed questionnaires to the students of Dilla University in Main Campus during 2014

The questionnaire that was distributed to students was correctly filled out and returned by 54% male and 46%

Table 2. Students' response to the following questionnaires at Dilla University, Ethiopia at 2014.

Types of questions to the Dilla University students, Main campus	Target answered questionnaires	Sex	
		Male (n,%)	Female (n,%)
Have you heard about HIV/AIDS?	Yes	25 (50.00)	23 (46.00)
	No	2 (4.00%)	-
From where you heard about IV/AIDS?	From parents	2 (4.00)	1 (2.00)
	From friends	1 (2.00)	1 (2.00)
	From health center	5 (10.00)	4 (8.00)
	From mass media	10 (20.00)	13 (26.00)
	From books	4 (8.00)	2 (4.00)
	From all the above	3 (6.00)	2 (4.00)
Knowledge about routs of HIV transmission	Unprotected sexual intercourse	19 (38.00)	20 (40.00)
	From infected mother to child	5 (10.00)	1 (2.00)
	Using sharp materials with others	2 (4.00)	2 (4.00)
	Through the bite of insect	1 (2.00)	-
Knowledge about HIV prevention method	Abstinence	11 (22.00)	13 (26.00)
	Condom	6 (12.00)	4 (8.00)
	Faithfulness	9 (18.00)	1 (2.00)
	I do not use any methods	1 (2.00)	5 (10.00)
Have you ever had sexual intercourse	Yes	16 (32.00)	8 (16.00)
	No	11 (22.00)	15 (30.00)
If your answer to question above is yes, with whom did you have sexual intercourse?	With my partner (husband or wife)	3 (12.50)	-
	With my beloved friend	13 (54.17)	8 (33.33)
	With prostitute	-	-
If your answer to questions above is yes, with how many people did you have casual sex?	Single person	12 (50.00)	5 (20.83)
	Two person	3 (12.50)	1 (4.17)
	Three and more	1 (4.17)	2 (8.33)
If your answer to question above is yes, did you use a condom during sexual intercourse?	Yes	9 (37.50)	5 (20.53)
	No	7 (29.17)	3 (12.50)
Have you ever used sharp materials with others?	Yes	1 (2.00)	2 (4.00)
	No	18 (36.00)	16 (32.00)
	Some times	5 (10.00)	2 (4.00)
	I don't remember	3 (6.00)	3 (6.00)

No, Number.

female. Most of the students (96.00%) has heard about HIV/AIDS whereas (4.00%) student respondents had never heard about HIV virus. As indicated in Table 2 the majority of students believed that the major route of HIV transmission was unprotected sex, from infected mothers to child, sharing sharp materials with other and through bite of insect: 78.00, 12.00, 8.00 and 2.00% respectively. On the other hand, our data further revealed that while, abstinence/ faithfulness and using a condom (48.00 and 20.00%) were thought of as majority ways of HIV prevention mechanisms and approaches, a few did not use any method.

Majority of students (about 70.83%) had sexual intercourse with a single person while some (16.67%) had sexual intercourse with two persons and three or more (12.50%) respectively.

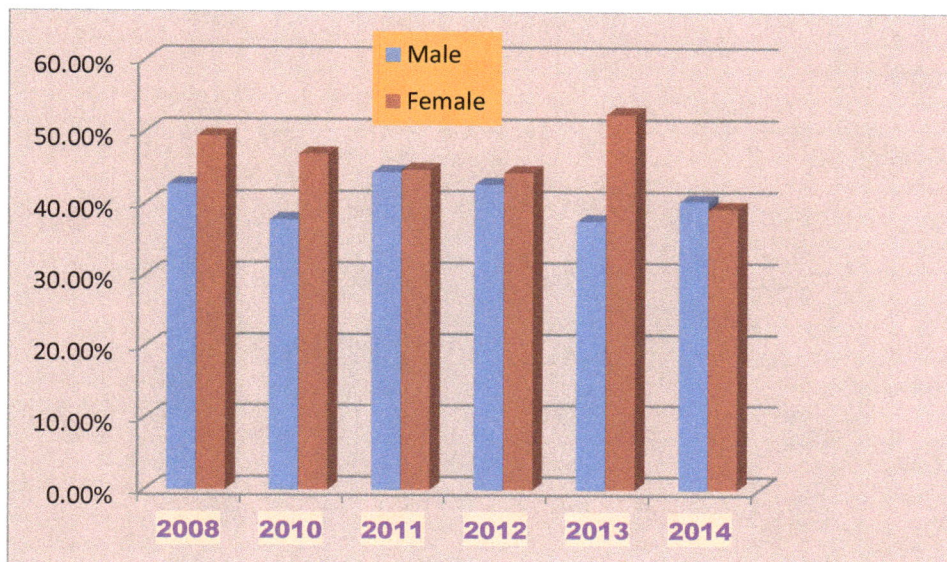

Figure 1. Prevalence of HIV infected individuals between the age of 15-49 at Dilla Referral Hospital.

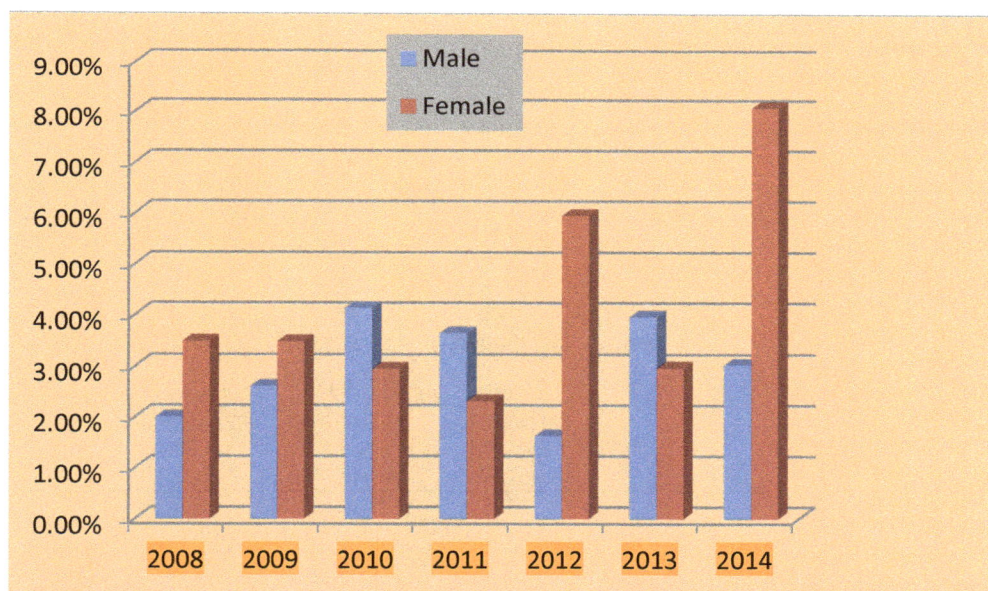

Figure 2. Prevalence of HIV infected individuals in the age group <15 at Dilla Referral Hospital.

Majority of the people that live with HIV virus belonged to the productive age (15-49 years) among both males and females. Among these, the number of females living with HIV virus were larger than male: 42.63, 44.35, 37.76, 44.37, 42.70, 37.62, and 40.40; and 49.38, 52.17, 46.90, 44.70, 44.32, 52.47 and 39.40% respectively for years 2008 to 2014. Next to the productive age were children (<15 age) infected (carrier) by HIV virus followed by old age (>49 age) (Figure 3).

Distribution of HIV infection between the age of 15-49 across the sex was more prevalent in females than males while, across the years 2008-2014, the prevalence was decreased as indicated in Figure 1.

Distribution of HIV infection in patients less than 15 years of age was across the gender; more prevalent in females than males while, across the year 2008-2014, the prevalence was increased in female as indicated in Figure 2.

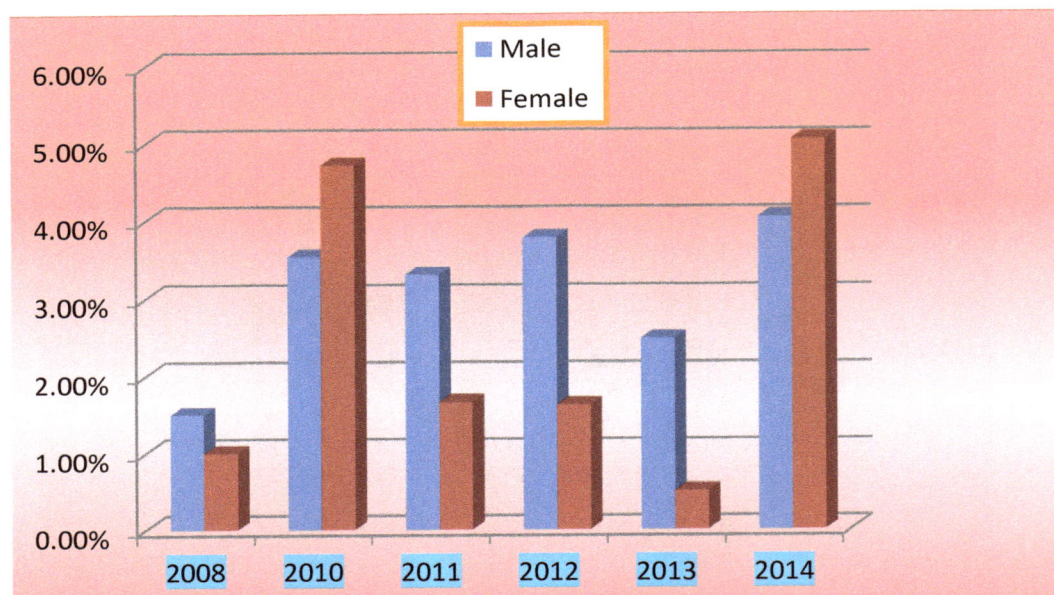

Figure 3. Prevalence of HIV infected individuals in the age group >49 at Dilla Referral Hospital. Distribution of HIV infection in patients greater than 49 years old was across the gender more prevalent on female than male while, across the year 2008-2014, the prevalence was increased in both males and females as indicted Figure 3.

Table 3. Prevalence of HIV/AIDS among voluntary counseling test for HIV virus from 2008-2014 at Dilla Referral Hospital, Dilla, Ethiopia at 2014.

Sex	Age category	Years						
		2008	2009	2010	2011	2012	2013	2014
Male	<15 age	8 (2.00)	9 (2.61)	14 (4.13%)	11 (3.64%)	3 (1.63%)	8 (3.96%)	3 (3.03%)
	15-49 age	171 (42.63%)	144 (41.74%)	128 (37.76%)	134 (44.37%)	79 (42.70%)	76 (37.62%)	40 (40.40%)
	>49 age	6 (1.50%)		12 (3.54%)	10 (3.31%)	7 (3.78%)	5 (2.48%)	4 (4.04%)
	Total	185 (46.13%)	153 (44.35%)	154 (45.43%)	155 (51.32%)	89 (48.11%)	89 (44.06%)	47 (47.47%)
Female	<15 age	14 (3.49%)	12 (3.48%)	10 (2.95%)	7 (2.32%)	11 (5.95%)	6 (2.97%)	8 (8.08%)
	15-49 age	198 (49.38%)	180 (52.17%)	159 (46.90%)	135 (44.70%)	82 (44.32%)	106 (52.47%)	39 (39.40%)
	>49 age	4 (1.00%)		16 (4.72%)	5 (1.66%)	3 (1.62%)	1 (0.50%)	5 (5.05%)
	Total	216 (53.87%)	192 (55.65%)	185 (54.57%)	147 (48.68%)	96 (51.89%)	113 (55.94%)	52 (52.53%)

DISCUSSION

As this study showed, among HIV positive individuals, the number of males and females were, 185, 153,154,155, 89, 89, 47 and 216, 192, 185, 147, 96, 52 respectively in the respective years as indicated in Table 3. In Ethiopia, in 2000, the median age of first sexual intercourse of women aged 20-49 was 16.4 years and for men it was 20.3 years, indicating the relatively greater vulnerability of teenage girls to HIV infection (CSA and ORCMacro, 2001). Ethiopia is classified along with Nigeria, China, India, and Russia as belonging to the "next wave countries" with large populations at risk from HIV infection, which will eclipse the current focal point of the epidemic in central and southern Africa (NIC, 2002).

As Table 1 shows 55.09% male and 35.78% female students believed using a condom can prevent HIV/ AIDS while 7.35% male and 11.76% female students believed that using a condom will not prevent HIV transmission from patient (carrier) to healthy individual. In 2003, the highest HIV infection rates in Ethiopia reportedly occurred in the 15-34 age groups. The highest rates in female Antenatal care (ANC) attendees were in the 15-24 age groups (8.6%). Children and adolescents have become increasingly exposed to HIV in recent years, with an estimated 14,000 new infections in the 0-14 age group in 2003 (MOH, 2004). Rates are higher in young females than males, apparently due to a combination of the earlier commencement of sexual activity of females, the older age of their partners, gender-based biological factors

(Quinn and Overbaugh, 2005), and prenatal and obstetric care/delivery exposures. In Ethiopia, in 2000, the median age of first sexual intercourse of women aged 20-49 was 16.4 years and for men 20.3 years, indicating the relatively greater vulnerability of teenage girls to HIV infection (CSA and ORCMacro, 2001).

However, a significant number of students 24 (11.8%) still favored keeping their HIV status as a secret although majority of respondents 180 (88.2%) disagreed with keeping HIV status a secret. Ethiopia is classified (along with Nigeria, China, India, and Russia) as belonging to the "next wave countries" with large populations at risk from HIV infection, which will eclipse the current focal point of the epidemic in central and southern Africa (NIC, 2002).

Majority of students came to know and got information about HIV from mass media followed by health center, books, and combination of all lists, parents and friends: 46.00, 18.00, 12.00, 10.00, 6.00 and 2.00% respectively. Worldwide, women now represent 50% of all adults living with HIV and AIDS and this proportion had been steadily increasing over time (UNAIDS, 2002). Perinatal transmission can occur in utero, during labor and delivery, or post-partum through breast-feeding (Gwinn and Wortley, 1996). Perinatal transmission rates average 25–30% (Blanche et al., 1989).

About 52 .00% of both male and female students had not made intercourse while others (48.00%) had sexual intercourse. Among those who had sexual intercourse, some of them had casual sex with their partner (husband or wife) and with their beloved friends (87.50 and 12.50% respectively). The dominant mode of transmission is through heterosexual contact (estimated to account for 87% of infections) and mother-to-child transmission (MTCT) (10% of infections) (GoE, 2004). Blood transfusion, harmful traditional practices, and unsafe injections are all recognized to be a small risk at present but require attention (GoE, 1998).

Majority of students (58.33%) used a condom correctly during their casual intercourse while few students (41.67%) did not use condom. Large number of student (68.00%) did not sharp materials with other, while 14.00% of students sometimes shared sharp materials with others and few of students (6.00%) always shared sharp materials with others but the remaining students (12.00%), did not remember whether they used it or not. Epidemiologic studies have demonstrated that HIV is transmitted by three primary routes: sexual, parenteral (blood-borne), and perinatal (Nancy et al., 2005). Soldiers, high-risk and mobile groups exposed to and spreading HIV through multi-partner sex contacts, were stationed in the 1980s and early 1990s in many Ethiopian towns in the war zone. Troops were also at risk of being infected during emergency blood transfusions (Eshete et al., 1993; Kloos, 1993).

HIV infection rates in soldiers were increased from 2.1% in 1985/1986 to 12.0% in 1989 (Gebretensae, 2003).

Factors that increase the risk of exposure to blood, such as genital ulcer disease (Cameron et al., 1989; Plummer et al., 1991), trauma during sexual contact (Marmor et al., 1986), and menstruation of an HIV-infected woman during sexual contact (European Study Group, 1992; Nair et al., 1993; St Louis et al., 1993) may all increase the risk of transmission.

The number of HIV infected individuals and the number of individuals of voluntary counseling test for HIV decreased from 2008 to 2014. Sexual transmission of HIV from an infected partner to an uninfected partner can occur through male-to-female, female-to-male, male-to-male, and female-to-female sexual contact. Worldwide, sexual transmission of HIV is the predominant mode of transmission (Quinn, 1996).

Parenteral transmission of HIV has occurred in recipients of blood and blood products, through transfusion of blood (estimated 95% risk of infection from transfusion of a single unit of HIV-infected whole blood) (CDC, 1998) or clotting factors, in intravenous or injection drug users through the sharing of needles (approximately 0.67% risk per exposure) (Kaplan and Heimer, 1992), in health care workers through needle sticks (approximately 0.3-0.4% risk per exposure, depending on the size and location of the inoculum) (Tokars et al., 1993; Updated PHS guidelines, 2001), and, less commonly, mucous membrane exposure (0.09% risk per exposure (Updated PHS guidelines, 2001; Hessol et al., 1989). Among cumulatively reported AIDS cases in U.S. women through December 2001, 39% had injection drug use as their exposure risk and 3% reported receipt of infected blood, blood products, or tissue (CDC, 2002).

Routes of HIV spread are unprotected vaginal, anal, or oral sex with an infected person, needles or drug equipment shared with injection drug users who have HIV, prenatal (before birth) and perinatal (during and right after birth) exposure of infants whose mothers are infected with HIV, breast-feeding by mothers with HIV, transfusion of blood products containing the virus, organ transplants from HIV-infected donors, penetrating injuries or accidents of health care workers (usually needle sticks) while caring for HIV-infected patients or handling their blood.

Conclusion

This study was focused on the students background on HIV virus, their attitude, knowledge and information. Large numbers of students heard about HIV from media. Most of the students had positive attitude towards someone who had HIV in his or her blood. In addition most of the students used condoms during sexual intercourse while other students had unsafe sex as well as multiple sexual partners among students. On the contrary, some students still had negative attitude towards use of condoms and lack adequate and correct information and hence are unable to practically utilize HIV/AIDS management services such as awareness raising, training, and peer conservations, and condom use. On the other hand, the result obtained from Dilla Referral Hospital from 2008 to 2014 indicated more females were infected with HIV virus than males. However, the prevalence of HIV virus across the year 2008 to 2014 decreased along with the number of infected individual with HIV virus.

Conflict of Interests

The author(s) have not declared any conflict of interests.

ACKNOWLEDGMENTS

I am grateful to the Dilla University, College of Natural and Computational Sciences, Department of Biology who gave the facilities to conduct this study. I extend also my thanks to Dilla University students who volunteered to participate in the study through giving relevant information as well as Dilla Referral Hospital which contributed by providing essential secondary data.

REFERENCES

Blanche S, Rouzioux C, Moscato ML, (1989). A prospective study of infants born to women seropositive for human immunodeficiency virus type 1. HIV Infection in Newborns French Collaborative Study Group. N. Engl. J. Med. 320:1643-1648.

Brenchley JM, Schacker TW, Ruff LE (2004). CD4+ T cell depletion during all stages of HIV disease occurs predominantly in the gastrointestinal tract. J. Exp. Med. 200:749-59.

Cameron DW, Simonsen JN, D'Costa LJ (1989). Female to male transmission of human immunodeficiency virus type 1: risk factors for sero-conversion in men. Lancet 2:403-407.

CDC (1998). Management of possible sexual, injecting-drug-use, or other non-occupational exposure to HIV, including considerations related to antiretroviral therapy. M.M.W.R. 17:1-14.

CSA, ORCMacro (2001). Ethiopia Demographic and Health Survey. Addis Ababa and Calverston, MA (USA), Central Statistical Authority and ORCMacro.

Elder G, Sever J (1988). Neurologic disorders associated with AIDS retroviral infection. Review of Infectious Diseases 10 (2):286-302.

Eshete H, Heast N, Lindan K, Mandel J (l993). Ethnic conflicts, poverty, and AIDS in Ethiopia. Lancet 341:1219.

European Study Group (1992). Comparision of female to male and male to female transmission of HIV in 563 stable couples. European Study Group on Heterosexual Transmission of HIV. Br Med. J. 304:809-813.

Gebretensae GT (2003). HIV/AIDS in the Ethiopian military: perceptions, strategies, and impacts. Draft working paper for the CSIS Task Force on HIV/AIDS Committee on Destabilizing Impacts of HIV/AIDS. Addis Ababa.

Government of Ethiopia (1998). Policy on HIV/AIDS. Addis Ababa, GoE.

Government of Ethiopia (2004). A comprehensive strategic plan to combat HIV/AIDS epidemic in Ethiopia (2004-2007), Final Report. Addis Ababa: GoE.

Gwinn M, Wortley PM (1996). Epidemiology of HIV infection in women and newborns. Clin. Obstet Gynecol. 39:292-304.

Hessol NA, Lifson AR, Rutherford GW (1989). Natural history of human immunodefi-ciency virus infection and key predictors of HIV disease progression. AIDS Clin. Rev. 69-93.

Hladik W (2005). HIV/AIDS in Ethiopia: Where is the Epidemic Heading? Sexually Transmitted Infections 8:32-l35.

Kaplan EH, Heimer R (1992). A model-based estimate of HIV infectivity via needle sharing. J. Acquir. Immune. Defic. Syndr. 5:1116-118.

Kloos H (l993). Health impact of war. In H. Kloos and Z.A. Zein (eds), The Ecology of Health and Disease in Ethiopia 121-132. Boulder and Oxford: Westview Press.

Marmor M, Weiss LR, Lyden M (1986). Possible female-to-female transmission of human immunodeficiency virus. Ann. Intern. Med. 105:969.

MOH (2004). AIDS in Ethiopia. 5th edition. Addis Ababa: Ministry of Health.

Nair P, Alger L, Hines S (1993). Maternal and neonatal characteristics associated with HIV infection in infants of seropositive women. J. Acquir Immune Defic Syndr 6:298-302.

Nancy A, Hessol MSPH, Monica, Gandhi MD, Ruth M, Greenblatt, MD (2005). A Guide to the Clinical care of women with HIV. Health Resources and Services Administration, HIV/AIDS Bureau.

National Intelligence Council (2002). The next wave of HIV/AIDS: Nigeria, Ethiopia, Russia, India and China.Washington, D.C.

Nelson J (1988). Human immunodeficiency virus detected in bowel epithelium from patients with gastrointestinal symptoms. Lancet 8580:259-262.

Plummer FA, Simonsen JN, Cameron DW (1991) Cofactors in male-female sexual transmission of human immunodeficiency virus type 1. J. Infect. Dis. 163:233-239.

Pomerantz R (1987). Infection of the retina by human immunodeficiency virus type I. N.E.J.M. 317(26):1643-1647.

Quinn T C, Overbaugh J (2005). HIV/AIDS in women: an expanding epidemic. Science 308:1582-1583.

Quinn TC (1996). Global burden of the HIV pandemic. Lancet 348:99-106.

St. Louis ME, Kamenga M, Brown C (1993). Risk for perinatal HIV-1 transmission according to maternal immunologic, virologic, and placental factors. J.A.M.A. 269:2853-2859.

Tokars JI, Marcus R, Culver DH, (1993). For the CDC Cooperative Needlestick Surveillance Group. Surveillance of HIV infection and zidovudine use among health care workers after occupational exposure to HIV-infected blood. Ann. Intern. Med. 118:913-919.

UNAIDS (2002). AIDS epidemic update: December 2002. Geneva, Switzerland, UNAIDS/WHO.

UNAIDS (2002). AIDS epidemic update: December 2002. Geneva, Switzerland, UNAIDS/WHO, 2002.

UNAIDS UNICEF and WHO (2004). Ethiopia, epidemiological fact sheets on HIV/AIDS and sexually transmitted infections. Geneva: WHO.

Updated US (2001). Public Health Service guidelines for the management of occupational exposures to HBV, HCV, and HIV and recommendations for postexposure prophylaxis. MMWR Rep. 50:1-52.

27

Comparison of immunoperoxidase monolayer assay, polymerase chain reaction and haemadsorption tests in the detection of African swine fever virus in cell cultures using Ugandan isolates

Mathias Afayoa[1], David Kalenzi Atuhaire[1,2], Sylvester Ochwo[1], Julius Boniface Okuni[1], Kisekka Majid,[1] Frank Norbert Mwiine[1], William Olaho-Mukani[3] and Lonzy Ojok[1]

[1]College of Veterinary Medicine, Animal Resources and Bio-security, Makerere University,P.O.BOX 7062, Kampala, Uganda.
[2]National Agricultural Research Organization, National Livestock Resources Research Institute, P.O. Box 96, Tororo, Uganda.
[3]African Union-InterAfrican Bureau of Animal Resources, P.O. Box 30786, Nairobi, Kenya.

African swine fever (ASF) is a devastating viral disease of pigs and is among the major hindrances to pig industry in sub-Saharan Africa including Uganda. The aim of this study was to compare immunoperoxidase monolayer assay (IPMA) to PCR in detection of ASF virus in infected macrophage cultures and to categorize ASF viral isolates in Uganda by haemadsorption assay. Field strains of ASF virus were isolated from infected pigs into swine alveolar macrophages culture. The effect of the inocula on the cell culture was monitored daily and the presences of ASF virus in the inoculated macrophages were detected using PCR and IPMA. The isolates were then categorized by haemadsorption assay. 58.8% of the samples had ASF virus DNA and ASF virus was isolated from 27% of the samples. IPMA detected ASF viral antigens in 80% of the inoculated macrophages culture 48 hours post infection compared to the 100% by PCR. 95% of the virulent ASF viral isolates from Uganda were haemadsorbing. This study makes the first attempt to use IPMA and haemadsorption assay for the detection of ASF virus and categorization of the African swine fever virus (ASFv) field isolates into haemadsorbing and non-heamadsorbing in Uganda, respectively. The study demonstrates that IPMA is an appropriate option to PCR and could be used to detect ASF virus in cell cultures. It is recommended that the genome of the non-haemadsorbing ASF viral isolates could be sequenced and compared with that of haemadsorption (HAD) isolates to identify molecular peculiarities and markers of these two categories of ASFv.

Key words: African swine fever (ASF), African swine fever virus (ASFv), Immunoperoxidase Monolayer Assay (IPMA), swine alveolar macrophages, haemadsorption (HAD) Polymerase Chain Reaction (PCR).

INTRODUCTION

African swine fever (ASF) is a highly contagious haemorrhagic disease of domestic pigs caused by a large icosahedral DNA virus that belongs to genus *Asfivirus* and family *Asfarviridae* (Dixon et al., 2005). The disease has devastating effect on pig industry in Africa and is the major setback to pig production particularly in Uganda.

The disease has been reported annually in different regions of Uganda during the last ten years (Rutebarika and Nantima, 2002; Atuhaire et al., 2013).

Currently, there is neither vaccine nor treatment for ASF; the only control strategy for the disease is early detection of the disease followed by instituting strict disease control measures such as quarantine measures and movement control (Solenne et al., 2009). Laboratory diagnostic methods for ASF include viral isolation, detection of ASF viral genomic DNA, detection of viral antigens in porcine tissues and detection of antibodies against ASF virus antigens in serum (Wilkinson, 2000).

In endemic areas, serologic diagnosis is frequently used while in regions where the disease is newly introduced, it is preferable to detect the virus DNA or antigens. Methods which have been used for ASF virus detection include the long-established haemadsorption test (Malmquist and Hay, 1960), immunofluorescence (Colgrove et al., 1969), polymerase chain reaction (PCR) (King et al., 2003; Aguero et al., 2003, 2004), and recently LAMP (James et al., 2010). In Uganda, however, standard confirmatory diagnostic techniques are not in place in many laboratories hence, suspected ASF cases are diagnosed principally on the basis of clinical signs and postmortem lesions. Molecular diagnostic techniques of ASF in many countries have been limited to research only (Gallardo et al., 2011; Tejlar, 2012; Atuhaire et al., 2013). This is due to lack of appropriate diagnostic facilities and reagents in these countries (Oura et al., 2012). Lack of readily available and reliable diagnostic tests in many developing countries often delays the institution of effective ASF control measures; hence farmers often incur enormous losses through pig mortalities and loss of market for pigs and pig products.

Immuno-assays are often used in diagnosis of ASF, however, the OIE recommended ELISA has been reported not to detect some of the East African strains of ASF virus (Gallardo et al., 2011), hence the need for a more sensitive and readily available techniques for the confirmation of ASF in this region. Both *in situ* hybridization and immunocytochemistry have been compared and evaluated for localization of ASF virus in infected cells as a prerequisite for their use in the diagnosis and studies of the pathogenesis of ASF in domestic pigs, warthogs and bush pigs (Oura et al., 1998). Recent immunohistochemical studies carried out in Uganda to determine the prevalence of African swine fever viral antigens in slaughter pigs at Wambizi abattoir, Kampala revealed that of the slaughtered pigs with lesions suggestive of ASF, 0.1% had ASF viral antigens

in their tissues (Ssajjakambwe et al., 2011).

Viral isolation is one of the sensitive diagnostic methods; however it requires special bio-containment facilities and a source of swine monocytes and macrophages (OIE Manuel for Terrestrial animals, 2012). The isolated ASF viruses can be categorized by HAD and characterized by molecular techniques such as PCR.

Pig macrophages and monocytes *in vitro* are cells suitable for cultivation of ASF virus as wild type ASFV isolates do not replicate in conventional cell cultures. This is because ASF virus naturally infects and replicate in mononuclear phagocytic cells (Malmquist and Hay, 1960; Sanchez–Torres et al., 2003). In mononuclear phagocytic cells *in vitro*, ASF virus mimic natural infection and most strains of the virus grow readily in monocytes and macrophages culture (Carolina et al., 2010). These cells are frequently used in ASF viral isolation and haemadsorption diagnostic tests. Use of other cell lines for ASF virus cultivation and plaque formation assays require cell culture adapted virus strains (Carolina et al., 2010). Pulmonary lavage can produce sufficient yields of alveolar macrophages that can be used for ASF viral culture and titration (Bustos et al., 2002). The culture medium should be supplemented with serum of a pig from which the alveolar macrophages were obtained (Bustos et al., 2002). In many laboratories, bovine fetal serum is added to cell culture medium to supplement macrophages, however, Bastos et al. (2002) reported that addition of bovine fetal serum reduces infective viral particles in macrophages to 10 - 20% when the cell culture is infected with several ASF virus isolates. The team further suggested that, the ability of each particular batch of pig serum to support the production of infective virus should be tested. This is because the viral yield could differ between the alveolar cell stocks.

Polymerase chain reaction (PCR) is a specific diagnostic test that detects genomic DNA in body fluids, tissue samples and can be used even when the samples are unsuitable for virus isolation and antigen detection (Aguero et al., 2003). Although PCR is a reliable diagnostic assay, it requires specialist equipments and the risk of cross contamination is high (Oura et al., 2012). There is therefore a need to evaluate other cheap, reliable, affordable and readily available diagnostic assays for detection of ASF.

The aim of this study was to compare IPMA to conventional PCR in detection of ASF virus in infected cell cultures and to categorise ASF viral isolates in Uganda based on their ability to cause haemadsorption to infected macrophages. This was because IPMA and

PCR can detect both haemadsorbing and non haemad-sorbing isolates as opposed to HAD test that only detect pathogenic haemadsorbing ASFv. IPMA is an immune-logical test that is based on the principle of the specific binding of antibodies to antigen. It detects ASF viral protein in fixed cells, tissues and is less expensive as it does not require specialist equipments unlike PCR.

MATERIALS AND METHODS

Study design

This was a prospective comparative study in which IPMA was compared against conventional PCR, (used as a reference test in this study) in detection of ASF virus in cell culture. ASF virus was isolated from field and experimental samples. Pig tissue samples were collected for viral isolation. Field ASF viral isolates were obtained from all the four major regions of the Uganda, namely, Northern, Eastern, Southern and Western regions. From each region at least 20 sets of tissue samples from dead or clinically sick pigs were collected depending on the frequency of outbreaks of ASF in each region. Stored samples from experimentally infected pigs in previous studies were also included in this study. The sample were then screen for ASF virus using conventional PCR, samples that had ASF virus were processed and filtrate used to inoculate swine macrophages culture. The cytopathic effects of the inoculums on the macrophages were monitored daily and presence of ASF virus in the culture was comparatively confirmed using PCR, IPMA and HAD assay. The details are as described under the subsequent subheadings.

Sample collection, preparation and ASF virus isolation in pig macrophages culture

Tissue samples for viral isolation were collected from 148 car-casses of pig suspected to have died of ASF or clinical cases. Of these, 136 samples were from the field cases and 12 were preserved samples from experimentally infected pigs. Samples collected for viral isolation included spleen, haemorrhagic lymph nodes, pharyngeal tonsils and kidneys. The samples were screened for ASF viral DNA using PCR and tissues that had ASF viral DNA were macerated, PBS then added. The suspension was sheaved through sterile gauze and filtered through 0.45 μ Whatman filter. Viral isolation was done in pig alveolar macrophages which were harvested from 4-5 months old healthy pigs based on procedure described elsewhere (Carrascossa et al., 1982). The viability of the harvested macrophages was evaluated using trypan blue dye exclusion technique (Strober, 2001) and cell viability of above 90% was used for viral isolation. The viable macrophage suspension was then transferred into six-well culture plates and incubated at 37°C in 5% CO_2 for 24 h. To the macrophage cultures, 100 μl of the sample filtrate (viral suspension) was added per well in six well plates. The inoculated plates were incubated at 37°C, 5% CO_2 for one hour to enable viral adsorption to take place. This was followed by addition of growth medium (EMEM) containing 5% pig serum to the inoculated macrophages and incubated under the same conditions. The effect of the inoculums on the cell culture was monitored daily for cytopathic effects (CPE). The success of ASF virus isolation was confirmed by PCR and IPMA. ASF virus isolates were characterised based on the ability of the virus to induce haemadsorption on the infected macrophages.

Immuno-peroxidase monolayer assay (IPMA) for detection of ASF virus

Immuno-peroxidase monolayer assay for detection of the ASF viral proteins in the infected macrophages was done using a modified protocol described by Direksin et al. (2001) and Liang et al. (2013). In brief, the culture medium in the wells discarded and 1 ml of 10% buffered formalin containing 1% Nonidet P40 (NP40) was added and the plates were incubated at room temperature for 30 min to allow thorough cell fixation. The fixed cells were then washed three times using 0.5% Tween 80 in PBS to remove excess fixative. The primary antibody used was lyophilised hyperimune pig serum raised against ASF viral antigens in experimental pigs (Kindly given by Dr. Gallardo from Spain). It was diluted 1:400 in PBS (PH 7.2) con-taining 2.5% pig serum and 600 μl of the diluted antibody was added to each well and incubated for 30 min at 37°C. The wells were then washed thrice followed by addition of 600 μl of anti pig IgG peroxidase conjugate diluted 1:600 in 0.5% Tween 80 in PBS and incubated at 37°C for 30 min. The wells were then washed with PBS (pH 7.2). Finally, 600 μl of substrate, 3-amino-9-ethyl carba-zole (AEC) solution was added to each well and incubated at room temperature for 30 min. The cells were then examined under inverted microscope to evaluate the IPMA results. Red intracytoplasmic staining of the macrophages was considered positive for ASF virus antigens. Non specific staining was considered doubtful and repeated, while unstained cells were considered negative for IPMA, hence not infected by ASF virus. PCR was also used to confirm ASF viral isolation alongside IPMA.

Haemadsorption (HAD) test

Haemadsorption test was done based on modification of the established protocol (Malmquist and Hay, 1960). In brief, the pig alveolar macrophages culture was prepared, seeded in six well plates and inoculated with ASF virus isolates as earlier described. To prevent non specific haemadsorption; the serum and the alveolar macrophages used were from the same pig. The inoculated plates were incubated at 37°C in 5% CO_2 atmosphere. 600 μl of 0.5% freshly prepared pig erythrocytes in buffered saline was added onto the inoculated cells per well and cultures were then examined for haemadsorption daily for 5 days. The result was considered positive when pig erythrocytes adhere to the surface of the infected macrophages forming a ring or clusters. It was declared negative when neither haemadsorption nor CPE occurred in inoculated cell cultured. Infection by non haemadsorbing ASF virus was suspected when CPC was observed but no haemadsorption occurred and this was confirmed by PCR and IPMA.

ASF viral DNA extraction and PCR

ASFv DNA was extracted from pig tissue samples and inoculated cell culture using QIAamp DNA mini kit for blood and tissues (www.qiagen.com/products/dna/qiaamp-dna-mini kit, QIAiamp® DNA and Blood mini handbook 2012). The DNA was eluded in 10 mM tris hydrochloride (PH 7.8, eluent volume of 200μl) and stored at -20°C until testing. PCR was performed using Go Tag® green Master Mix PCR kit (Promega Corporation USA2012). The primers designed against vp 72 region of the ASF virus genome and recommended for detection of ASF DNA by OIE manual for diagnostic tests and vaccines for terrestrial animals (2009) was used, thus (F 5' – ATG GAT ACC GAG GGA ATA GC – 3', R 5' CTT ACC GAT GAA AAT GAT AC – 3'), (Wilkinson 2000). The reaction mixture consisted of 12.5 μl 1x GoTaq®green Master mix,

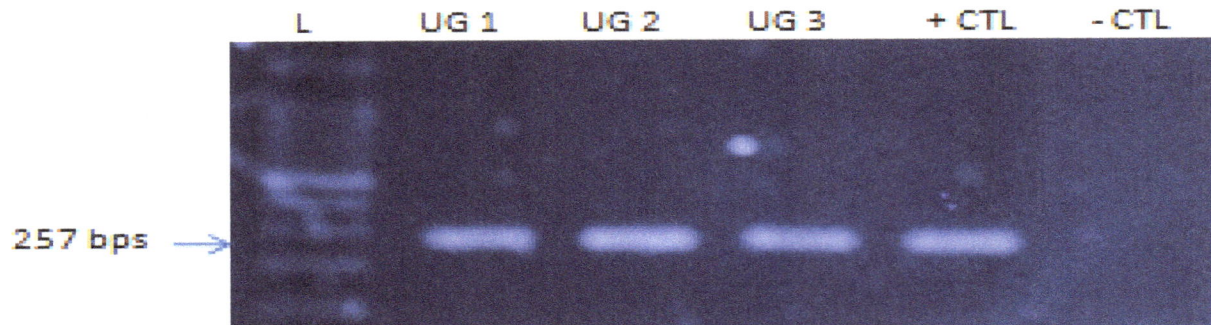

Figure 1. Detection of ASF viral DNA from cell culture by PCR. Culture 1 to 3 (Ug1, UG 2 and Ug 3) had ASF viral DNA, hence were all positive, + CTL is positive control culture, − CTL is negative control and L is the DNA ladder.

1 µl of 0.8 µM forward and reverse primers each, 2 µl of DNA template and 8.5 µl of double distilled water in a total reaction volume of 25 µl. The mixture was loaded into and run using MultigeneTM Optimax labnet thermal cycler (Multigene Labnet 99.9°C and temperature accuracy of ±0.5/±0.5. Thermal cycling started at 94°C denaturation 35 cycles each for 30 s, followed by annealing at 50°C and extension at 72°C using reaction volume of 25 µl. PCR products obtained were electrophoresed in 1.2% agarose gel (Nacalai Tesque, inc Kyoto, Japan lot № M2K1602), illuminated by UV system and images were photographed.

Ethics statement

Full ethical clearance was obtained from the Uganda National Council for Science and Technology (UNCST) and the College of Veterinary Medicine, Animal Resources and Bio-security of Makerere University under reference number VAB/REC/11/110. Animal welfare and care was ensured in accordance with the International Guideline on Animal Welfare and Euthanasia. Any experimental animal in pain or moribund was immediately euthanized to relieve it from further suffering. Clean water and commercial feed were provided *ad libitum* to all pigs during study period.

RESULTS

Sample collection, screening and ASF virus isolation

Out of the 148 pigs autopsied, 12 carcasses were from experimentally infected pigs and 136 were field cases. 69.6% (n =103) of the carcasses had lesions suggestive of ASF and diagnostic PCR done on all the tissue samples revealed that 58.8% (n= 87) of the samples had ASF viral DNA (Figure 1). Of the 87 samples with ASF viral DNA, 82 samples were from pig carcasses that had lesions suggestive of ASF while 5 PCR positive samples were from pigs without lesions suggestive of the disease. More still of the 75 pig tissue samples obtained from the field that had ASF viral DNA, ASF virus was successfully isolated from only 28 samples. On the other hand, ASF virus was isolated from tissues of all the 12 (100%) experimentally infected pigs, making total isolates to be 40 (Table 1).

International inc.), with programmable temperature range of 4 to

Comparative detection of ASF virus in cell culture by PCR, IPMA and HAD test

To detect ASF virus in the infected cell culture with time, we used conventional PCR and IPMA, comparatively. Viral isolates were then categorized based on their ability to cause haemadsorption in infected macrophages. PCR detected all the 40 (100%) isolates while IPMA detected 32 (80%) of the isolates in macrophages culture within 24 h post infection. However, HAD-test did not show distinct result in the first day post infection in the infected cells. In the second to the fifth day post infection PCR and IPMA detected all ASF virus isolates 100% (n = 40) while HAD detected 95% (n = 38) of the isolates (Tables 2 and 3, and Figure 2).

Categorization of Ugandan ASF virus isolates using haemadsorption (HAD) test

Of the 40 isolates confirmed by IPMA and PCR, two isolates (5%) were non-haemadsorbing, although they caused clinical disease in experimental pigs and CPE in macrophages cell cultures similar to the rest of the isolates (Figure 3E). Erythrocytes distinctly adhered to the surfaces of the infected macrophages by 48 h post-infection (PI) and by 72 h PI erythrocytes clumped on the remaining intact macrophages (Figure 3F).

Morphological changes in infected macrophages post inoculation

In the first two days post infection (48 h PI), there were no observable changes in the infected macrophages. However, at 56 h post inoculation (slide C of Figure 4),

Table 1. Total number of pigs autopsied and the different tests results.

Sources of samples	Carcasses autopsied and sampled	Carcasses with ASF-like lesions	PCR positive cases		Viral isolation
			Carcasses with ASF-like lesions	Carcasses without ASF-like lesions	
Field pigs samples	136	91	70	5	28
Experimental pigs samples	12	12	12	0	12
Total	148	103 (69.6%)	82 (55.4%)	5 (3.4%)	40 (27%)

Table 2. Detection of ASF virus in cell culture by PCR, immunoperoxidase monolayer assay (IPMA) and haemadsorption test (HAD) 73 h post inoculation (PI).

Diagnostic test	No. of isolates	No. positive	Percentage
PCR	40	40	100
IPMS	40	40	100
HAD	40	38	95

Table 3. Progressive detection of ASF virus in cell culture by PCR, IPMA and HAD

Diagnostic Assay	Days post inoculation				
	1	2	3	4	5
PCR	++++	++++	++++	++++	++++
IPMA	++	+++	++++	++++	++++
HAD	-/+	++	++	+++	+++

PCR = Polymerase chain reaction, IPMA = immunoperoxidase monolayer assay, HAD = haemadsoption test. + (plus) = Positive test result indicating week detectable ASF virus infection in the cell culture and ++++ strongest positive test result. - (minus) = negative test result, viral DNA, and proteins not detected by a given test at that period

ASF virus infected macrophages were enlarged and more rounded (balloon degeneration). By 72 h post-inoculation (slide D Figure 4) about 60% of the infected cells detached from the surface of the culture plates and were lysed and the intact macrophages were rounded and swollen. The changes are as shown in Figure 4.

DISCUSSION

In the present study, 69.5% of the sampled pigs had lesions suggestive of ASF; 58.9%, of the samples had ASF DNA. However, lesions due to ASF are not pathognomonic and are often confused with lesions due to other haemorrhagic swine diseases such as classical swine fever, septicaemic salmonellosis, acute trypanosomiasis due to *Trypanosoma Simiae* infection and thrombocytopaenic parpura (Kleiboecker, 2002). To differentiate these haemorrhagic diseases, the use of laboratory test is a prerequisite (Radostits et al., 1995; Aguero et al., 2003). Therefore, pigs that had lesions similar to those of ASF and were diagnosed negative for the disease which could have died of other swine haemorrhagic diseases. This finding therefore emphasises the limitation of relying on clinical and pathologic diagnosis of ASF which happen to be the common practice in developing countries where laboratory diag-nostic services are not readily available. It therefore calls for confirmatory diagnostic capacity to be established to address such suspected cases.

Isolation of ASF virus in this study was done from both field and experimental pig samples. Viral isolation and detection is one of the key diagnostic tests for ASF and in this study, ASF virus was successfully isolated from only 27% (n = 40) of the collected samples, though 58.9% of the total samples collected had ASF viral DNA. This could be attributed to the state at which samples were obtained from the field. Some tissue samples from field were obtained from pigs that had died in the previous day and hence the tissues were partly autolysed. This limited the possibility of isolating ASF virus from such samples. However, PCR was able to detect ASFv DNA in some of the autolysed samples from which viral isolation was not successful. This is contrary to the notion that ASFv is resistant to a number of physical conditions and as it is known to persist in putrefying tissues for several days. On the other hand ASFv were isolated successfully from all samples obtained from carcasses of experimental pigs (n =12) and this was likely due to the controlled environmental conditions in which the experiment was done and samples were collected immediately after death and stored at -20°C awaiting the process of viral isolation. This minimised the chances of viral inactivation, hence good result of viral isolation obtained.

In addition to viral isolation and detection, several techniques have been used to diagnose ASF; these include pathologic diagnosis, immunoassays, PCR and haemadsorption test (HAD). The commercially available diagnostic technologies have varying sensitivities and specificities. In this study, we compared conventional PCR and IPMA which are among sensitive, rapid and

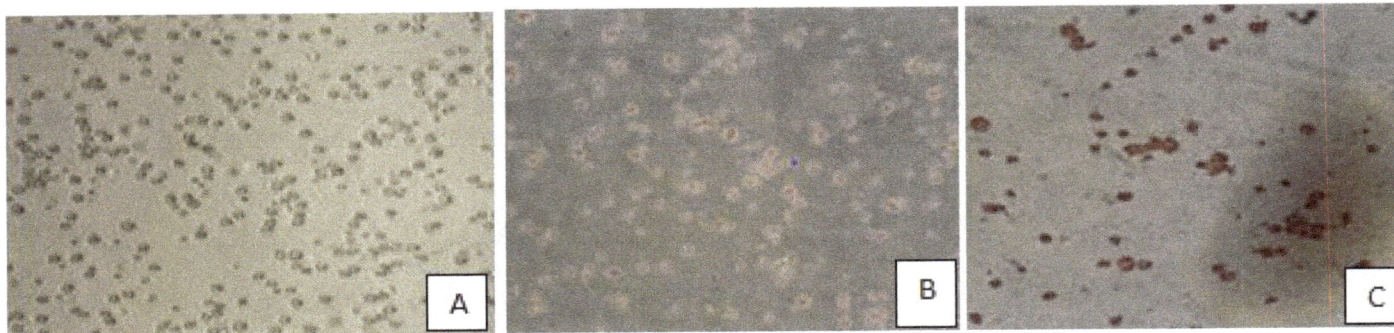

Figure 2. Detection of ASF viral antigen from cell culture by IPMA at different time intervals. Culture A is non infected macrophages 72 h PI (negative control), B is ASF virus infected macrophages 48 h PI and C is ASF virus infected macrophages 72.

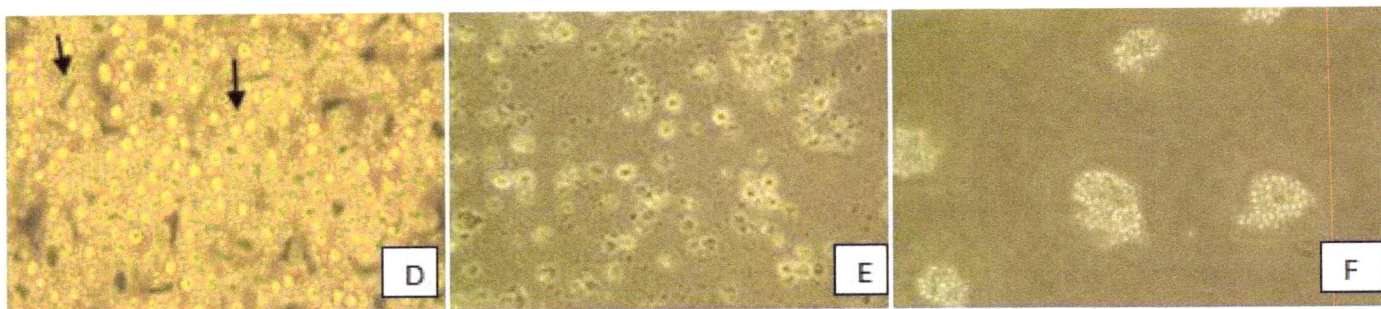

Figure 3. Haemadsorption of pig erythrocytes to ASF virus infected macrophages. Slides D and F indicate positive HAD test result 48 and 72 h post–infection, respectively. Slide E, shows negative HAD test of macrophages infected with non-HAD ASF.

specific diagnostic tests (Oura et al., 2002) to detect ASF viral DNA and antigen in ASF virus infected macrophages at different durations post-infection. Previous studies showed that PCR was able to detect all known ASF virus genotypes including that of non-haemadsorbing and less pathogenic isolates. Furthermore it detected ASF genome in degraded or inactivated samples (Oura et al., 2002). Despite the enormous advantages of using PCR as diagnostic test for ASF, it is prone to cross contamination; hence false positive results may occur (Oura et al., 2002). Oura et al. (1998) detected ASF DNA in ASFv infected IBRS2 cells 12 h post infection using in situ hybridisation employing an35S labelled DNA probe. In this study, we found that ASF viral DNA and proteins (antigen) in infected macrophages were at detectable levels within one day post-infection by conventional PCR and IPMA, respectively. PCR detected all the 40 (100%) isolates while IPMA detected 32 (80%) isolates in macrophages culture within 24 h post infection. However, HAD assay did not show distinct result in the first day post infection in some of the infected cells. From the second to the fifth day post infection all the three assays used (PCR, IPMA and HAD) clearly detected ASF viral DNA, antigen and

surface adhesive molecules on the infected macrophages, respectively. By the 48 h post infection, PCR and IPMA detected all ASF virus isolates 100% (n = 40) while HAD detected 95% (n = 38) of the isolates. This shows that IPMA can be highly specific in detection of ASF virus in cell cultures. However, PCR was in this case more sensitive than IPMA as the latter detected ASF virus in all the 40 isolates in macrophages culture within 24 h post infection unlike IPMA that detected 80% of the isolates in the same period of culture. Oura et al. (1998) reported that detection of ASFviral protein vp73 using immunocyto-chemistry was as sensitive as the use of DNA hybridisation and the team noted that, immunocytochemistry assay is a quick, safe and easy diagnostic technique that allows morphological detection of the antigen. Oura et al. (1998) further reported that, attenuated ASF virus isolate infected a high percentage of endothelial cells, alveoli and bone marrow derived macrophages. In this study, we demonstrated that within 24 h post-infection, macrophages infected with Ugandan isolates of ASF virus had detectable amounts of ASF virus proteins and the viral protein concentration increased with time post inoculation. The detection of ASF antigen in infected cells within24 h on this study is in agreement with what was reported by Gallardo et al. (2012), though the later first detected ASF virus 48 h PI in infected COS – 1 cells. The difference in the period of first detection of ASF viral antigen in the infected cells could be due to the different cell types used in the two studies, more so macrophages being the natural target cells for ASF virus unlike COS – 1 cells.

Although earlier studies reported that the OIE recommended ELISA could not detect some of East African strains of ASF virus (Gallardo et al., 2011), the hyper immune serum raised against European (Spain) isolate of ASF virus was able to detect all Ugandan isolates of the virus using IPMA. This was probably because direct IPMA was used to detect ASF viral proteins in cell culture in this study. More so, the viral antigen concentration in infected macrophages culture was probably higher as compared to that in naturally infected pigs.

Majority (95%, n = 38) of ASF virus isolates from Uganda during this study caused adsorption of erythrocytes to infected cells, however 5% of the isolates were non haemadsorbing (non-HAD). Haemadsorption test (HAD) is one of the oldest diagnostic tests for ASF (Malmquist and Hay, 1960) and is based on the fact that porcine erythrocytes adhere to the surface of the infected swine monocytes and macrophages. Most of virulent ASF virus isolates are known to be haemadsorbing and a very small percentage of the isolates are non haemadsorbing (Boinas et al., 2004). Previous studies showed that most of non haemadsorbing ASF strains were a virulent. How-ever, it is well known that a small proportion of non-haemadsorbing ASFv do cause acute disease (Gonzague et al., 2001). This property of haemadsorption could be used as diagnostic assay for haemadsorbing strains of ASF virus in cell culture (Boinas et al., 2004). Specific protein CD2v is responsible for haemadsorbing infected leukocytes (Kay-jackson et al., 2004). CD2v is encoded by a gene EP402R and this protein is similar to adhesion receptor or CD2 on T – lymphocytes (Rodriquez et al., 1993). EP402R gene is reported to be responsible for the adhesion of swine erythrocytes to the infected leukocytes, while EP153R encodes for protein that stabilizes the adhesion molecules (SOP/CISA/ASF/VI/I/2008). In addition to causing HAD, CD2v is also associated with impairment of lymphocytes replication in response to mitogens (Barca et al., 1998). Contrary to what we found in this study, Kay-Jackson et al. (2004) reported that non HAD ASF virus did not cause disease in experimental pigs but induced antibody production that persisted throughout their study period (49 days post infection). The pigs infected with non pathogenic, non-HAD ASF virus isolates were protected against pathogenic HAD virus isolated from the same farm. However, in this study we found that the non-HAD ASF virus isolates were virulent and caused cytopathic effect (CPE) in infected cells by 56 - 72 h post inoculation and they caused clinical disease in infected

pigs similar to that due to HAD ASF virus. Our findings were in conformity with what was reported by Gonzague et al. (2001) that showed that some of the non-HAD ASF viral isolates caused high mortalities of 80-90% in domestic pigs in Southern Africa and Madagascar. This shows that phenotypic characteristic of haemad-sorption or non-haemadsorption of ASF virus is not exclusively the determinant of the pathogenicity of ASF isolates. The finding of Gonzague et al. (2001) was similar to what we reported in this study where the non HAD ASF virus isolates were virulent in domestic pigs and caused clinical disease in experimentally infected pigs. This limits the sensitivity of HAD assay in detection of ASF virus in cell culture as false negatives test result can be generated in case of non haemadsorbing ASF virus isolates. Although, viral isolation and detection by HAD is internationally accepted diagnostic test, the result should be confirmed by other tests such as immuno-assays or PCR (Oura et al., 2002).

Previous studies revealed that non-pathogenic non-HAD ASF virus often cause sporadic viremia in infected pigs and small amounts of the virus occur in various organs of the infected pigs (Kay-Jackson et al., 2004). This makes isolation of non-pathogenic and non-HAD ASF virus more difficult than HAD virus. This could explain why probably, we were unable to isolate non HAD avirulent strains of ASF virus in this study. The non-haemadsorbing avirulent ASF virus isolates have genomic deletions unlike the HAD virus isolated (Kay-jackson et al., 2004). The low pathogenicity of some of the non-HAD isolates may be related to loss of virulence factors associated with the deleted genes which is probably associated with mutation in the gene that encode adhesion protein CD2v (Zsak et al., 2001). In Portugal and Spain, non-HAD ASF virus were isolated in many pig tissue samples during the period of attempted vaccination of pigs against ASF (Vivagio et al., 1974). Nevertheless in the Ugandan case, there were no documentation showing vaccination attempt to control or prevent ASF outbreak in the country in the previous years. The emergence of pathogenic non-HAD ASF virus isolates was probably due to natural phenomenon (mutation).

Cytopathic effect in ASF virus infected macrophages was clearly observed by the third day post infection (56 -72 h PI) in this study. The infected cells were distended and many of them were detached from the surface of the culture plates and lysed. Greig et al. (1967) pointed out that it is difficult to distinguish between the true cytopathic effect due to ASF virus and that as a result of cell degeneration caused by other factors especially in commercial cell lines. For example in vero cells, true cytopathic effect is clearer after several passages, especially between the 4th and 8th passages and it reaches its maximum between 8th and 20th passages (Greig et al., 1967). Cytopathic effect reflects the quantity of virus production and the state of adaptation of the virus in a given cell type.

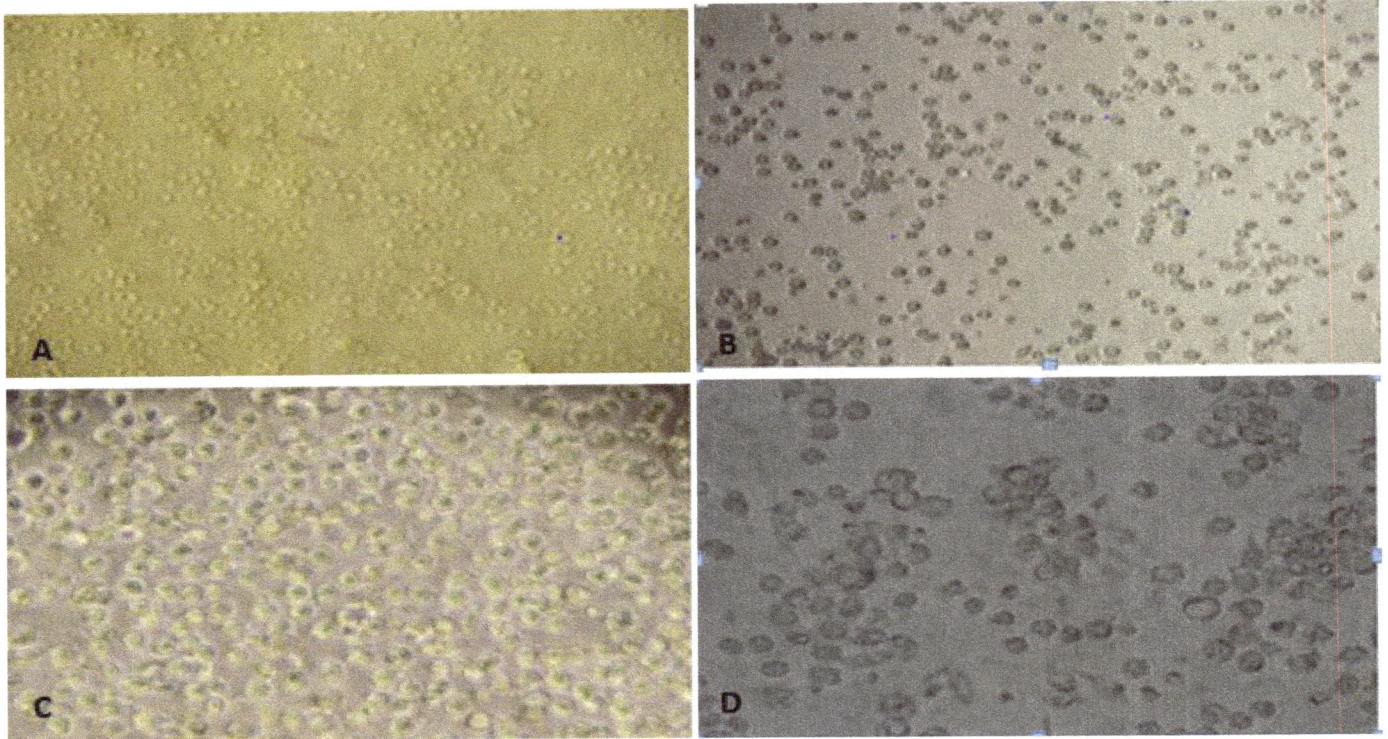

Figure 4. Changes observed at x10 (A and B) and x20 magnification in macrophages infected with Ugandan isolates of ASF virus. (A): non infected macrophages 72 h post infection, no changes noted in cell sizes, (B): infected macrophages 48 h PI no significant morphological change was noted, (C): infected macrophages 56 h PI, the cells were swollen and rounded. (D): Infected macrophages 72 h PI, majority of the infected cells were lysed and intact cells swollen.

Success of adapting a virus isolate to a foreign host (cell) depends on the degree of virus selection. It is usually those members of the viral population that are best suited to grow in foreign cell line that eventually become dominant in the cell culture. ASF virus generally adapts slowly to grow in pig kidney cells and cytopathic effect takes a longer time as compared to other viruses (Greig et al., 1967). Unlike what has been reported in conventional cell lines, in this study, CPE was first observed by the 56th h post inoculation in infected macrophages and by 72 h post infection majority of the cells were infected by ASF virus and had CPE (Figure 4D and B. To confirm that the observed CPE was due to ASF virus, we detected the presence of ASF viral DNA and antigens in the infected macrophages by conven-tional PCR and IPMA. The cytopathic effects of ASF virus on infected macrophages at different durations post infection were as shown in Figure 4A to D. Sanchez–Torres et al. (2003) noted that pig macrophages and monocytes *in vitro* are cells of choice for cultivation of ASF virus and field isolates of ASF virus do not replicate in conventional cell cultures. This is because ASFv naturally infects and replicates in mononuclear phagocytic cells (Sanchez–Torres et al., 2003, Malmquist and Hay, 1960). In mononuclear phago-

cytic cells *in vitro*, ASF virus mimics natural infection and most strains of the virus grow readily in monocytes and macrophages culture (Carolina et al., 2010). This probably could explain the early CPE observed in this study and the high infectivity of the isolates used might also contribute to the short time of the observable effect of the virus on the cell culture. The viral suspension in the sample filtrate used in this study was evaluated in terms of haemadsorbing units and the actual viral load per unit volume was not titrated. Macrophages cultures that showed early CPE (54 hours PI) probably had higher viral load than others where CPE appeared 72 h PI.

Challenges and limitations of the study

Samples for this study were obtained from pigs reported to have died of swine haemorrhagic diseases, hence probably only virulent strains of ASF virus were isolated. Other less virulent or avirulent strains of ASF virus were not used in this study.

The study was limited to strains of ASF virus obtained from domestic pigs in Uganda, though it is known that wild swine species and soft ticks (*Ornithodoros moubata*) which have no boundaries are the reserviour hosts and

Comparison of immunoperoxidase monolayer assay, polymerase chain reaction and haemadsorption tests in the...

209

vectors of ASF virus, respectively.

CONCLUSIONS AND RECOMMENDATIONS

Majority (95%) of the ASF virus isolates from Uganda during this study were haemadsorbing though 5% of the isolates were non haemadsorbing. Virulent African swine fever virus isolates from Uganda caused noticeable cytopathic effect in infected macrophages as early as 56 h post inoculation.

Imunoperoxidase monolayer assay was able to detect ASF viral antigens in the macrophages culture as early as 48 h post infection similar to PCR though the sensitivity of PCR is more than the former at this time. Therefore, IPMA is an appropriate option to PCR technique, which could be used to detect ASF viral antigens in cell culture especially in less established laboratories as it detects both haemadsorbing and non haemadsorbing strains of ASF virus.

We recommend that a survey should be conducted to investigate occurrence of classical swine fever and other pig haemorrhagic diseases in Uganda as only 58.8% of the pigs that had lesions suggestive of ASF had ASF viral DNA. More so, the genome of the non-haemadsorbing virulent ASF viral isolates should be sequence to see if there is nucleotide sequence variation as compared to HAD isolates. A study could also be conducted to determine the virulence factor in ASF virus isolates and explain why some non-HAD ASF virus were also pathogenic.

Declaration of non conflict of interest

The authors declare that there is no conflict of interest in this study

ACKNOWLEDGEMENTS

This study was funded by the Millennium Science Initiative, under the Uganda National Council of Science and Technology through a grant to Prof. Lonzy Ojok, Dr. William Olaho-Mukani and Dr. Julius Boniface Okuni of the Appropriate Animal Diagnostic Technologies project. The authors acknowledge the technical guidance of Dr. Yoshikazu Iritani, a Japanese virologist and technical expert to Uganda on JICA project for improvement of diagnostic services in the country, for giving technical guidance during ASF virus isolation and pathogenicity study. Special regard to JICA, the management and staff members of Central Diagnostic Laboratory, College of Veterinary Medicine, Animal Resources and Biosecurity, Makerere University, Uganda, for providing conducive environment for accomplishing this research.

REFERENCES

Aguero M, Fernandez J, Romer L, Sanchez Mascaraque C, Arias M, and Sanchez Vizcaino JM (2003). Highly sensitive PCR assay for routine diagnosis of African Swine fever virus in clinical samples. J. Clin. Microbiol. 41(9):4431-4434.

Ana LR, Parkhouse R.M.E, Ana RP, Carlos M and, Alexandre L (2007). Systematic analysis of longitudinal serological responses of pigs infected experimentally with African swine fever virus. Portugal. Ave. Univ. Tec. Lisbon.

Anderson EC (1986). African swine fever, current concepts on its pathogenesis and immunology. Revised scientific technique for OIE, 5 (2): 477-486. Depat. Experiment. Pathol. animal virus Res. institute, pirbright, Surrey GU 24 ONF, UK.

Atuhaire DK, Afayoa M, Ochwo S, Mwesigwa S, Okuni J.B, Olaho-Mukani W, and Ojok L (2013). Molecular characterisation and phylogenetic study of African swine fever virus isolates from recent outbreaks in Uganda (2010 – 2013). J. Virol.10:247-248.

Boinas FS, Hutchings GH, Dixon LK and, Wilkinson PJ (2004). Characterisation of pathogenic and non –pathogenic African swine fever virus isolated from Ornithodoros eraticus inhabitic pig premises in Portugal. J. Gen. virol. 85(8):2117-2187.

Boinas FS., Cruz B, Portugal FC, Portugal R, Mendes s, Leitao A, Matin SC and, Rosinha A (2001). Evaluation of the role of ornithodorus erraticus as a reservoir of African swine fever in Alentego – Portugal. Report on the annual meeting of national swine fever laboratories. 97-98.

Borca MC, Carrillo C, Zsak L, Laegreid WW, Kutish GF, Neihan JG, Burraje TG and, Rock DL (1998). Deletion of a CD2 like gene, 8 DR from African swine fever virus affects viral infection in domestic swine. J. virol. 72:2881-2889.

Carolina urtado, Maria Jose Bustos, Angle L. Carrascosa (2010). The use of Cos – 1 Cells for studies of field and Laboratory African swine Fever Virus samples. Elsevier editorial system (tm) J. Virol. methods.

Carrascossa AL, Santeren JF and, Vinuela E (1982). Production and titration of African swine fever virus in porcine alveolar macrophages. J. Virol. methods 3:303-310.

Casal I, Enjuanes L, Vinuela E, (1984). Porcine leukocytes cellular subsets sensitive to African swine fever virus in vitro. J. Virol. 52:37-46.

Colgrove GS, Haelterman EO and, Coggins L (1969). Pathogenesis of African swine fever in young pigs. American J. Vet. Res. 30: 1343-1359.

Direksin K, Joo H and, Goyal MS (2002). An immunoperoxidase monolayer assay for detection of antibodies against swine influenza virus. J. Vet. Diagnos. investig. 14:169.

Dixon LK, Escribano JM, Martins C, Rock DL, Salas ML, Wilkinson PJ (2005). Asfarviridae in : Fauquent CM, Mayo MA, Maniloff J and Ball LA (eds). Virus Taxonomy, VIIIth Report of International Committee on Taxonomy of Viruses. London, UK. Elsev. Academ. Press: 135-143.

Gallardo C, Ademun AR, Nieto R, Nantima N, Arias M, Martin E, Pelayo V, Bishop RP (2011). Genotyping of African swine fever virus isolates associated with disease outbreaks in Uganda in 200. Afr. J Biotechnol. 10(17):3488-3497.

Gallardo C, Solar A, Nieto R, Carraishop RP, Martin C, Fasine FO, Couacy- Hymman E, Heath L, Pelayo V, Martin E, Simon A, Martin R, and Arias M (2012). Comparative evaluation of noval African swine fever virus antibody detection techniques derived from specific African swine fever viral genotypes with the OIE Internationally prescribed serological tests. Vet. microbiol.

Go Tag® green Master Mix 2012, Promega Corporation USA. Product information sheet.

James HE, Ebert K, Mc Gonigle R, Reid SM, Boonham N, Tomlinson JA, hutchings GH, Denyer M, oura CA, Dukes JP and , King DP (2010). Detection of African swine fever by loop mediated isothermal amplification. J. Virol. Methods. 1640:68-74.

Kay- Jackson PC, Goatley LC, Cox L, Miskin JE, Parkhouse RME, Wienands J And, Dixon LK (2004). The CD2v protein of African swine fever virus interact with the action binding adapter protein SH 3P7. J. Gen. Virol. 85:119-130.

Kleiboecker SB (2002). Swine fever: classical swine fever and African swine fever. Vet. Clin. Food anim. 18:431-451.

Liang H, Wang H, Zhang L, Gu H and, Zhang G (2013). Development of a novel Immuno-Peroxidase Monolayer Assay for detection of Swine hepatitis E virus antibodies based on stable cell lines expressing the ORFS protein.

Malmquist, WA, and Hay D (1960). Haemadsoption and cytopathic effect produced by African Swine fever Virus in bone marrow and buffy coat cultures. Ame. J. Vet. Research. 21:104-108.

Mebus CA (1998). African swine fever In: Foreign animal diseases. United States animal Health Association, Richmound. 52-61.

OIE Terrestrial manual (2012). African swine fever; Chapter 2.8.1 section 2.8 SUIDAE. Version adopted by the World assembly of delegates of the OIE in May 2012.

Oura CA, powell PP, Parkhouse RM, (1998). Detection of African swine fever virus in infected pig tissues by immunocytochemistry and in situ hybridisation. J. Virol. Methods. 72: 205-217.

Oura CAL, Edwards L, and balten CA (2012). Virological diagnosis of African swine fever. Comparative study of available tests.

Pan IC and Hess WR (1984). Virulence in African swine fever: its measurement and implication. Ame. J. Vet. Res. 45:361-366.

Ruteberika C and ,Nantema (2002), Uganda and neighbouring under serious threat of African swine fever. Emergency prevention system for transboundary Animal Pests and Diseases. Htt://www.fao.org/AG/AGAInfo/programmes/en/early warning.

Sajjakambwe P, Okee-Acai J and, Ojok L (2012). Prevalence of African swine fever viral antigens in slaughter pigs at Nalukolongo abattoir, Kampala. ROAVS. www.roavs.com.

Solenne C, Barbara W, Willian de Glanville, Ferran J, Rebecca R, Wilna V, Francois R, Dirk UP and, Linda KD (2009). African Swine fever, How can Global Spread be Prevented, philosophical transactions. UK, J.Royal soc. Pub. Biol. Sci. 364(1530):2683-2696.

Strobe (2001). Trypan blue exclusion test for cell viability. Current protocol in immunology. US National Library of Medicine, National institute of Health. Pubmed, PMD 18432654.

Tejler Emma (2012). Outbreak of African swine fever in domestic pigs in Gulu distric- Uganda. Swedish university of Agricultural sciences (SLU). ISSN 1652-8697. http:/epsilon.slu.se

Vigario JD, Terrinha AM and, Moura Nunes JF (1974). Antigenic relationships among strains of African swine fever virus. Arch Gesamte virus forsch 5:272-277.

Wilkinson PJ (2000). African swine fever manual of standards for diagnostic test and vaccines. 4th Edition, International Office for Epizootics, Paris France. 189 -198. www.qiagen.com/products/dna/qiaamp-dna-mini kit, QIAiamp® DNA and Blood mini handbook (2012)

Zsak L, Lu Z, Burrage TG, Neiland JG, kutish GF, Moore DM and, Rock DL (2001). African swine fever virus multigene family 360 and 530 genes are novel macrophage host range determinants. J. Virol. 75: 3066-3076.

Permissions

List of Contributors

Kamal Sharma
International Institute of Tropical Agriculture, PMB 5320, Ibadan, Nigeria
Central Tuber Crops Research Institute, Thiruvananthapuram, Kerala -695017, India
International Institute of Tropical Agriculture, PMB 5320, Ibadan, Nigeria

Raj Shekhar Misra
International Institute of Tropical Agriculture, PMB 5320, Ibadan, Nigeria

M. Tavasoli
Division of Microbiology, College of Science, Alzahra University, Tehran, Iran

N. Shahraeen
Plant Virus Research Department, Iranian Research Institute for Plant Protection (IRIPP), Tehran, P.O. Box-19395-1454, Iran

SH. Ghorbani
Division of Microbiology, College of Science, Alzahra University, Tehran, Iran

Mohd Apandi Yusof
Virology Unit, Infectious Disease Research Centre, Institute for Medical Research, Kuala Lumpur, Malaysia

Lau Sau Kuen
Virology Unit, Infectious Disease Research Centre, Institute for Medical Research, Kuala Lumpur, Malaysia

Norfaezah Adnan
Virology Unit, Infectious Disease Research Centre, Institute for Medical Research, Kuala Lumpur, Malaysia

Nur Izmawati Abd Razak
Virology Unit, Infectious Disease Research Centre, Institute for Medical Research, Kuala Lumpur, Malaysia

Liyana Ahmad Zamri
Virology Unit, Infectious Disease Research Centre, Institute for Medical Research, Kuala Lumpur, Malaysia

Khairul Izwan Hulaimi
Virology Unit, Infectious Disease Research Centre, Institute for Medical Research, Kuala Lumpur, Malaysia

Zainah Saat
Virology Unit, Infectious Disease Research Centre, Institute for Medical Research, Kuala Lumpur, Malaysia

Mohd Apandi Yusof
Virology Unit, Infectious Disease Research Centre, Institute for Medical Research, Kuala Lumpur, Malaysia

Lau Sau Kuen
Virology Unit, Infectious Disease Research Centre, Institute for Medical Research, Kuala Lumpur, Malaysia

Norfaezah Adnan
Virology Unit, Infectious Disease Research Centre, Institute for Medical Research, Kuala Lumpur, Malaysia

Nur Izmawati Abd Razak
Virology Unit, Infectious Disease Research Centre, Institute for Medical Research, Kuala Lumpur, Malaysia

Liyana Ahmad Zamri
Virology Unit, Infectious Disease Research Centre, Institute for Medical Research, Kuala Lumpur, Malaysia

Khairul Izwan Hulaimi
Virology Unit, Infectious Disease Research Centre, Institute for Medical Research, Kuala Lumpur, Malaysia

Zainah Saat
Virology Unit, Infectious Disease Research Centre, Institute for Medical Research, Kuala Lumpur, Malaysia

Patrick F. Sullivan
Department of Genetics, University of North Carolina at Chapel Hill, NC, USA

Franzcp
Department of Genetics, University of North Carolina at Chapel Hill, NC, USA
Department of Medical Epidemiology and Biostatistics, Karolinska Institutet, Stockholm, Sweden

Tobias Allander
Laboratory for Clinical Microbiology, Department of Microbiology, Tumor and Cell Biology, Karolinska University Hospital, Karolinska Institutet, Stockholm, Sweden

Cecilia Lindau
Laboratory for Clinical Microbiology, Department of Microbiology, Tumor and Cell Biology, Karolinska University Hospital, Karolinska Institutet, Stockholm, Sweden

Kristina Fahlander
Laboratory for Clinical Microbiology, Department of Microbiology, Tumor and Cell Biology, Karolinska University Hospital, Karolinska Institutet, Stockholm, Sweden

Andreas Jacks
Department of Medical Epidemiology and Biostatistics, Karolinska Institutet, Stockholm, Sweden

Birgitta Evengård
Department of Clinical Microbiology, University of Umeå, Umeå, Sweden

Nancy L. Pedersen
Department of Medical Epidemiology and Biostatistics, Karolinska Institutet, Stockholm, Sweden

Björn Andersson
Department of Cell and Molecular Biology, Science for Life Laboratory, Karolinska Institutet, Stockholm, Sweden

Rakib A. Al-Ani
Department of Plant Protection, College of Agriculture, University of Baghdad, Iraq

Mustafa A. Adhab
Department of Plant Protection, College of Agriculture, University of Baghdad, Iraq

Kareem A. H. Ismail
Department of Plant Protection, College of Agriculture, University of Baghdad, Iraq

Y. Apandi
Virology Unit, Institute for Medical Research, Kuala Lumpur, Malaysia

W. A. Nazni
Medical Entomology Unit, Institute for Medical Research, Kuala Lumpur, Malaysia

Z. A. Noor Azleen
Medical Entomology Unit, Institute for Medical Research, Kuala Lumpur, Malaysia

I. Vythilingam
Parasitology Unit, Institute for Medical Research, Kuala Lumpur, Malaysia

M. Y. Noorazian
Parasitology Unit, Institute for Medical Research, Kuala Lumpur, Malaysia

A.H. Azahari
Medical Entomology Unit, Institute for Medical Research, Kuala Lumpur, Malaysia

S. Zainah
Virology Unit, Institute for Medical Research, Kuala Lumpur, Malaysia

H. L. Lee
Medical Entomology Unit, Institute for Medical Research, Kuala Lumpur, Malaysia

H. Akbar
National Center of Excellence in Molecular Biology University of Punjab, Lahore-Pakistan

M. Idrees
National Center of Excellence in Molecular Biology University of Punjab, Lahore-Pakistan

S. Manzoor
National Center of Excellence in Molecular Biology University of Punjab, Lahore-Pakistan

I. ur Rehman
National Center of Excellence in Molecular Biology University of Punjab, Lahore-Pakistan

S. Butt
National Center of Excellence in Molecular Biology University of Punjab, Lahore-Pakistan

M. Z. Yousaf
National Center of Excellence in Molecular Biology University of Punjab, Lahore-Pakistan

S. Rafique
National Center of Excellence in Molecular Biology University of Punjab, Lahore-Pakistan

Z. Awan
National Center of Excellence in Molecular Biology University of Punjab, Lahore-Pakistan

B.Khubaib
National Center of Excellence in Molecular Biology University of Punjab, Lahore-Pakistan

M. Akram
National Center of Excellence in Molecular Biology University of Punjab, Lahore-Pakistan

M. Aftab
National Center of Excellence in Molecular Biology University of Punjab, Lahore-Pakistan

A. A. Ogun
Department of Epidemiology, Medical Statistics and Environmental Health, Faculty of Public Health, College of Medicine, University of Ibadan, Ibadan, Nigeria

I. O. Okonko
Department of Virology, Faculty of Basic Medical Sciences, College of Medicine, University of Ibadan, University College Hospital (UCH) Ibadan, Nigeria

A. O. Udeze
Virology Unit, Department of Microbiology, Faculty of Sciences, University of Ilorin, Ilorin, Nigeria

I. Shittu
Department of Viral Research, National Veterinary Research Institute, P.M.B. 01, Vom, Plateau State, Nigeria

K. N. Garba
Department of Virology, Faculty of Basic Medical Sciences, College of Medicine, University of Ibadan, University College Hospital (UCH) Ibadan, Nigeria

A.Fowotade
Department of Medical Microbiology and Parasitology, University of Ilorin Teaching Hospital, Ilorin, Kwara State, Nigeria

O. G. Adewale
Department of Biochemistry, Olabisi Onabanjo University, Ago-Iwoye, Ogun State, Nigeria

E. A. Fajobi
Department of Basic Sciences, Federal College of Wildlife Management, New Bussa, Niger State, Nigeria

B. A. Onoja
Department of Virology, Faculty of Basic Medical Sciences, College of Medicine, University of Ibadan, University College Hospital (UCH) Ibadan, Nigeria

E. T. Babalola
Department of Veterinary Microbiology and Parasitology, Faculty of Veterinary Medicine, University of Ibadan, Ibadan,Nigeria

A. O. Adedeji
Department of Veterinary Microbiology and Parasitology, Faculty of Veterinary Medicine, University of Ibadan, Ibadan,Nigeria

Vaishali S. Tatte
Enteric Viruses Group, National Institute of Virology, Pune, India

D. Shobha D. Chitambar
Enteric Viruses Group, National Institute of Virology, Pune, India

S. Anbazhagi
National Environmental Engineering Research Institute Chennai Zonal Laboratory, CSIR Madras Complex, Taramani, Chennai – 600 113, India

S. Kamatchiammal
National Environmental Engineering Research Institute Chennai Zonal Laboratory, CSIR Madras Complex, Taramani, Chennai – 600 113, India

D. Jayakar Santhosh
National Environmental Engineering Research Institute Chennai Zonal Laboratory, CSIR Madras Complex, Taramani, Chennai – 600 113, India

Ali K. Ageep
Department of Pathology, Faculty of Medicine, Red Sea University, Port Sudan, Sudan

Behzad Esfandiari
Pasteur Institute of Iran-Amol Research Center, Iran

Mohammad Reza Youssefi
Department of Veterinary Parasitology, Islamic Azad University, Babol – Branch, Iran

Ahmad Fayaz
Rabies Section in Pasteur Institute of Iran

Seyed Hamidreza Monavari
Department of Virology and Antimicrobial Resistance Research Center, Tehran University of Medical Sciences,Tehran, Iran

Hossein Keyvani
Department of Virology and Antimicrobial Resistance Research Center, Tehran University of Medical Sciences,Tehran, Iran

Hamidreza mollaie
Department of Medical Virology, Tehran University of Medical Sciences, Tehran, Iran

Mehdi fazlalipour
Department of Medical Virology, Tehran University of Medical Sciences, Tehran, Iran

Farzin sadeghi
Department of Medical Virology, Tehran University of Medical Sciences, Tehran, Iran

Mostafa Salehi-Vaziri
Department of Virology and Antimicrobial Resistance Research Center, Tehran University of Medical Sciences,Tehran, Iran

Roghaeh mollaie
Department of Medical Technology, Hazrat rasool Hospital, Tehran University of Medical Sciences, Tehran, Iran

Farah Bokharaei-Salim
Department of Virology and Antimicrobial Resistance Research Center, Tehran University of Medical Sciences,Tehran, Iran

A. A. Fajinmi
Department of Crop Protection, COLPLANT, University of Agriculture Abeokuta, P.M.B. 2240 Alabata, Ogun State, Nigeria

O. B. Fajinmi
National Institute of Horticultural Research and Training, Idi-Ishin, Ibadan, Oyo state, Nigeria

Nussieba A. Osman
Department of Pathology, Parasitology and Microbiology, College of Veterinary Medicine and Animal Production, Sudan/Sudan University of Science and Technology, P.O.Box 204, Khartoum North, Sudan

A. S. Ali
Department of Preventive Medicine and Veterinary Public Health, Faculty of Veterinary Medicine, University of Khartoum, Post code 13314, Khartoum North, Sudan

M. E. A/Rahman
Department of Virology, Central Veterinary Research Laboratories, Soba, P.O.Box 8067, Khartoum, Sudan

M. A. Fadol
Viral Vaccine Production Unit, Central Veterinary Research Laboratories, Soba, P.O.Box 8067, Khartoum, Sudan

Emad-Aldin Ibrahim Osman
Department of Haematology, Faculty of Medical Laboratory Sciences, Elrazi College of Medical and Technological Sciences, Khartoum, Sudan
Department of Clinical Laboratories, Al-Shaab Teaching Hospital, Federal Ministry of Health, Khartoum, Sudan

Nagwa Ahmed Abdulrahman
Department of Clinical Laboratories, Al-Shaab Teaching Hospital, Federal Ministry of Health, Khartoum, Sudan

Osman Abbass
Department of Clinical Laboratories, Al-Shaab Teaching Hospital, Federal Ministry of Health, Khartoum, Sudan

Waleed Hussein Omer
Al-Neelain Medical Research Center, Faculty of Medicine and Health sciences, Al-Neelain University, Khartoum, Sudan

Hafi Anwer Saad
Department of Community Medicine, Faculty of Medicine, University of Shendi, Shendi, Sudan

Muzamil Mahdi Abdel Hamid
Department of Molecular Biology, Institute of Endemic Diseases, University of Khartoum, P. O. Box 102, Khartoum, Sudan

Mary Bridget Nanteza
Uganda Virus Research Institute, Plot 51-59 Nakiwogo Road, P. O. Box 49, Entebbe, Uganda

David Yirrell
MRC/UVRI Uganda Research Unit on AIDS, c/o Uganda Virus Research Institute, Plot 51-59 Nakiwogo Road, P. O.Box 49, Entebbe, Uganda
Department of Medical Microbiology, Ninewells Hospital, Dundee, DD1 9SY United Kingdom

Benon Biryahwaho
Uganda Virus Research Institute, Plot 51-59 Nakiwogo Road, P. O. Box 49, Entebbe, Uganda

Natasha Larke
MRC Tropical Epidemiology group, London School of Hygiene and Tropical Medicine, Keppel Street, London WC1E7HT, United Kingdom

Emily Webb
MRC Tropical Epidemiology group, London School of Hygiene and Tropical Medicine, Keppel Street, London WC1E7HT, United Kingdom

Frances Gotch
Department of Immunology, Imperial College, Chelsea and Westminster Hospital 369 Fulham Rd, London SW10 9NH,United Kingdom

Pontiano Kaleebu
MRC/UVRI Uganda Research Unit on AIDS, c/o Uganda Virus Research Institute, Plot 51-59 Nakiwogo Road, P. O.Box 49, Entebbe, Uganda
Uganda Virus Research Institute, Plot 51-59 Nakiwogo Road, P. O. Box 49, Entebbe, Uganda

Y Mohd Apandi
Virology Unit, Infectious Disease Research Centre, Institute for Medical Research, Jalan Pahang, 50588 Kuala Lumpur, Malaysia

M. Z Zarina
Virology Unit, Infectious Disease Research Centre, Institute for Medical Research, Jalan Pahang, 50588 Kuala Lumpur, Malaysia

O Khairul Azuan
Virology Unit, Infectious Disease Research Centre, Institute for Medical Research, Jalan Pahang, 50588 Kuala Lumpur, Malaysia

A. R Nur Ismawati
Virology Unit, Infectious Disease Research Centre, Institute for Medical Research, Jalan Pahang, 50588 Kuala Lumpur, Malaysia

S Zainah
Virology Unit, Infectious Disease Research Centre, Institute for Medical Research, Jalan Pahang, 50588 Kuala Lumpur, Malaysia

Endale Tadesse
Department of Medical Laboratory Science, College of Medicine and Health Sciences, Hawassa University, P.O. Box1560, Hawassa, Ethiopia

Lobna Metwally
The Department of Medical Microbiology and Immunology, Faculty of Medicine, Suez Canal University, Egypt

Alaa EL- Din Saad Abd-El Hamid
The Department of Clinical Pathology, Faculty of Medicine, Suez Canal University, Egypt

W. A. Nazni
Medical Entomology Unit, Infectious Diseases Research Centre, Institute for Medical Research, Jalan Pahang, 50588 Kuala Lumpur, Malaysia

M. Y Apandi
Virology Unit, Infectious Diseases Research Centre, Institute for Medical Research, Jalan Pahang, 50588 Kuala Lumpur, Malaysia

M Eugene
Medical Entomology Unit, Infectious Diseases Research Centre, Institute for Medical Research, Jalan Pahang, 50588 Kuala Lumpur, Malaysia

A. H Azahari
Medical Entomology Unit, Infectious Diseases Research Centre, Institute for Medical Research, Jalan Pahang, 50588 Kuala Lumpur, Malaysia

M. K Shahar
Medical Entomology Unit, Infectious Diseases Research Centre, Institute for Medical Research, Jalan Pahang, 50588 Kuala Lumpur, Malaysia

S Zainah
Virology Unit, Infectious Diseases Research Centre, Institute for Medical Research, Jalan Pahang, 50588 Kuala Lumpur, Malaysia

I Vthylingam
Department of Parasitology, Faculty of Medicine, University of Malaya, 50603 Kuala Lumpur, Malaysia

H. L. Lee
Medical Entomology Unit, Infectious Diseases Research Centre, Institute for Medical Research, Jalan Pahang, 50588 Kuala Lumpur, Malaysia

B. Mware
Department of Plant Science and Crop Protection, Faculty of Agriculture, University of Nairobi. P. O. Box 30197-00200 Nairobi, Kenya
Kenya Agricultural Research Institute/National Agricultural Research Laboratories, P. O. Box 14733, 00800 Nairobi,Kenya

R. Narla
Department of Plant Science and Crop Protection, Faculty of Agriculture, University of Nairobi. P. O. Box 30197-00200 Nairobi, Kenya

R. Amata
Kenya Agricultural Research Institute/National Agricultural Research Laboratories, P. O. Box 14733, 00800 Nairobi,Kenya

F. Olubayo
Department of Plant Science and Crop Protection, Faculty of Agriculture, University of Nairobi. P. O. Box 30197-00200 Nairobi, Kenya

J. Songa
Kenya Agricultural Research Institute/National Agricultural Research Laboratories, P. O. Box 14733, 00800 Nairobi,Kenya

S. Kyamanyua
Makerere University Kampala-Uganda

E. M. Ateka
Department of Horticulture, Jomo Kenyatta University of Agriculture and Technology. P. O. Box 62000-00200 Nairobi,Kenya

Mohd Apandi Yusof
Virology Unit, Infectious Disease Research Centre, Institute for Medical Research, Kuala Lumpur, Malaysia

Hariyati Md Ali
Virology Unit, Infectious Disease Research Centre, Institute for Medical Research, Kuala Lumpur, Malaysia

Hamadah Mohammad Shariff
Virology Unit, Infectious Disease Research Centre, Institute for Medical Research, Kuala Lumpur, Malaysia

Noor Khairunnisa Ramli
Virology Unit, Infectious Disease Research Centre, Institute for Medical Research, Kuala Lumpur, Malaysia

Zarina Mohd Zawawi
Virology Unit, Infectious Disease Research Centre, Institute for Medical Research, Kuala Lumpur, Malaysia

Syarifah Nur Aisyatun
Virology Unit, Infectious Disease Research Centre, Institute for Medical Research, Kuala Lumpur, Malaysia

Jasinta Anak Dennis
Virology Unit, Infectious Disease Research Centre, Institute for Medical Research, Kuala Lumpur, Malaysia

Zainah Saat
Virology Unit, Infectious Disease Research Centre, Institute for Medical Research, Kuala Lumpur, Malaysia

P. A. Asare
Department of Crop Science, University of Cape Coast, Cape Coast, Ghana

I. K. A. Galyuon
Department of Molecular Biology and Biotechnology, University of Cape Coast, Ghana

E. Asare-Bediako
Department of Crop Science, University of Cape Coast, Cape Coast, Ghana

J. K. Sarfo
Department of Biochemistry, University of Cape Coast, Cape Coast, Ghana

J. P. Tetteh
Department of Crop Science, University of Cape Coast, Cape Coast, Ghana

S Jeyanthi
Virology Unit, Institute for Medical Research, Jalan Pahang, 50588 Kuala Lumpur, Malaysia

T. A. R Tengku Rogayah
Virology Unit, Institute for Medical Research, Jalan Pahang, 50588 Kuala Lumpur, Malaysia

R Thayan
Virology Unit, Institute for Medical Research, Jalan Pahang, 50588 Kuala Lumpur, Malaysia

A Az-UlHusna
Virology Unit, Institute for Medical Research, Jalan Pahang, 50588 Kuala Lumpur, Malaysia

A Aruna
Virology Unit, Institute for Medical Research, Jalan Pahang, 50588 Kuala Lumpur, Malaysia

B. V Khebir
Disease Control Division, Ministry of Health Malaysia, Putrajaya. Malaysia

B Thevarajah
Kumpulan Perubatan Johor Hospital,Kota Kinabalu, Sabah, Malaysia

S Maria
Sabah Health Department, Kota Kinabalu, Sabah Malaysia

S Zainah
Virology Unit, Institute for Medical Research, Jalan Pahang, 50588 Kuala Lumpur, Malaysia

Fekadu Alemu
Department of Biology, College of Natural and Computational Sciences, Dilla University, P.O.Box. 419, Dilla, Ethiopia

Mathias Afayoa
College of Veterinary Medicine, Animal Resources and Bio-security, Makerere University,P.O.BOX 7062, Kampala,Uganda

David Kalenzi Atuhaire
College of Veterinary Medicine, Animal Resources and Bio-security, Makerere University,P.O.BOX 7062, Kampala, Uganda
National Agricultural Research Organization, National Livestock Resources Research Institute, P.O. Box 96, Tororo, Uganda

Sylvester Ochwo
College of Veterinary Medicine, Animal Resources and Bio-security, Makerere University,P.O.BOX 7062, Kampala,Uganda

Julius Boniface Okuni
College of Veterinary Medicine, Animal Resources and Bio-security, Makerere University,P.O.BOX 7062, Kampala,Uganda

Kisekka Majid
College of Veterinary Medicine, Animal Resources and Bio-security, Makerere University,P.O.BOX 7062, Kampala,Uganda

Frank Norbert Mwiine
College of Veterinary Medicine, Animal Resources and Bio-security, Makerere University,P.O.BOX 7062, Kampala,Uganda

William Olaho-Mukani
African Union-InterAfrican Bureau of Animal Resources, P.O. Box 30786, Nairobi, Kenya

Lonzy Ojok
College of Veterinary Medicine, Animal Resources and Bio-security, Makerere University,P.O.BOX 7062, Kampala,Uganda

www.ingramcontent.com/pod-product-compliance
Lightning Source LLC
Chambersburg PA
CBHW080630200326

41458CB00013B/4580